Matthias Rögner (Ed.)
Photosynthesis

I0047326

Also of Interest

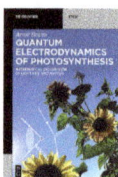

Quantum Electrodynamics of Photosynthesis.
Mathematical Description of Light, Life and Matter
Artur Braun, 2020
ISBN 978-3-11-062692-6, e-ISBN 978-3-11-062994-1

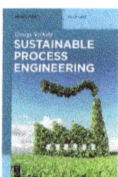

Sustainable Process Engineering
Gyorgy Szekely, 2021
ISBN 978-3-11-071712-9, e-ISBN 978-3-11-071713-6

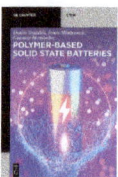

Polymer-based Solid State Batteries
Daniel Brandell, Jonas Mindemark, Guiomar Hernández, 2021
ISBN 978-1-5015-2113-3, e-ISBN 978-1-5015-2114-0

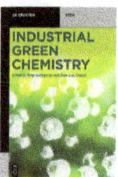

Industrial Green Chemistry
Edited by Serge Kaliaguine, Jean-Luc Dubois, 2021
ISBN 978-3-11-064684-9, e-ISBN 978-3-11-064685-6

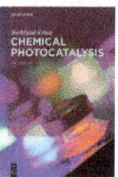

Chemical Photocatalysis
Edited by Burkhard König, 2020
ISBN 978-3-11-057654-2, e-ISBN 978-3-11-057676-4

Photosynthesis

Biotechnological Applications with Microalgae

Edited by
Matthias Rögner

DE GRUYTER

Editor
Prof. Dr. Matthias Rögner
Ruhr-Universität Bochum
Fakultät für Biologie & Biotechnologie
Lehrstuhl für Biochemie der Pflanzen
Universitätsstr. 150
44780 Bochum
Germany
matthias.roegner@rub.de

ISBN 978-3-11-071691-7
e-ISBN (PDF) 978-3-11-071697-9
e-ISBN (EPUB) 978-3-11-071704-4

Library of Congress Control Number: 2021939323

Bibliographic information published by the Deutsche Nationalbibliothek
The Deutsche Nationalbibliothek lists this publication in the Deutsche Nationalbibliografie;
detailed bibliographic data are available on the Internet at http://dnb.dnb.de.

© 2021 Walter de Gruyter GmbH, Berlin/Boston
Cover image: Photography by Andreas Heddergott; in the picture: Andreas Apel
Graphical processing: Dieter Wunsch (Ruhr-University Bochum)
Typesetting: Integra Software Services Pvt. Ltd.
Printing and binding: CPI books GmbH, Leck

www.degruyter.com

Preface

Photosynthesis is regarded as the most important process on the Earth. It is the energetic basis of evolution and of our development and living. Fundamental is especially the process that releases oxygen by the light-powered splitting of water, which gradually transformed our atmosphere into an oxygenating one, that is, the basis of respiration and the development of higher life. Also, photosynthesis is the main source of our everyday energy: regarding the fact that the world energy use of oil per year corresponds to the amount which was created in algae and plankton by photosynthesis within 450,000 years is worrying and should remind us to switch our way of thinking energy from consumers to producers. This is especially mandatory as these energy resources are limited and also combined with massive generation of CO_2 causing serious environmental problems. Solvation of these energy crises is possible if we try to understand the principles of photosynthesis as to how to harness the free sun energy efficiently and use it for our purposes: here and now!

The aim of this book is to provide an overview of the state-of-the-art photosynthesis research with focus on its biotechnological applications and to combine this with an outlook on its future potential. While application of photosynthetic principles for food production with higher plants is established since ages, transfer of the basic knowledge of photosynthesis into application with microalgae is just starting. The focus of this book on microalgae is based on the fact that they grow much faster than the land plants, reaching productivities that exceed higher plants by up to a factor of 10. Also, their growth in saltwater and freshwater is not in competition with farmland and could even be established in special photobioreactors in deserts, requiring just water, some salts and sunlight. Especially challenging is the now available knowledge on genetic transformation of many microalgae species, enabling the generation of design cells tailored for the production of required compounds and suitable for mass production. In contrast to bacterial-based mass production as already established in industrial scale, microalgae supply their own energy requirement by photosynthesis which should help to make these processes substantially cheaper.

This book shows that an intelligent mix of energy and high value products may be an attractive model for the future application of microalgae even under an economic point of view: prerequisite is the ongoing development and investment into the production of next-generation large-scale production facilities, that is, photobioreactors, including sophisticated strain improvements. This approach may be complemented by most recent findings that cyanobacteria may also serve as a basis for biological life-support systems relying on local materials on Mars. Interestingly, by combination with biotechnology, photosynthesis may come back to its cyanobacterial roots in evolution – now called "Astrobiology."

https://doi.org/10.1515/9783110716979-202

Contents

Part 1: Cell design/metabolic engineering

Part 2: Environment and photobioreactor design

Part 3: **Emerging technologies**

List of authors

Hitesh Medipally
Plant Biochemistry
Faculty of Biology and Biotechnology
Ruhr-University Bochum
44780 Bochum
Germany
Hitesh.Medipally@rub.de

Marc M. Nowaczyk
Plant Biochemistry
Faculty of Biology and Biotechnology
Ruhr-University Bochum
44780 Bochum
Germany
Marc.M.Nowaczyk@rub.de

Matthias Rögner
Plant Biochemistry
Faculty of Biology and Biotechnology
Ruhr-University Bochum
44780 Bochum
Germany
matthias.roegner@rub.de

Xufeng Liu
Microbial Chemistry
Dept. of Chemistry-Ångström
Uppsala University
SE-751 20 Uppsala
Sweden
Xufeng.Liu@kemi.uu.se

Hao Xie
Microbial Chemistry
Dept. of Chemistry-Ångström
Uppsala University
SE-751 20 Uppsala
Sweden
Hao.Xie@kemi.uu.se

Stamatina Roussou
Microbial Chemistry
Dept. of Chemistry-Ångström
Uppsala University
SE-751 20 Uppsala
Sweden
Stamatina.Roussou@kemi.uu.se

Rui Miao
Department of Protein Science
Science for Life Laboratory
KTH - Royal Institute of Technology
SE-171 21, Stockholm
Sweden
rui.miao@scilifelab.se

Peter Lindblad
Microbial Chemistry
Dept. of Chemistry-Ångström
Uppsala University
SE-751 20 Uppsala
Sweden
Peter.Lindblad@kemi.uu.se

Hanna C. Grimm
Institute for Molecular Biotechnology
TU Graz
8010 Graz
Austria
h.buechsenschuetz@tugraz.at

Robert Kourist
Institute for Molecular Biotechnology
TU Graz
8010 Graz
Austria
kourist@tugraz.at

Oliver Lampret
Plant Biochemistry
AG Photobiotechnology
Faculty of Biology and Biotechnology
Ruhr-University Bochum
44780 Bochum
Germany
Oliver.Lampret@rub.de

Claudia Brocks
Plant Biochemistry
AG Photobiotechnology
Faculty of Biology and Biotechnology
Ruhr-University Bochum
44780 Bochum
Germany
Claudia.Brocks@rub.de

https://doi.org/10.1515/9783110716979-204

Martin Winkler
Elektrobiotechnologie
TUM Campus Straubing
Schulgasse 22
94315 Straubing
Germany
martin-h.winkler@tum.de

Tobias Erb
Max Planck Institute for Terrestrial
Microbiology
35043 Marburg
Germany
toerb@mpi-marburg.mpg.de

Jan Zarzycki
Max Planck Institute for Terrestrial
Microbiology
35043 Marburg
Germany
zarzycki@mpi-marburg.mpg.de

Marieke Scheffen
Max Planck Institute for Terrestrial
Microbiology
35043 Marburg
Germany
marieke.scheffen@mpi-marburg.mpg.de

Christian Wilhelm
Sen.Prof. Algal Biotechnology
Faculty of Life Science
Institute of Biology
University of Leipzig
04318 Leipzig
Germany
cwilhelm@rz.uni-leipzig.de

Heiko Wagner
Sen.Prof. Algal Biotechnology
Faculty of Life Science
Institute of Biology
University of Leipzig
04318 Leipzig
Germany
hwagner@rz.uni-leipzig.de

Rosa Rosello Sastre
Institute of Process Engineering in Life
Sciences - Bioprocess Engineering
KIT Karlsruhe Institute of Technology
76131 Karlsruhe
Germany
rosa.rosello@kit.edu

Clemens Posten
Institute of Process Engineering
in Life Sciences - Bioprocess
Engineering
KIT Karlsruhe Institute of Technology
76131 Karlsruhe
Germany
clemens.posten@kit.edu

Simon MoonGeun Jung
Green Carbon Research Center
Korea Research Institute of Chemical
Technology
Daejeon 34114
South Korea
mgjung@krict.re.kr

Jong-Hee Kwon
Dept. of Food Science and Technology
College of Agriculture and Life Science
Gyeongsang National University
South Korea
jhkwon@gnu.ac.kr

Felix Melcher
Werner Siemens-Chair of Synthetic
Biotechnology
Department of Chemistry
Technical University of Munich
85748 Garching
Germany
felix.melcher@tum.de

Michael Paper
Werner Siemens-Chair of Synthetic
Biotechnology
Department of Chemistry
Technical University of Munich
85748 Garching
Germany
michael.paper@tum.de

Thomas Brück
Werner Siemens-Chair of Synthetic
Biotechnology
Department of Chemistry
Technical University of Munich
85748 Garching
Germany
brueck@tum.de

Lauri Nikkanen
Molecular Plant Biology
Dept. of Biochemistry
University of Turku
Turku
Finland
lenikk@utu.fi

Michal Hubacek
Molecular Plant Biology
Dept. of Biochemistry
University of Turku
Turku
Finland
michal.hubacek@utu.fi

Yagut Allahverdiyeva-Rinne
Molecular Plant Biology
Dept. of Biochemistry
University of Turku
Turku
Finland
allahve@utu.fi

Katja Bühler
Dept. Solar Materials / Catalytic Biofilms
Helmholtz Centre for Environmental
Research – UFZ GmbH
04318 Leipzig
Germany
katja.buehler@ufz.de

Bruno Bühler
Dept. Solar Materials / Applied Biocatalysis
Helmholtz Centre for Environmental
Research – UFZ GmbH
04318 Leipzig
Germany
bruno.buehler@ufz.de

Stephan Klähn
Dept. Solar Materials / Molecular Biology of
Cyanobacteria
Helmholtz Centre for Environmental
Research – UFZ GmbH
04318 Leipzig
Germany
stephan.klaehn@ufz.de

Jens O. Krömer
Dept. Solar Materials / Systems
Biotechnology
Helmholtz Centre for Environmental
Research – UFZ GmbH
04318 Leipzig
Germany
jens.kroemer@ufz.de

Christian Dusny
Dept. Solar Materials / Microscale Analysis
and Engineering
Helmholtz Centre for Environmental
Research – UFZ GmbH
04318 Leipzig
Germany
christian.dusny@ufz.de

Andreas Schmid
Dept. Solar Materials and MIKAT-Center for
Biocatalysis
Helmholtz Centre for Environmental
Research – UFZ GmbH
04318 Leipzig
Germany
andreas.schmid@ufz.de

Hitesh Medipally, Marc M. Nowaczyk, Matthias Rögner

1 Parameters of photosynthesis relevant for a biotechnological application

Abstract: This chapter introduces natural photosynthesis and the potential of its redesign in living organisms. Due to their high application potential in biotechnology, microalgae are the main focus of this chapter. We do not cover the potential of semi-artificial and artificial photosynthesis as this was recently laid out in a detailed report of Acatech [1]. Instead, we emphasize the significance of such systems for the development and improvement of modified natural systems and show their potential and limitations. Besides the design of such self-reproducing organisms and the state-of-the-art methods involved, we also focus on the design of the environment, especially on the optimization of photobioreactors and new developments. Overall, this chapter serves as an introduction for the more specialized in-depth analyses presented in the following chapters of this book.

1.1 Photosynthesis in microalgae – evolution and limitations

1.1.1 What are "microalgae"?

Microalgae comprise both pro- and eukaryotic single cell organisms living either in seawater or freshwater (Fig. 1). Their sizes range between about 1 and 50 μm with some colonies consisting of cell clusters. As phototrophs, they are capable to survive through photosynthesis, synthesizing high-energy organic compounds from inorganic materials using light as an energy source. Notably, microalgae are primary producers in aquatic environments with tens of thousands of species growing far more quickly than land plants (Fig. 2). As they can easily be (mass) cultivated, they have a great potential to be used in bio-industry, provided they are purified and isolated. Under stable conditions, their energy production efficiency is reportedly around ten times higher than for land plants [2, 3].

1.1.2 Which premises are relevant for picking a suitable model organism?

For biotechnological applications – depending on the desired product – it may be mandatory that the organism is genetically transformable, has a low doubling time,

https://doi.org/10.1515/9783110716979-001

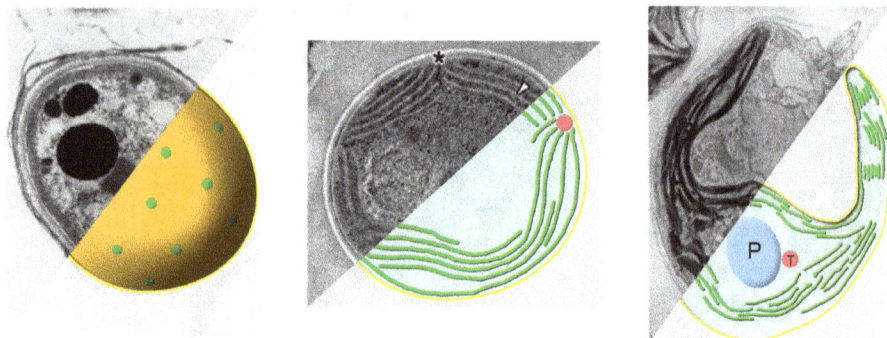

Fig. 1: Three microalgae species representing various levels of complexity: evolutionarily *Gloeobacter represents* one of the most simple structured cells (left, lacking an internal membrane system), followed by *Synechocystis* PCC 6803 (middle, with internal membrane system consisting of various layers). In contrast to these two bacterial species, *Chlamydomonas reinhardtii* (right) belongs to the eukaryotes and provides a chloroplast as organelle. Common to all three species is the presence of thylakoid membranes (green), which harbor the components of the photosynthetic electron transport chain (Fig. acc. to ref. 4).

is genetically stable and is robust. Additionally, the large range of microalgae existing in nature (most still being uncharacterized) also allows to choose species adapted to extreme environmental conditions such as psychrophilic algae (<15 °C) (tolerating low temperatures) and thermophiles (45–70 °C) which may be advantageous for various biotechnological applications. In this context, it may be useful to mention that the upper limit of microalgae performing oxygenic photosynthesis may be around 70–75 °C, while in case of thermophiles individual components like isolated photosystems (i.e., PS1) can tolerate much higher temperatures. With respect to future mass cultivation and also to the final economic evaluation of the process costs it may be important whether a freshwater or a seawater/marine organism is chosen: In case of a location close to the sea, choosing a marine organism may save considerable costs, especially as freshwater is becoming more and more limiting (current global percentages of seawater vs. freshwater being approx. 97.5%/2.5%).

Here we have chosen *Synechocystis* PCC 6803 (see above, hereafter *Synechocystis*) as a model organism to illustrate the internal composition of the photosynthetic machinery and the most decisive parameters for future biotechnological applications. However, in principle, most of these parameters are valid for all kinds of microalgae (and even – with some minor deviations – for higher plants).

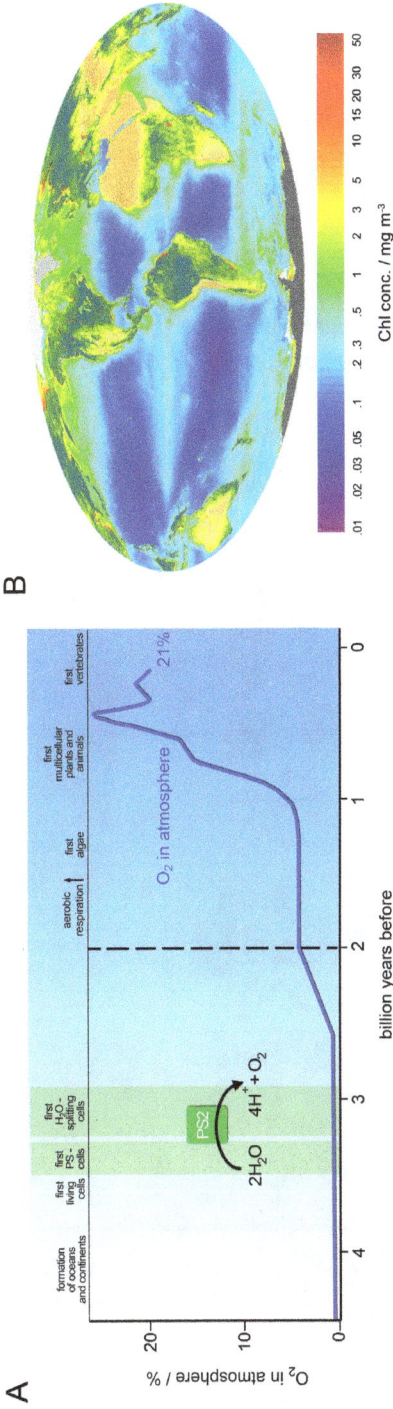

Fig. 2: (A) Evolution of water-splitting photosynthesis goes in line with a transformation of the atmosphere. (B) Distribution of photosynthetic organisms worldwide according to chlorophyll measurements from space (figures from ref. 5, 6).

1.1.3 Which are the rate-limiting aspects of photosynthetic efficiency?

Figure 3A shows the major electron transport (ET) routes in cyanobacterial membranes: while the outer cytoplasmic membrane provides the respiratory ET chain but lacks both photosystems, the focus of biotechnological engineering should be the thylakoid membrane (TM) which contains the complete photosynthetic ET chain, starting with the light-triggered water-splitting reaction at PS2 up to the acceptor side of PS1, which plays a central role in energy distribution, depending on physiological requirements. Figure 3B, based on structural investigations on the molecular level, reveals that all three intrinsic membrane protein complexes (PS2, Cyt. b_6f complex

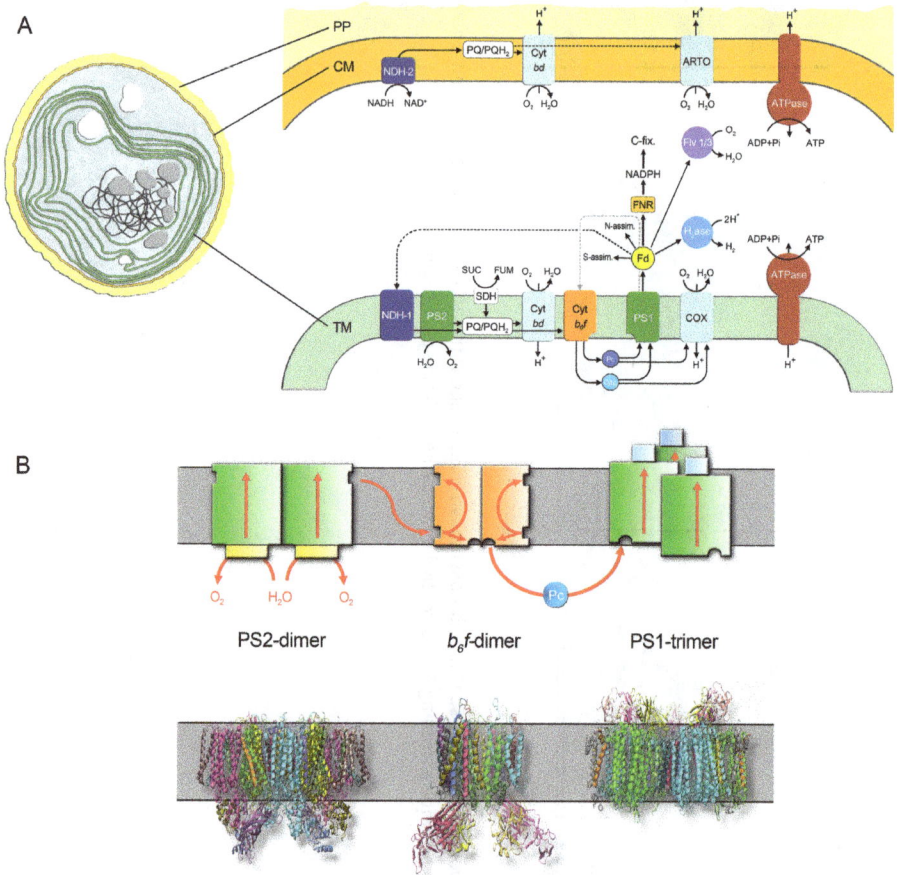

Fig. 3: (A) Membrane systems in the cyanobacterial *Synechocystis* cell and major electron flows in cytoplasmic (CM) and thylakoid (TM) membrane. (B) Crystal structures of the three major photosynthetic electron transport complexes (PS2, b_6f and PS1 [7–9]) and schematic view of their electron transport routes.

and PS1) form oligomeric complexes, which in case of PS1 and PS2 are also active as (isolated) monomers, whereas the b_6f complex requires a dimeric state to function. Within the TM, depending on the light conditions, the oligomeric complexes of PS2 and PS1 are in equilibrium with the respective monomers. This reflects the high adaptability of the cellular system to changing light conditions which is difficult/impossible to imitate in a semiartificial system.

Figure 4 shows the kinetics of electron flow through this light-powered ET chain. Apparently, PQ reduction at PS2 is the rate-limiting step for the light reaction [10]. In the absence of this step and other limitations, the light-triggered reactions of PS2 and

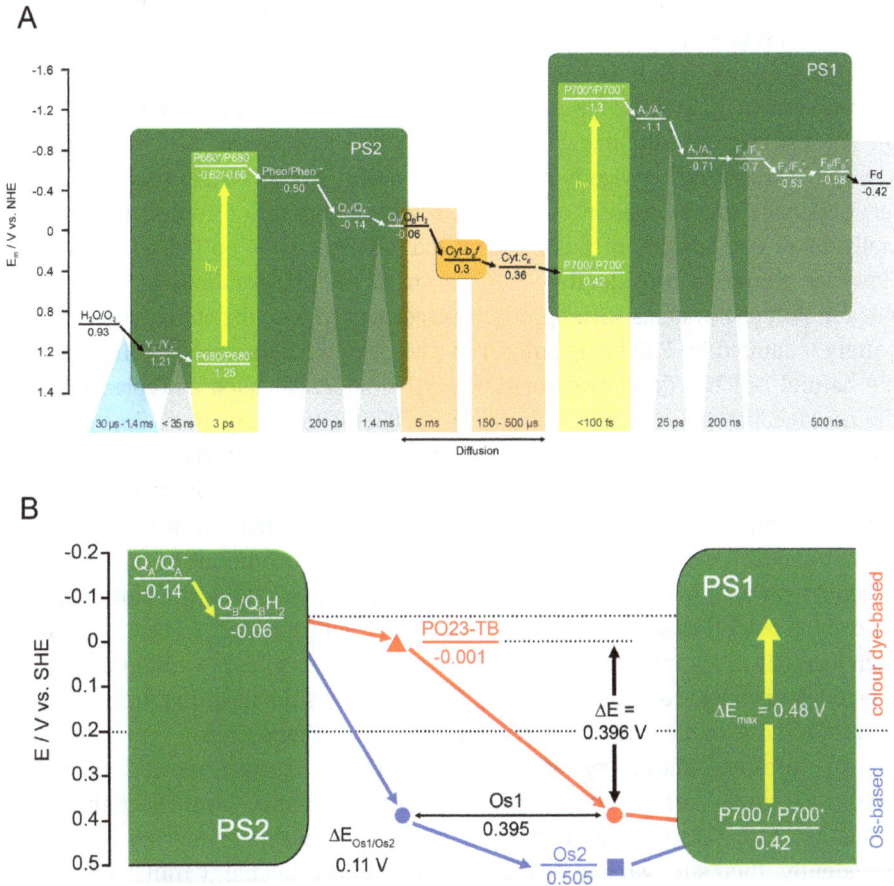

Fig. 4: (A) Redox components and kinetics of the photosynthetic light reactions in the thylakoid membrane, involving PS2, Cyt. b_6f and PS1, arranged in sequence for linear electron transport [4]. (B) Redox situation in a semiartificial minimal device supplemented by synthetic components ("biobattery"):PS2-based photoanode and PS1-based photocathode, being connected by Os polymers (Os1 and Os2) or, alternatively, by a phenothiazine-modified polymer in combination with Os1, realizing a semiartificial Z-scheme [12].

PS1 could proceed with much higher speed, which is decisive for an evaluation of the capacity of photosynthetic systems and their potential for biotechnological applications. This can be convincingly demonstrated with a semiartificial system, consisting of isolated photosystems which are immobilized on gold electrodes (Figs. 4 and 5, respectively) [11]. In this case, the function of the Cyt. b_6f complex is replaced by an artificial tailor-made redox polymer, that is, osmium polymers and/or phenothiazine, respectively, which enable to immobilize photosystems on the electrode surfaces [12]. They are not rate limiting and guarantee an optimized downhill electrode transport (Fig. 4B). It could be shown that in case of PS1 the ET rate under these conditions could be extended by at least a factor of 6 [13]. In case of PS2, the water-splitting reaction occurring in a time range of about 1.4 ms is rate limiting, but under the conditions of the semiartificial device this could be compensated by using larger amounts of PS2 at the anode side than PS1 at the cathode side. In contrast, natural *Synechocystis* cells contain up to 10-fold more PS1 than PS2 [14]; thus, the ET is not balanced and the much smaller amount of PS2 is limiting the whole ET reaction.

The semiartificial device is an extremely useful model system to simulate the most effective transformation of harvested light energy into electric energy, especially as all components of this minimalized system can be optimized separately. However, despite a considerable stabilization of the photosystems by embedding them in polymer systems with appropriate redox polymers, the stability of this bio-battery is limited by PS2 being subject to photooxidative damage [15]. In contrast, the natural cell has developed some protection strategies such as replacement of the damaged parts (especially subunit D1 of PS2) within very short time periods, that is, 20–30 min. Such an efficient biogenesis and repair system [16] is difficult or impossible to realize in an engineered or semiartificial system but very important for a biotechnological application of photosynthesis. Another advantage of the cellular system is the dynamic adjustment to external light conditions by switching between D1 variants with different properties [17]: under low light, the more efficient D1 copy is used (PsbA1), being however not very light tolerant. In contrast, under high light (HL) a slightly less efficient D1 copy (PsbA3) enables more light protection due to faster charge recombination [18]. Figure 6 shows the resulting different reactions of PS2 due to the exchange of D1 copies under HL conditions.

An extremely efficient system was recently found in algae growing on desert sand crusts under harsh environmental conditions (filamentous cyanobacteria *Microcoleus* sp., Fig. 7A) which are immediately lethal for "normal" (mesophilic) algae. Apparently, their survival strategy involves alterations in energy transfer from the antennae to the reaction centers in dependence of the diurnal desiccation/rehydration cycles and the activation of a nonradiative charge recombination due to a smaller redox gap between P_{680}^+ and Q_A^- (Fig. 7B) [20]. In consequence, an alternative pheophytin-independent recombination may minimize the damaging $^1O_2$ production associated with radiative recombination [21].

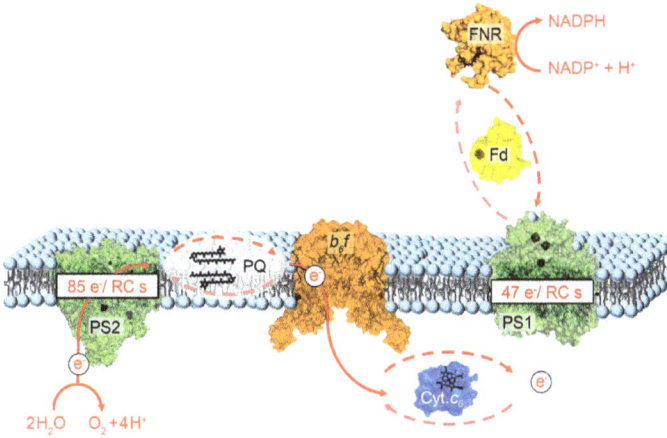

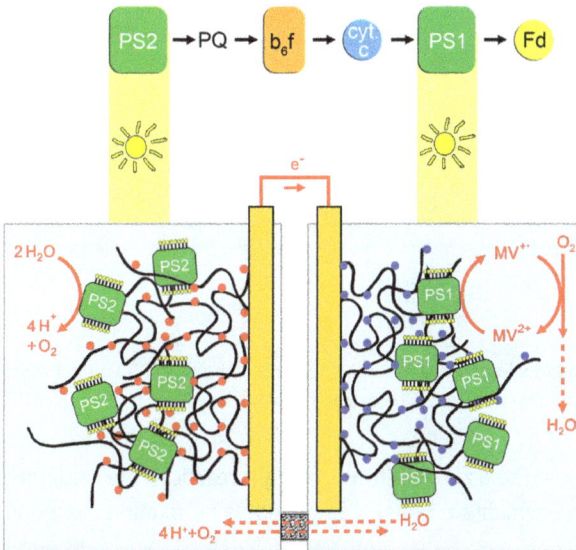

Fig. 5: Kinetics of photosynthetic electron flow in native thylakoid membranes (A) in comparison with a semiartificial device ("biobattery," (B) with isolated photosystems embedded and immobilized into a matrix of two osmium polymers with appropriate redox potentials at the acceptor side of PS2 (red) and at the donor side of PS1 (blue), respectively [11].

These desert organisms could be the blueprint for strategies to overcome damage by extreme HL intensities and dryness, especially as deserts provide unlimited space for mass cultivation of algae under HL conditions (for more details, see chapter 6).

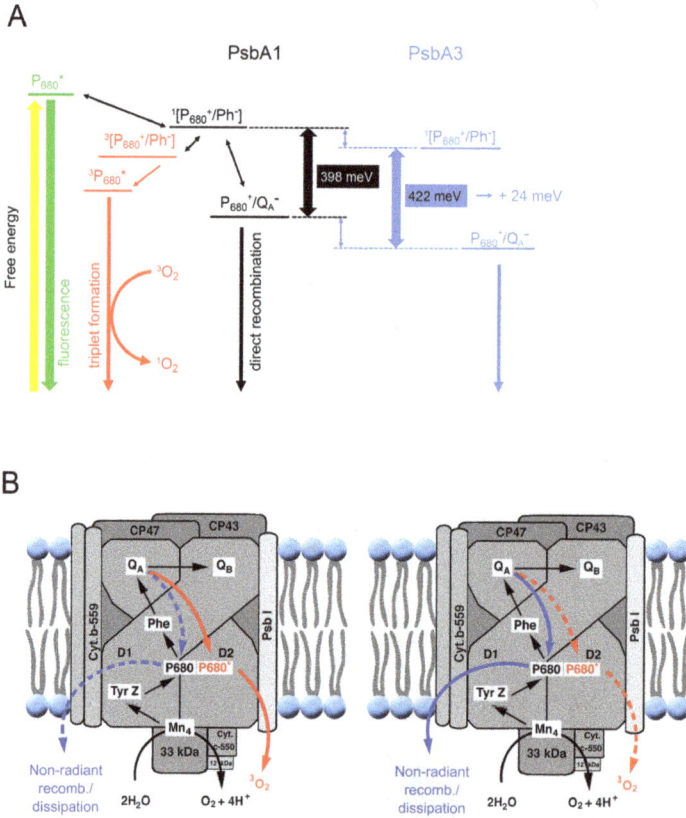

Fig. 6: (A) High light (PsbA1) and low light (PsbA3) copies of the D1 subunit in PS2 on the redox scale [19] and (B) with correlated reactions of PS2 electron flow [17].

In the cellular environment, the above-introduced light reactions are combined with the so-called dark reactions, which are better designated as "C-fixation reactions." Essentially, they use energy (ATP) and redox equivalents (NADPH) generated during the light reaction and CO_2 (from the air) to synthesize sugar compounds as energy storage for periods without light (i.e., nighttime, shading, etc.). Considering the efficiency of photosynthesis-dependent biotechnological processes, it would be reasonable to rate "quantum efficiency" as the efficiency of each light quantum which hits the TM to trigger a primary reaction in PS1 and PS2 – especially as (sun) light is available in large excess and for free. Following this definition, the efficiency of PS1 and PS2 is beyond 99% according to a measurable charge separation across the TM occurring within a few nanoseconds. However, if compared with technical systems such as solar cells [22], efficiency has to be based on the full solar spectrum of which the photosynthetically active radiation (PAR) corresponds to approximately 45%. For this reason, primary photosynthetic events start with an efficiency of max. 45%, which decreases with the following

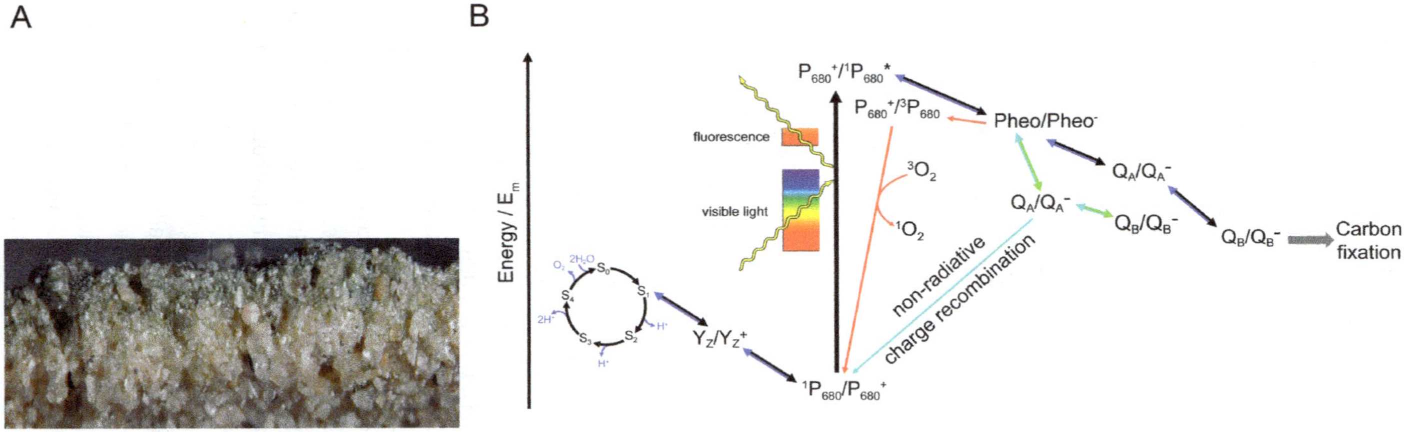

Fig. 7: (A) Side view perpendicular to the surface of a natural sand crust with a cyanobacterial filament layer on top (*Microcoleus* sp. [20]). (B) Non-radiative charge recombination in *Microcoleus* PS2 due to changes in the redox potential of Q_A and Q_B sites (blue arrows: forward ET; green arrows: protective route) (Fig. taken from ref. 21); reactions generating singlet oxygen marked in red.

ET chain between PS2 and PS1 to theoretically max. 17% (Fig. 8). Due to loss of energy during each electron transfer step, the efficiency of one of the first measurable final products, "sugar," is typically less than 1%. However, for microalgae in bubbled bioreactors a maximum of 5–7% under optimal conditions has been estimated [23].

The major gap in efficiency occurs between light and dark reactions, and the major general reason is the fundamental difference in the kinetics of the reactions involved (see time range in Fig. 8).

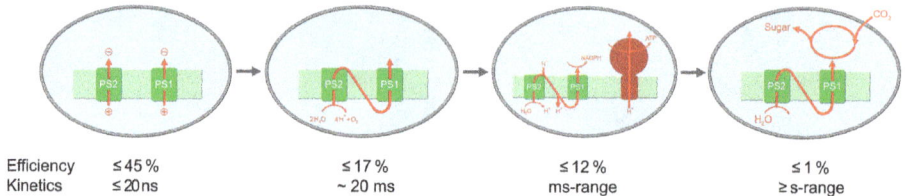

| Efficiency | ≤45% | ≤17% | ≤12% | ≤1% |
| Kinetics | ≤20 ns | ~20 ms | ms-range | ≥s-range |

Fig. 8: Efficiency of major steps in photosynthetic reactions starting from 100% incident sunlight: processes of the light reactions in the thylakoid membrane [4].

While all light reactions occur in the pico- and (lower) ms range, the CO_2-fixation reaction catalyzed by ribulose-bisphosphate-carboxylase (Rubisco) in microalgae has a turnover of 0.3 per second, which is combined with a general inefficiency of the CO_2-fixing enzymes and pathways (for details see chapter 5). This is the reason for the 99% loss of the harvested light energy. In consequence, using algae for the production of biomass via photosynthesis is energetically an extreme waste of energy. Biotechnologically, the best gain of the harvested light energy would be to use the electrons directly from the acceptor sites of the two photosystems, that is, PS2 and PS1. Which portion of the available electrons can be harvested without seriously compromising downstream reactions relevant for the survival of the cell (C-fixation reactions, etc.)? Efficiency measurements of whole cells under HL conditions show that a considerable amount of the harvested light energy under these conditions cannot be used due to a blockage of electron flow at the slow dark reactions. To prevent HL-induced damage via photooxidative effects, the cell releases this surplus energy as fluorescence and heat, which of course is also a considerable waste of energy: as pointed out in more detail in chapter 6, the energy released as heat and/or fluorescence can amount to about 70% of the absorbed radiation energy. Preliminary studies have shown that a considerable part of this energy may be used if an appropriate sink is available in the cells (see Fig. 9). There are also indications that more than two-thirds of the so-called carbon partitioning could be used to create other products. Examples are a more than 80% carbon partitioning rate into ethanol reported by Kopka et al. [24] and close to 70% by Savakis et al. [25]. However, optimization toward such goals normally requires various steps of metabolic engineering involving also approaches of synthetic biology (see especially chapters 2, 3, 5 for such example).

1.2 Tuning of intrinsic PS parameters relevant for application

Figure 9 provides an overview on the possible targets for manipulating the ET on a cellular level. Most of these parameters have been experimentally tested and some of the results will be presented as examples in the following section.

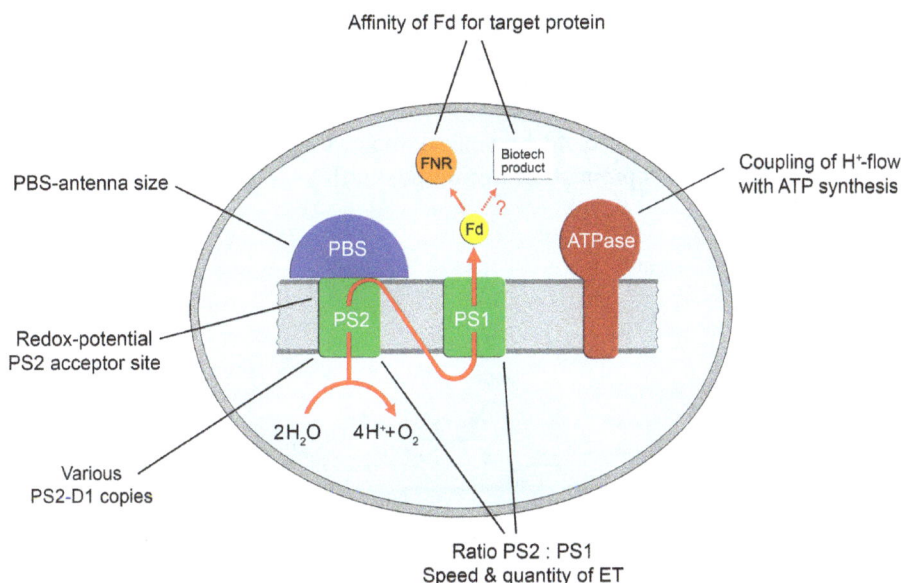

Fig. 9: Impact sites for manipulation of cellular photosynthetic electron transport (ET): uncoupling ATPase yields a rate increase of ET by about a factor of 2 [26], switching to a different PS2-D1 copy yields stabilization [18], truncation of PBS increases the ratio of PS2/PS1 with corresponding speeding up of ET [27] and engineering of Fd/FNR, for example, increases H_2 photoproduction yields by a factor of 18, while decreasing FNR activity [28].

1.2.1 Manipulation of light harvesting

Reduction of the phycobilisome (PBS) antenna size promotes biotechnological application of photosynthesis in many ways:

– PBS constitute up to 63% [29, 30] of the soluble cellular proteins, that is, PBS reduction can save tremendous energy for the synthesis of other products.
– Reduction or even deletion of PBS increases the PS2/PS1 ratio of the cell from approximately 1:5 to up to 1:1 (F. Mamedov, pers. com.). As the PS2 content is the bottleneck for electron supply from the water-splitting reaction, this in parallel increases the linear ET up to a factor of 5.5 as can be shown by light-dependent

oxygen evolution measurements [31]. This additional electron supply can be used for the generation of biotechnological products.

- A smaller PBS antenna size also enables to tolerate significantly higher light intensities such as direct exposure to bright sunlight which is relevant for biotechnological mass cultivation under extreme conditions as known from deserts [14].
- Due to self-shading effects, reduction of PBS antenna size also enables higher cell densities, that is, more product formation, in mass cultures [27].

However, the complete removal of PBS as in case of the PAL mutant (Fig. 10) apparently causes major metabolic problems as shown by quantitative proteomics [14, 30]. For this reason, a balance between positive and negative effects is crucial in optimizing a design cell for biotechnological applications. The olive mutant seems to represent the best compromise for this purpose [31].

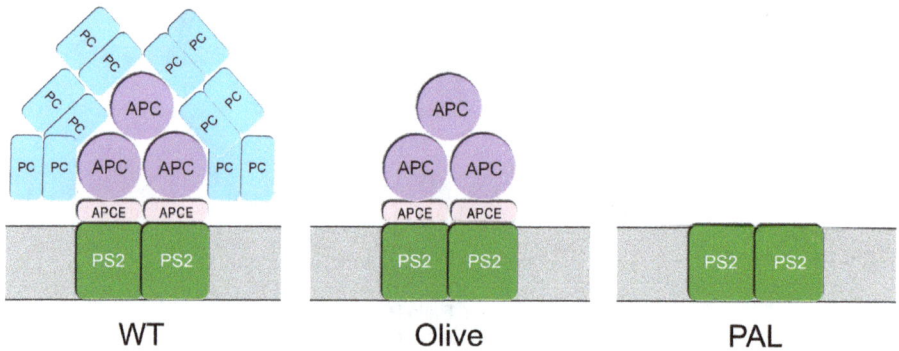

Fig. 10: Phycobilisome antenna sizes of *Synechocystis* PCC 6803 wild-type (WT) cells and mutants Olive and PAL.The olive mutant lacks the PC (phycocyanin) –subunits, while in the PAL mutant phycobilisomes are abolished entirely [27].

Another approach to optimize light harvesting could be an antenna extension into the infrared (IR) range (Fig. 11). Such an antenna already exists in photosynthetic bacteria which contain only one photosystem (PS1 or PS2 precursor without water-splitting ability). If integrated into the oxygenic photosynthesis of cyanobacteria, such IR-ranged antenna could increase the quantum efficiency by using a range of the sunlight spectrum up to now completely unused, especially if both photosystems would be equipped with different antennae [33].

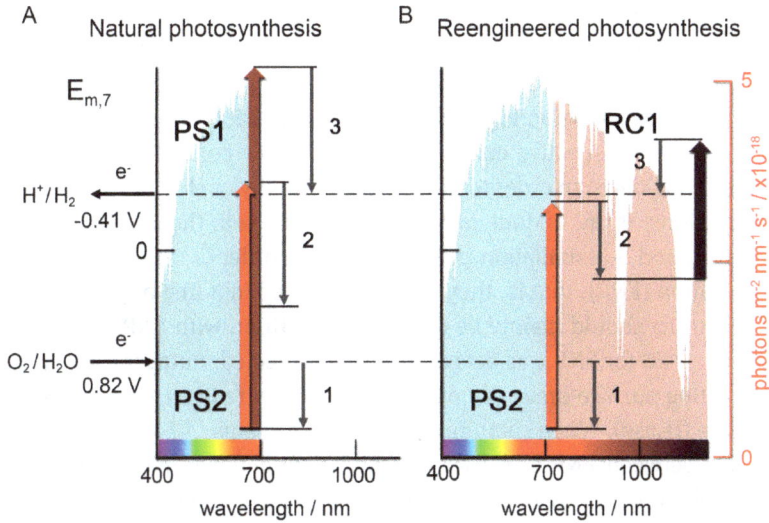

Fig. 11: Photoelectrochemical energy capture diagram for natural (A) and engineered (B) photosynthesis, according to the thermodynamic principles for improved efficiency. In (B) PS1 is replaced by RC1, a new reaction center with farther-red-absorbing Chl pigments, which increases the photosynthetic efficiency by approximately doubling the solar photon capture. For further details, see [33] (figure acc. to ref. 33).

1.2.2 Product-oriented design of photosynthetic electron flow

Figure 12 shows potential sites to modify the photosynthetic electron flow for a biotechnological application. Branching off parts of these electrons at the level of the light reaction ensures a high quantum efficiency for the products that were estimated to range between 10% and 14% [33, 34], thus being about 10–20-fold higher than the quantum efficiency of "classical" photosynthesis with a sugar component as a product. As mentioned earlier, branching off electrons for the generation of a desired product should not endanger the "maintenance" of the cell, that is, the minimal required energy for a cell to maintain the most fundamental metabolic processes without dying.

Of course, optimization does not only concern the design cell itself but also the respective environment, that is, growth management and so on (see below). Especially for cyanobacteria, achieved values for C-partitioning seem to be very high for desired products, and for this reason these species are better suited for biotechnology than eukaryotic microalgae.

Potential sites for branching off electrons from the light reaction are P1 at PS2 (Fig. 12, for instance, via a fusion with H_2ase, see below and also chapter 11), P2 at PS1 (fusion with H_2ase, see [35, 36]), P3 at Fd [37] and P4 at NADPH (reduction of externally added components [38]).

In reality, the situation is more complicated than shown in Fig. 12A, as PS1 via its electron acceptor ferredoxin (Fd) is an important switching point for the cellular metabolism, funneling photosynthetically mobilized electrons in many different biochemical pathways as shown in Fig. 12B. Practically, for each of these reactions, the affinity of Fd for the respective enzyme and the enzyme concentration within the cell have to be considered in order to predict the hierarchy of electron supply for all competing reactions. If this is done for a cyanobacterial cell, the significance of N-assimilation (NIR) and S-assimilation (SIR) is about one order of magnitude lower compared to C-fixation (FNR), that is, they can be neglected in a first approach and branching off electrons should mainly be done in competition with FNR. As an example, the affinity of Fd for FNR was decreased by site-directed mutagenesis (SDM) within the interacting surface areas of both Fd and FNR to redirect electrons to a target hydrogenase (H$_2$ase). When combining such SDM variants of Fd and FNR in a competition assay between FNR and H$_2$ase for photosynthetically mobilized electrons, 73% of the photohydrogen evolution activity measured in the absence of FNR could be recovered, while H$_2$ase only reached 4% of this control activity when using wildtype proteins of Fd and FNR (see Fig. 13). This demonstrates the high potential of engineering metabolic pathways and cells for new products at the expense of photosynthetic electrons [28, 39]. An alternative approach has been published by Appel et al. [35] by fusing hydrogenase directly to PS1 [35]. In this case, there is a direct electron transfer from PS1 to hydrogenase which is presently only hampered by the fact that this hydrogenase is inhibited by the O$_2$ molecules generated in the water-splitting reaction at PS2, that is, presently, in the living cell this reaction can only occur under anaerobic conditions. However, gaining a detailed understanding of the inactivation mechanisms occurring at the active centers of [NiFe] and [FeFe] hydrogenases and establishing effective molecular tolerance or resistance strategies will hopefully allow photohydrogen production to occur under oxygenic conditions which will assure a high quantum yield (see chapter by Winkler!) [40]. Considering most of the possible improvements by cellular design outlined above, model calculations for a potential H$_2$ production rate by cyanobacteria predict an approximately 100-fold increase in comparison with typical production yields by green algae [32].

Another example is the use of photosynthetic electrons originating from water splitting to provide NADPH for recombinant oxidoreductases [38]. It could be shown that small hydrophobic substrates can be taken up by the cell from the surrounding medium; reduction in the cell is followed by product excretion into the environment. This facilitates harvesting of the desired product considerably [38] (see also chapter 3).

These selected examples show that directed re-routing of electrons in specially designed cells is possible. However, it should also be considered that reactions in the test tube with selected reaction partners are different from conditions in the living cell. For instance, comparison with selected mutants of Fd, FNR and H$_2$ase under test tube and physiological conditions shows severe differences. In case of *Synechocystis*, experiments have shown that cells do not survive with FNR mutants

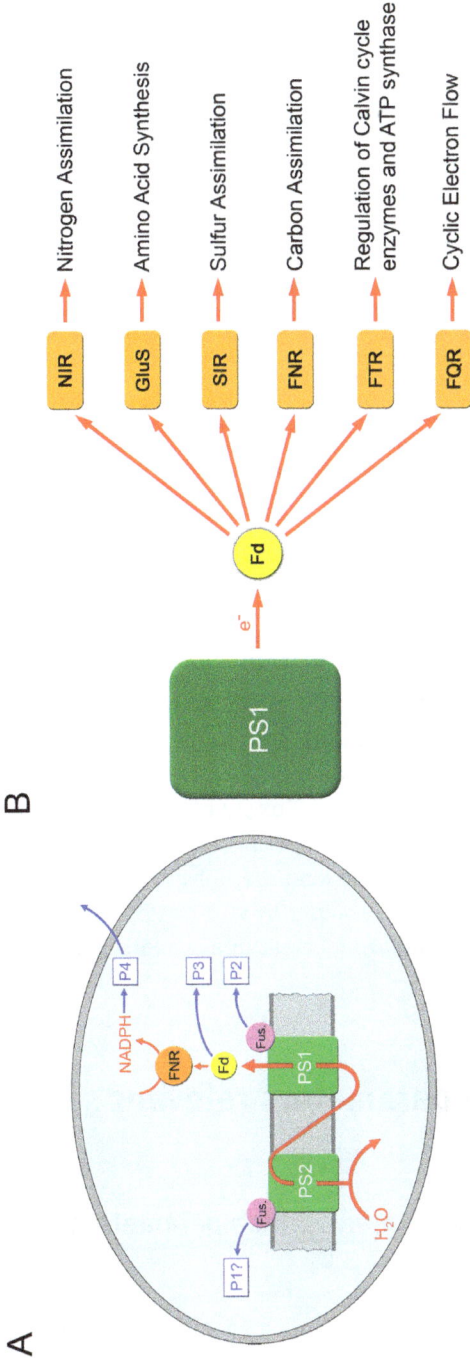

Fig. 12: Potential sites for branching off photosynthetic electrons for biotechnological product generation (A) and significance of PS1/Fd branching point for electron distribution in the cyanobacterial cellular metabolism (B); for detailed explanations, see text.

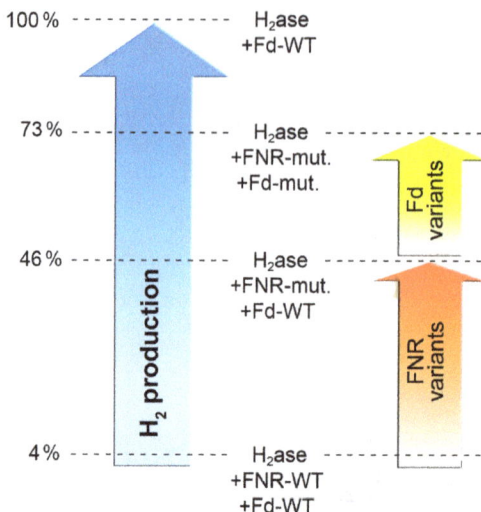

Fig. 13: Competition of FNR and H$_2$ase for photosynthetic electrons from Fd: comparison of H$_2$ generation in the test tube after mixing isolated H$_2$ase with ferredoxin (Fd) (left) and H$_2$ase with Fd in competition with FNR (wild type and mutants of Fd and FNR) (right) [28].

being severely affected in their affinity for Fd, thus preventing a transfer of the most impactful combination of protein variants measured in vitro to the living cell [39]. However, in this mutant strain, a functional electron sink (i.e., an engineered hydrogenase) did not exist which may change the whole redox situation. This exemplifies that in vitro experiments are just an indication for the results that may be achieved in vivo with corresponding mutant strains and that for gradual improvements each individual mutant needs to be tested under in vivo conditions. Establishing reliable in vitro simulations of all cellular processes in a cyanobacterial cell is not yet possible (Ralf Steuer, pers. comm.).

1.3 Tuning of extrinsic PS parameters relevant for application

1.3.1 Artificial versus natural environment: which parameters can be optimized?

Microalgae have a relatively low cell density compared to other microbes. In most cases, light is limiting in their natural environment which requires a sophisticated arrangement of additional antennae – PBS in case of cyanobacteria and LHC in case of green algae. In contrast, industrial conditions normally provide sufficient light which permits a considerable reduction of the native antenna size (see above). This in turn allows to reach much higher cell densities – combined with correspondingly higher product yields – in photobioreactors (PBR) as surface-exposed cells with a decreased antenna size show a much smaller shading effect for internal cells than wild-type cells [27] (see Fig. 14). Additionally, light intensity and light quality can now be adjusted tailor-made, and even different sunlight conditions equivalent to region-dependent variations can now be simulated due to advanced LED technology (see chapter 9).

Example: Reduction of the PBS antenna size can increase the dry cell weight per liter culture to a factor of 1.57 [41] and – due to more cells with more PS2 – the rate of linear photosynthetic ET from water splitting increases up to 5.5-fold [31]. However, this maximal performance only applies under conditions when cell density, illumination and nutrient supply are optimal. For this reason, all those parameters need to be maintained at a constant level, which can only be realized by a continuous cultivation system as introduced in chapters 7 and 8. This type of cultivation under turbidostatic process control eliminates the relation between cultural growth and cell density which is kept constant by controlled dilution with fresh growth media. If this approach is complemented by a detailed proteome analysis, bottlenecks of metabolism and stress conditions can be indicated and prevented. Upscaling of such a system is possible and has been shown in a 100 L flat plate reactor in cooperation with an industrial partner (for details, see Fig. 15).

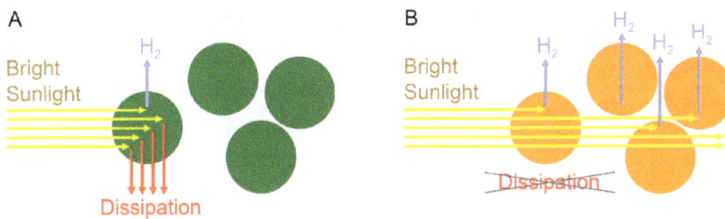

Fig. 14: Impact of PBS size on self-shading effect in PBR with high cell densities. WT cell (A) and cells with truncated PBS antenna (B) [27].

1.3.2 Product-oriented design of the photosynthetic electron flow

The highly optimized photosynthetic ET chain of cyanobacteria can be used to drive redox biotransformations for the conversion of externally supplied substrates into desired products. Cyanobacteria are particularly interesting host organisms because their metabolism is "over-reduced" compared to heterotrophic bacteria. This surplus of photosynthetic redox power can be harnessed for biotechnological applications, particularly to supply oxidoreductases with NADPH (see chapter 3).

Oxidoreductases play an important role for the industrial production of fine chemicals and pharmaceuticals. As redox enzymes, they rely on electron supply by redox co-

Fig. 15: A 100 L flat plate PBR as a pilot example for scaling up for future mass fermentation of cyanobacteria (coop. with KSD Co., Hattingen) [27].

factors, which are often more expensive than the product (e.g., NADPH). Therefore, co-factor regeneration systems were developed that depend on other organic co-substrates like glucose or isopropanol. Their application led to other problems like a poor atom efficiency of the process and the stoichiometric production of co-products like gluconic acid in case of glucose. An ideal system would regenerate the redox co-factor just from water, like photosynthesis does. However, photosynthetic co-factor regeneration is very complex and not fully understood in all details. At least two consecutive light-driven reactions, catalyzed by PS2 and PS1, are necessary to bridge the potential difference between H_2O and NADPH. Water oxidation releases oxygen as a by-product, which is another advantage of the system, as many oxidoreductases (e.g., monooxygenases or dioxygenases) depend on oxygen. Cyanobacteria are the simplest organisms performing oxygenic photosynthesis and several model organisms like *Synechocystis* have been extensively studied. In principle, wild-type cyanobacteria can be employed for redox biotransformations but the productivity and space-time yield of these native systems are fairly limited due to the low expression level of native oxidoreductases.

However, the availability of various tools for the genetic manipulation of cyanobacteria enables the expression of recombinant oxidoreductases with optimized genetic elements (e.g., promotors, ribosome-binding sites and riboswitches) to increase productivity. In a proof-of-concept study, Königer et al. introduced the gene of the ene-reductase YqjM from *B. subtilis* under the control of the light-induced psbA promotor into *Synechocystis* [38]. With this system, several prochiral substrates were converted with a maximum specific activity of more than 100 U per gram cell dry weight (U g_{DCW}^{-1}) into the optical pure products (99% ee). The success of the concept of light-driven whole-cell biotransformation was later verified with other oxidoreductases, for example, with a CYP450 monooxygenase from *Acidovorax* sp. CHX100 with a specific activity of 39 U g_{DCW}^{-1} [42]. Further optimization of the YqjM system was achieved by exchanging the promotor P_{psbA2} to P_{cpc} (1.3-fold increase of activity) and especially by increasing the NADPH supply. Changes in the cellular NADPH pool were monitored by PAM fluorescence spectroscopy, which led to the conclusion that the photosynthetic supply of NADPH was not sufficient to compensate for the additional consumption by YqjM. Cyanobacteria feature a very versatile electron transfer network that enables a quick adaptation to changing environmental conditions, particularly to changing light intensities, as overreduction of the photosynthetic electron transfer chain (imbalance of, e.g., NADPH production and consumption) leads to the generation of reactive oxygen species and damage of the photosynthetic apparatus. Therefore, additional electron sinks, like flavodiiron proteins (FLVs), are important to enable the efficient dissipation of excess energy in a so-called water–water cycle [43] (see chapter by A. Rinne). FLVs are essential for cyanobacteria to survive under natural conditions but they are expendable under controlled laboratory conditions. A knockout of Flv1 and Flv3 in the *Synechocystis* YqjM overexpression strain almost doubled the maximum product formation rate [44]. This example illustrates the potential of the general streamlining approach to increase productivities. Future strategies will include controlled inactivation of other competing electron sinks and particularly address the low space-time yield of cyanobacterial cultures. Usually, the volumetric production of heterotrophic bacteria can be easily optimized by increasing the cell density of the culture. This is currently not possible with cyanobacteria, as the maximum cell density and particularly the productivity of redox-driven biotransformations are limited by the ability of light to penetrate into the culture. High-density cultures are usually not performing better, as the substrate light becomes less available. Novel PBR designs and techniques of cell immobilization may help solve this fundamental problem (see below).

1.4 State of the art and future potential

1.4.1 Future design cells

Bioengineering of cyanobacterial cells for the generation of special products can now make use of the experiences gained from various mutants. Most importantly, all these effects are usually not additive and the respective mutants have to be introduced step by step with detailed metabolic analysis. Simulation of such effects on the cellular level by metabolic programming would be highly interesting in the future.

Especially the source–sink dynamics should be analyzed carefully and individually for each product: if there is no balance between electrons generated by water splitting and electrons used for products, aggressive side products will be formed like oxygen radicals which limit the yield by destroying components of the ET chain.

The following is a toolbox for tentative approaches toward a design cell. They may be mixed/combined according to specific requirements:

- Individual mutants, some of which already introduced above, should be combined in one single design cell: they involve stabilization of PS2 (subunits from thermophiles or high light-resistant species), increasing the number of photosynthetic electrons (increasing PS2 concentration by PBS reduction), speeding up electrons (by partial ATPase uncoupling) and re-routing Fd-gated electrons (by Fd and FNR mutants).
- Introduction of light-regulated genes, which may switch on special product generation under HL, while cell maintenance could profit under normal/low light conditions, allowing cells to proliferate. This also could save energy under growth conditions.
- Designing fusion proteins with subunits at the acceptor side of the photosystems to minimize interfering reactions and optimize product yields. This has recently been shown for PS1 with hydrogenase [35]. Energetically superior would be the direct fusion or the wiring of the PS2 acceptor side with hydrogenase as shown in Fig. 16 – possibly as proof of principle for other enzymes. Here, the challenge would be to divert the electrons from pheophytin despite the rapid kinetics of competitive redox reactions. In addition, due to the water-splitting reaction at PS2 an oxygen-tolerant hydrogenase is required (for strategies see chapter by Winkler). However, the combination of light-powered water-splitting and hydrogen production within a single engineered enzyme complex would be unique. As outlined in [12], the capacity of such systems could be prechecked and optimized in electrode-based semiartificial devices before engineering them in living cells.

1.4.2 Environmental design

The commercial production of bio-hydrogen, biomass and biomass-based products requires low cost and efficient production systems – in case of microalgae especially efficient PBR. Key factors for the design of PBR are temperature, light and agitation [45] (see also chapter 7). In detail:

Temperature: Tolerance for light intensity and optimal behavior of metabolism in microalgae PBR compared to open ponds. This problem can be addressed by passive evaporative heating systems such as freshwater sprays or heat exchangers but they require a large amount of water. Major problem is that 95% of the sunlight is outside of the PAR

A

B

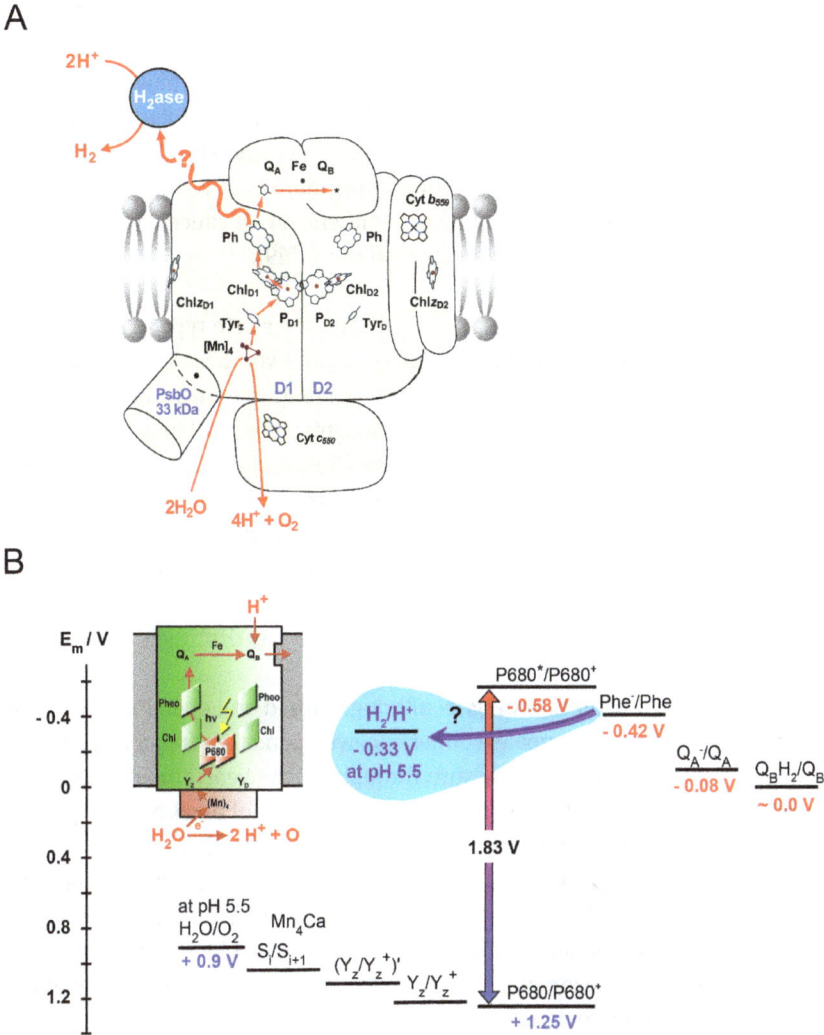

Fig. 16: (A) Example for a fusion construct between PS2 and hydrogenase, which is thermodynamically favored but depends on the availability of an oxygen-tolerant enzyme, maybe in combination with oxygen-scavenging reactions. (B) Energetics of the coupling represented by the redox potential of the components, that is, PS2 and fusion partner hydrogenase..

which involves heat dissipation and an increase in temperature [46]. This can be minimized by engineering the photonic spectrum and by use of IR spectrum filters in PBR.

Light: Light plays a crucial role in photosynthetic efficiency, especially as factors like atmospheric scattering and latitude change of the light source limit its availability. This spatiotemporal behavior of light which impacts the light capture can be optimized by proper positioning of the PBR – for instance by east–west orientation at latitude ≥35°N or north–south orientation for latitudes ≤35°N [47]. Additionally, synchronizing technologies help (1) automatic movement of the PBR with solar tracking devices and angling the reactor to the trajectory of the light angle [48]; (2) wireless light emitters for uniform light distribution and increase of overall light intensity; and (3) using wavelength shifters (quantum dots and fluorescent or phosphorescent compounds) to convert the spectral range of light with less photosynthetic significance, such as green gap, to a more photosynthetically efficient spectral range [49].

Also, the design of PBR depends on specific needs and the strain type.

- **Production of lipids from microalgae:** presently, a two-phase two-step system is most promising. In the first phase, microalgae are grown in closed PBR under controlled conditions (pH, light, cell density, no contaminations) to achieve higher growth rates in less time. In the second phase (lipid induction), microalgae are grown in open raceway ponds under nutrient deprivation or deliberate environmental stress or alternatives, for instance, for the production of astaxanthin [50].
- **Bio-hydrogen production from microalgae:** hydrogen gas requires a closed PBR, which also helps in microbial cultivation. The diversity of PBR for this purpose – tubular, vertical, column and flat panel – is discussed in detail in chapter 7.
- **Biomass production from microalgae:** although closed and open PBRs are used for biomass production of microalgae, open raceway ponds are frequently utilized [51]. Due to their high sunlight exposure throughout the year, tropical countries could be a strategic region for the large-scale production of algal-based products or biomass [52]. However, recently small-scale PBR have also extensively improved. The so-called high-density cultivation systems enabled a 10-fold increase in cell density by modifying both the dimension and the aeration of the system [53]. With this system, cell densities similar to those reached by *E. coli* cultures are within the reach [54]. An example for a successful application is the production of cyanophycin [55]. For more details, see chapter 7.

A completely new approach is the use of biofilms for the generation of products as outlined in chapters 10 and 11.

1.4.3 Semiartificial systems

While semiartificial systems involving PS2 always face the problem of lability due to the missing repair function, PS1 is much more stable – especially if isolated from thermophilic strains [56, 57]. As even mesophilic complexes are quite stable and also easy to manipulate genetically, PS1 from *Synechocystis* PCC 6803 has been used as a light-powered electron pump as illustrated in Fig. 18.

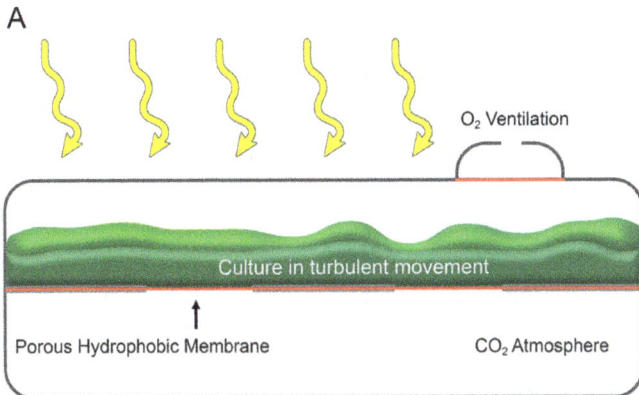

Fig. 17: High-density cultures (HDC system, patent DE 10 2013 015 969 B4), based on a membrane which is impermeable for water and supplies CO_2 from below. This system achieves very high cell densities (up to 30 g DW/L for some strains) without negative effects due to photoinhibition even at very high light intensities (>2000 uE/m^2/s) (R. Steuer, HU Berlin, pers. com.).

In this case, PS1–carbon nanotube conjugates in solution have been shown to transfer photoelectrons from PS1 to the carbon nanotube, generating directed photocurrents [58]. Such a principle may be very useful for an application in artificial photosynthesis, for instance, as a nano-optoelectronic device. Usage of ITO (Indium tin oxide) electrodes provides an opportunity to load high amounts of PS1 and PS2 on electrode surfaces which enables the use of biophotoelectrochemical devices for solar fuel production [59].

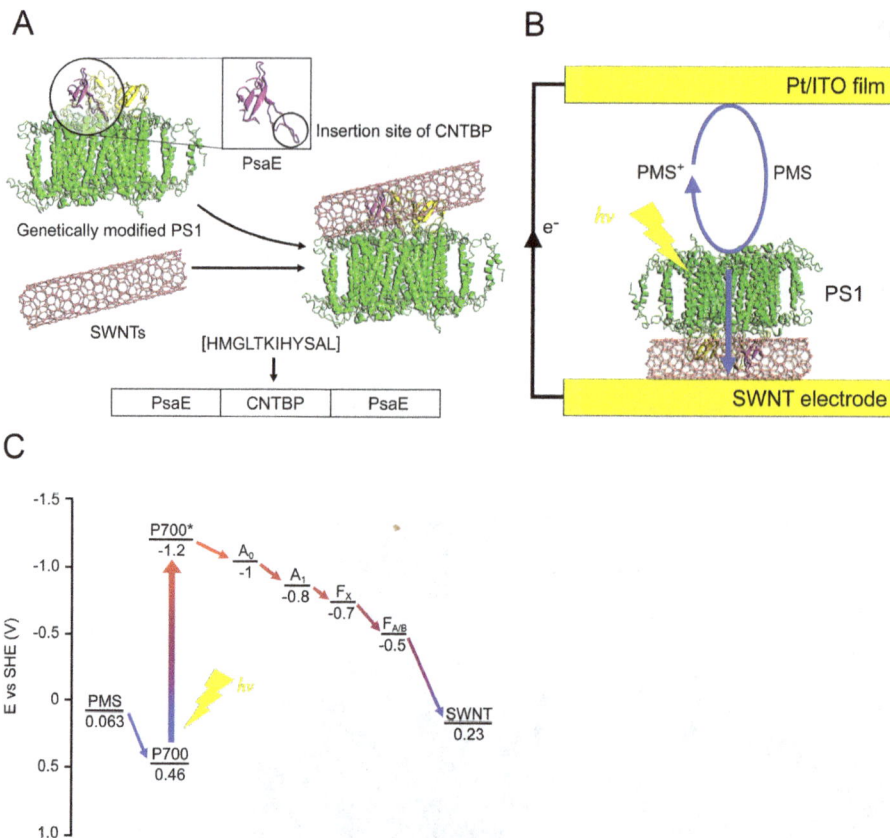

Fig. 18: (A) PS1 complexes with engineered carbon nanotube binding peptide, conjugated on a single-walled carbon nanotube; (B) electron flow from PS1 to the carbon nanotube (SWNT) electrode as part of a device with PMS as a redox mediator, which generates current during photoirradiation; and (C) electron flow diagram representing the redox potential of each component. (Figures acc. to ref. 58 modified).

However, one of the major limitations in PS1-based electrochemical systems is their stability. In recent years, a significant improvement could – for instance – be achieved by encapsulating PS1 in organic microparticles: this retained the 100% activity of PS1 even after 10 days [60]. However, this system is incompatible with PS1 deposition on electrode surfaces. In future, the design of an electrode setup considering the stability will make the system even more efficient for large-scale applications.

Another example is the spatial organization and immobilization of algal/bacterial cell communities in dextran-in-PEG emulsion microdroplets [61]. Enclosing the photosynthetic cells with a shell of bacterial cells undergoing aerobic respiration enables photohydrogen production in daylight under air due to localized oxygen depletion (Fig. 19). This is also a proof of principle for utilizing aqueous two-phase separated droplets as vectors for controlling algal cell organization and photosynthesis in synthetic microspaces. Such photosynthetic microbial microreactors may generally be useful for multiple functionalities of photosynthesis in the future.

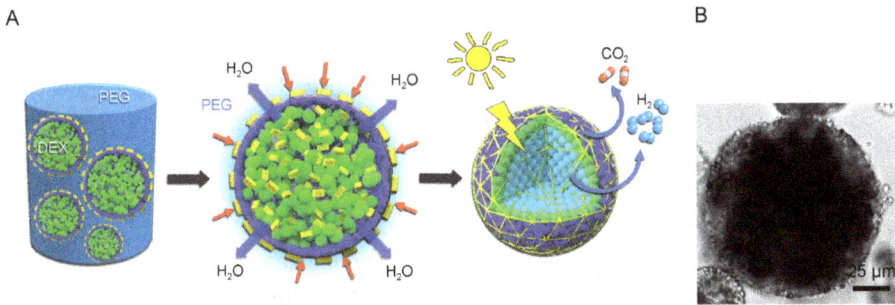

Fig. 19: (A) Multicellular droplet-based living microreactors, based on algal/bacterial hybrid spheroids: droplets consist of mixtures of *Chlorella* cells (inside, green) and PEGylated *E. coli* cells (outside, blue), that is, the photosynthetic algal cells are enclosed by a thin oxygen-depleting layer of respiratory bacterial cells. Hyperosmotic compression results in consolidation and immobilization of the two cellular microniches. Hydrogen production is enhanced by hypoxic photosynthesis in the core, consisting of a closely packed aggregate of algal cells immobilized in a dextran/BSA (yellow rectangles) hydrogel matrix. (B) Bright-field image of a single dextran-in-PEG emulsion droplet with captured algal and surface-adsorbed bacterial cells (Figures from ref. 61, modified).

1.4.4 Modeling

In future, modeling will help considerably in predicting and engineering of photosynthesis-based products [62]. As cyanobacteria are the most simple organisms performing oxygenic photosynthesis, and as thousands of characterized mutants are available – for instance, from *Synechocystis sp.* PCC 6803 – they could be ideal "simple" model organisms. The final aim is to optimize the metabolic engineering of cells to generate products by photosynthesis without compromising vital reactions that are the basis for survival (and reproduction) of the cell. This way, highest efficiency could be achieved without testing many mutants by "trial and error." The involvement of synthetic biology in such approaches can be seen in chapter by Erb, although – on the whole cell level – this is still a future scenario and will require more knowledge about all the fundamental reactions in a cell, cyanobacteria may be the ideal choice due to their simplicity, the possibility of mass cultivation and genetic engineering. Such a modeling will also help optimally distribute the harvested light energy between basic cellular ("housekeeping") and specific biotechnological reactions. Overall, the combination of modeling and synthetic biology will help overcome the natural limitations of photosynthetic cells.

1.5 Conclusions and outlook

Figure 20 summarizes the potential use of photosynthetic electrons for a selection of "solar" products. Strategies for a realization and an outlook on future perspectives of photosynthesis are as follows:

- Semiartificial systems (biophotovoltaic cells) can be extremely useful to develop and optimize new approaches for harnessing light energy with the help of photosynthesis. In comparison with "natural" photosynthesis they lack, however, dynamic adaptations such as the reversible mono-/trimerization of PS1 and the state transitions to optimize light harvesting or the repair function of PS2 (see below). Future will show whether advanced (semi-)artificial photosynthesis-based systems will be able to imitate such dynamics.
- Due to rational, energetic, ecologic and economic reasons electrons, have to come from water, which is available nearly everywhere worldwide. For this process, water splitting at PS2 is mandatory. However, PS2 is the weakest point of the whole photosynthetic ET chain and – due to photooxidative damage – has to be repaired every 20–30 min by a very sophisticated cellular repair system which (up to now) cannot be realized in a semiartificial system. For this reason, for the time being, most or all photosynthesis-based "applied" systems have to be modified/optimized cellular systems, that is, "design cells."

Using photosynthesis for product generation is not primarily a problem of efficiency, but – due to cost-free availability of sunlight – will finally be decided by the price of this product. For this purpose, all parameters have to be considered, including the efficiency of the design cell, costs of PBR generation, its lifetime, its running costs and its disposal. Also the amount of product generated per time is decisive. For this reason, a life cycle assessment of the environmental impacts and cumulative energy demand is mandatory before starting such a project, as is a comparison with competitive procedures [63] (see chapter 2). The use of seawater versus freshwater organisms could also be an important point to evaluate.

- Due to their robustness and versatility, cyanobacteria have recently been shown to be the basis for biological life-support systems during Mars exploration programs [64]. Apparently, they can use N_2 and C from the atmosphere and mineral nutrients from the Martian regolith and also grow under low pressure as demonstrated with *Anabaena* PCC 7938 in a specifically designed low-pressure PBR. Via photosynthesis, they could regenerate atmospheric gases, especially O_2, supporting the growth of other organisms and also feed secondary consumers like *E. coli*. In future, cyanobacteria may play an important role in such bioregenerative life-support systems for spaceflight and planetary outposts.

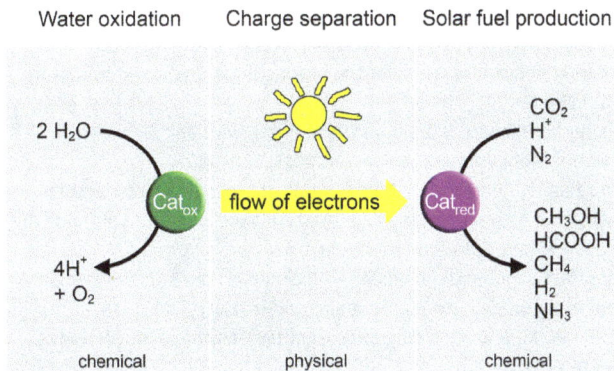

Fig. 20: Summary of the potential use of photosynthetic electrons for "solar" products (selection). Solar energy has an outstanding potential as a cheap, clean and sustainable energy source but it has to be captured and transformed into chemical energy in the form of chemical bonds as in natural photosynthesis. One of the chemical products that can be synthesized by photosynthetic organisms from basic components and yet exhibits huge potential as a sustainable energy carrier is hydrogen gas. Its combustion produces water and allows to use CO_2 as primary source for generating chemicals such as methanol and various acids, which could be used in manufacturing and processing industries [65].

References

[1] Acatech – National Academy of Science and Engineering, German National Academy of Sciences Leopoldina, Union of the German Academies of Sciences and Humanities (Eds.). Artificial Photosynthesis. Munich, 2018, 74.

[2] Kim SK. Microalgae, a biological resource for the future. In: Kim SK, ed. Essentials of Marine Biotechnology ed. Cham Switzerland, Springer, 2019, 197–225.

[3] Kim SK. Marine bioenergy production. In: Kim SK, ed. Essentials of Marine Biotechnology ed. Cham Switzerland, Springer, 2019, 297–341.

[4] Rexroth S, Nowaczyk MM, Rögner M. Cyanobacterial photosynthesis: The light reactions. In: Hallenbeck P, ed. Modern Topics in Phototrophic Prokaryotes ed. Cham Switzerland, Springer, 2017, 163–91.

[5] Ocean color and land vegetation data from SeaWiFS. Reveals the dynamic seasonal pattern of distribution of primary production (Accessed April 4, 2021 at https://oceancolor.gsfc.nasa.gov/SeaWiFS/BACKGROUND/Gallery/biosphere_egg_caption.jpg).

[6] The SeaWiFS Project. NASA/Goddard Space Flight Center and ORBIMAGE. (Accessed April 4 2021 at https://oceancolor.gsfc.nasa.gov/SeaWiFS/BACKGROUND/Gallery/index.html).

[7] Umena Y, Kawakami K, Shen JR, Kamiya N. Crystal structure of oxygen-evolving photosystem II at a resolution of 1.9 Å. Nature 2011, 473, 55–60.

[8] Kurisu G, Zhang H, Smith JL, Cramer WA. Structure of the cytochrome b_6f Complex of oxygenic photosynthesis: Tuning the cavity. Science 2003, 302, 1009–14.

[9] Fromme P, Jordan P, Witt HT, Klukas O, Saenger W, Krauss N. Three-dimensional structure of cyanobacterial photosystem I at 2.5 angstrom resolution. Nature 2001, 411, 909–17.

[10] Baker NR, Harbinson J, Kramer DM. Determining the limitations and regulation of photosynthetic energy transduction in leaves. Plant Cell Environ 2007, 30, 1107–25.

[11] Rögner M. Metabolic engineering of cyanobacteria for the production of hydrogen from water. Biochem Soc Trans 2013, 41, 1254–9.

[12] Kothe T, Schuhmann W, Rögner M, Plumeré N. Semi-artificial photosynthetic Z-scheme for hydrogen production from water. Biohydrogen, De Gruyter, 2015, 189–210.

[13] Kothe T, Pöller S, Zhao F, Fortgang P, Rögner M, Schuhmann W, et al. Engineered electron-transfer chain in photosystem 1 based photocathodes outperforms electron-transfer rates in natural photosynthesis. Chem – A Eur J 2014, 20, 11029–34.

[14] Rexroth S, Wiegand K, Rögner M. Cyanobacterial design cell for the production of hydrogen from water. In: Rögner M, ed. Biohydrogen. Berlin, Germany, De Gruyter, 2015, 1–18.

[15] Badura A, Kothe T, Schuhmann W, Rögner M. Wiring photosynthetic enzymes to electrodes. Energy Environ Sci 2011, 4, 3263–74.

[16] Zabret J, Bohn S, Schuller SK, Arnolds O, Möller M, Meier-Credo J, et al. Structural insights into photosystem II assembly. Nat Plants 2020, 7, 524–38.

[17] Sander J, Nowaczyk M, Buchta J, Dau H, Vass I, Deák Z, et al. Functional characterization and quantification of the alternative PsbA copies in *Thermosynechococcus elongatus* and their role in photoprotection. J Biol Chem 2010, 285, 29851–6.

[18] Hartmann V, Ruff A, Schuhmann W, Rögner M, Nowaczyk MM. Analysis of photosystem II electron transfer with natural PsbA-variants by redox polymer/protein biophotoelectrochemistry. Photosynthetica 2018, 56, 229–35.

[19] Sugiura M, Azami C, Koyama K, Rutherford AW, Rappaport F, Boussac A. Modification of the pheophytin redox potential in *Thermosynechococcus elongatus* Photosystem II with PsbA3 as D1. Biochim Biophys Acta – Bioenerg 2014, 1837, 139–48.

[20] Ohad I, Nevo R, Brumfeld V, Reich Z, Tsur T, Yair M, et al. Inactivation of photosynthetic electron flow during desiccation of desert biological sand crusts and *Microcoleus* sp.-enriched isolates. Photochem Photobiol Sci 2005, 4, 977–82.

[21] Ohad I, Berg A, Berkowicz SM, Kaplan A, Keren N. Photoinactivation of photosystem II: Is there more than one way to skin a cat? Physiol Plant 2011, 142, 79–86.

[22] Chen M, Blankenship RE. Expanding the solar spectrum used by photosynthesis. Trends Plant Sci 2011, 16, 427–31.

[23] Janssen M, Tramper J, Mur LR, Wijffels RH. Enclosed outdoor photobioreactors: Light regime, photosynthetic efficiency, scale-up, and future prospects. Biotechnol Bioeng 2003, 81, 193–210.

[24] Kopka J, Schmidt S, Dethloff F, Pade N, Berendt S, Schottkowski M, et al. Systems analysis of ethanol production in the genetically engineered cyanobacterium *Synechococcus* sp PCC 7002. Biotechnol Biofuels 2017, 10, 1–21.

[25] Savakis PE, Angermayr SA, Hellingwerf KJ. Synthesis of 2,3-butanediol by *Synechocystis* sp. PCC6803 via heterologous expression of a catabolic pathway from lactic acid- and Enterobacteria. Metab Eng 2013, 20, 121–30.

[26] Imashimizu M, Bernát G, Sunamura EI, Broekmans M, Konno H, Isato K, et al. Regulation of F_0F_1-ATPase from *Synechocystis* sp. PCC 6803 by γ and ε subunits is significant for light/dark adaptation. J Biol Chem 2011, 286, 26595–602.

[27] Nowaczyk M, Rexroth S, Rögner M. Biotechnological potential of cyanobacteria. Biotechnology. In: Rögner M, ed. Biohydrogen. Berlin, Germany, De Gruyter, 2015, 141–64.

[28] Wiegand K, Winkler M, Rumpel S, Kannchen D, Rexroth S, Hase T, et al. Rational redesign of the ferredoxin-NADP$^+$-oxido-reductase/ferredoxin-interaction for photosynthesis-dependent H_2-production. Biochim Biophys Acta – Bioenerg 2018, 1859, 253–62.

[29] Moal G, Lagoutte B. Photo-induced electron transfer from photosystem I to NADP+: Characterization and tentative simulation of the in vivo environment. Biochim Biophys Acta – Bioenerg 2012, 1817, 1635–45.

[30] Liberton M, Chrisler WB, Nicora CD, Moore RJ, Smith RD, Koppenaal DW, et al. Phycobilisome truncation causes widespread proteome changes in *Synechocystis* sp. PCC 6803. PLoS One 2017, 12, 1–18.

[31] Bernát G, Waschewski N, Rögner M. Towards efficient hydrogen production: The impact of antenna size and external factors on electron transport dynamics in *Synechocystis* PCC 6803. Photosynth Res 2009, 99, 205–16.

[32] Waschewski N, Bernát G, Rögner M. Engineering photosynthesis for H_2 production from H_2O: Cyanobacteria as design organisms. In: Vertes AA, Editor Nasib Qureshi (Co-Editor), Hideaki Yukawa (Co-Editor), Hans P. Blaschek (Co-Editor). Biomass to Biofuels-Strategies for Global Industries. Chichester, UK, John Wiley & Sons, 2010, 387–401.

[33] Blankenship RE, Tiede DM, Barber J, Brudvig GW, Fleming G, Ghirardi M, et al. Comparing photosynthetic and photovoltaic efficiencies and recognizing the potential for improvement. Science 2011, 332, 805–9.

[34] Dau H, Zaharieva I. Principles, efficiency, and blueprint character of solar-energy conversion in photosynthetic water oxidation. Acc Chem Res 2009, 42, 1861–70.

[35] Appel J, Hueren V, Boehm M, Gutekunst K. Cyanobacterial in vivo solar hydrogen production using a photosystem I–hydrogenase (PsaD-HoxYH) fusion complex. Nat Energy 2020, 5, 458–67.

[36] Kanygin A, Milrad Y, Thummala C, Reifschneider K, Baker P, Marco P, et al. Rewiring photosynthesis: A photosystem I-hydrogenase chimera that makes H_2: In vivo. Energy Environ Sci 2020, 13, 2903–14.

[37] Eilenberg H, Weiner I, Ben-Zvi O, Pundak C, Marmari A, Liran O, et al. The dual effect of a ferredoxin-hydrogenase fusion protein in vivo: Successful divergence of the photosynthetic electron flux towards hydrogen production and elevated oxygen tolerance. Biotechnol Biofuels 2016, 9, 1–10.

[38] Köninger K, Gómez Baraibar Á, Mügge C, Paul CE, Hollmann F, Nowaczyk MM, et al. Recombinant cyanobacteria for the asymmetric reduction of C=C bonds fueled by the biocatalytic oxidation of water. Angew Chemie – Int Ed 2016, 55, 5582–5.

[39] Kannchen D, Zabret J, Oworah-Nkruma R, Dyczmons-Nowaczyk N, Wiegand K, Löbbert P, et al. Remodeling of photosynthetic electron transport in *Synechocystis* sp. PCC 6803 for future hydrogen production from water. Biochim Biophys Acta – Bioenerg 2020, 1861, 148208.

[40] Winkler M, Duan J, Rutz A, Felbek C, Scholtysek L, Lampret O, et al. A safety cap protects hydrogenase from oxygen attack. Nat Commun 2021, 12, 1–10.

[41] Kirst H, Formighieri C, Melis A. Maximizing photosynthetic efficiency and culture productivity in cyanobacteria upon minimizing the phycobilisome light-harvesting antenna size. Biochim Biophys Acta – Bioenerg 2014, 1837, 1653–64.

[42] Hoschek A, Toepel J, Hochkeppel A, Karande R, Bühler B, Schmid A. Light-dependent and aeration-independent gram-scale hydroxylation of cyclohexane to cyclohexanol by CYP450 harboring *Synechocystis* sp. PCC 6803. Biotechnol J 2019, 14, 1–10.

[43] Allahverdiyeva Y, Isojärvi J, Zhang P, Aro EM. Cyanobacterial oxygenic photosynthesis is protected by flavodiiron proteins. Life 2015, 5, 716–43.

[44] Assil-Companioni L, Büchsenschütz HC, Solymosi D, Dyczmons-Nowaczyk NG, Bauer KKF, Wallner S, et al. Engineering of NADPH supply boosts photosynthesis-driven biotransformations. ACS Catal 2020, 10, 11864–77.

[45] Huang Q, Jiang F, Wang L, Yang C. Design of photobioreactors for mass cultivation of photosynthetic organisms. Engineering 2017, 3, 318–29.

[46] Nwoba EG, Parlevliet DA, Laird DW, Alameh K, Moheimani NR. Light management technologies for increasing algal photobioreactor efficiency. Algal Res 2019, 39, 101433.

[47] Sierra E, Acién FG, Fernández JM, García JL, González C, Molina E. Characterization of a flat plate photobioreactor for the production of microalgae. Chem Eng J 2008, 138, 136–47.

[48] Hindersin S, Leupold M, Kerner M, Hanelt D. Irradiance optimization of outdoor microalgal cultures using solar tracked photobioreactors. Bioprocess Biosyst Eng 2013, 36, 345–55.

[49] Delavari Amrei H, Nasernejad B, Ranjbar R, Rastegar S. Spectral shifting of UV-A wavelengths to blue light for enhancing growth rate of cyanobacteria. J Appl Phycol 2014, 26, 1493–500.

[50] Zhang BY, Geng YH, Li ZK, Hu HJ, Li YG. Production of astaxanthin from *Haematococcus* in open pond by two-stage growth one-step process. Aquaculture 2009, 295, 275–81.

[51] Acién FG, Molina E, Reis A, Torzillo G, Zittelli GC, Sepúlveda C et al. Photobioreactors for the production of microalgae. In: Gonzalez-Fernandez C, Muñoz R, ed. Microalgae-Based Biofuels and Bioproducts. Duxford, UK, Elsevier, 2017, 1–44.

[52] Ugwu CU, Aoyagi H, Uchiyama H. Photobioreactors for mass cultivation of algae. Bioresour Technol 2008, 99, 4021–8.

[53] Bähr L, Wüstenberg A, Ehwald R. Two-tier vessel for photoautotrophic high-density cultures. J Appl Phycol 2016, 28, 783–93.

[54] Dienst D, Wichmann J, Mantovani O, Rodrigues JS, Lindberg P. High density cultivation for efficient sesquiterpenoid biosynthesis in *Synechocystis* sp. PCC 6803. Sci Rep 2020, 10, 5932.

[55] Lippi L, Bähr L, Wüstenberg A, Wilde A, Steuer R. Exploring the potential of high-density cultivation of cyanobacteria for the production of cyanophycin. Algal Res 2018, 31, 363–6.

[56] Nguyen K, Bruce BD. Growing green electricity: Progress and strategies for use of Photosystem I for sustainable photovoltaic energy conversion. Biochim Biophys Acta – Bioenerg 2014, 1837, 1553–66.

[57] Badura A, Guschin D, Kothe T, Kopczak MJ, Schuhmann W, Rögner M. Photocurrent generation by photosystem 1 integrated in crosslinked redox hydrogels. Energy Environ Sci 2011, 4, 2435–40.

[58] Nii D, Miyachi M, Shimada Y, Nozawa Y, Ito M, Homma Y, et al. Conjugates between photosystem I and a carbon nanotube for a photoresponse device. Photosynth Res 2017, 133, 155–62.

[59] Bobrowski T, Conzuelo F, Ruff A, Hartmann V, Frank A, Erichsen T, et al. Scalable fabrication of biophotoelectrodes by means of automated airbrush spray-coating. ChemPlusChem 2020, 85, 1396–400.

[60] Cherubin A, Destefanis L, Bovi M, Perozeni F, Bargigia I, De La Cruz Valbuena G, et al. Encapsulation of Photosystem I in organic microparticles increases its photochemical activity and stability for Ex Vivo photocatalysis. ACS Sustain Chem Eng 2019, 7, 10435–44.

[61] Xu Z, Wang S, Zhao C, Li S, Liu X, Wang L, et al. Photosynthetic hydrogen production by droplet-based microbial micro-reactors under aerobic conditions. Nat Commun 2020, 11, 1–10.

[62] Zavřel T, Faizi M, Loureiro C, Poschmann G, Stühler K, Sinetova M, et al. Quantitative insights into the cyanobacterial cell economy. Elife 2019, 8, e42508.

[63] Rosner V. Analysis and assessment of current photobioreactor systems for photobiological hydrogen production. In: Rögner M, ed. Biohydrogen. Berlin, Germany, De Gruyter, 2015, 19–40.

[64] Verseux C, Heinicke C, Ramalho TP, Determann J, Duckhorn M, Smagin M, et al. A low-pressure, N_2/CO_2 atmosphere is suitable for cyanobacterium-based life-support systems on Mars. Front Microbiol 2021, 12, 611798.

[65] El-Khouly ME, El-Mohsnawy E, Fukuzumi S. Solar energy conversion: From natural to artificial photosynthesis. J Photochem Photobiol C Photochem Rev 2017, 31, 36–83.

Part 1: **Cell design/metabolic engineering**

Xufeng Liu, Hao Xie, Stamatina Roussou, Rui Miao, Peter Lindblad

2 Engineering cyanobacteria for photosynthetic butanol production

Abstract: Cyanobacteria are photoautotrophic microorganisms that can be engineered to convert CO_2 and water into fuels and chemicals via photosynthesis using solar energy in direct processes. Based on knowledge and progress in fermentative heterotrophic biobutanol production, cyanobacteria have been engineered to produce photosynthetic butanol from sunlight, water and CO_2. This chapter discusses the present status of engineering cyanobacteria for photosynthetic isobutanol and 1-butanol production. Special focus is on recent advances in introducing enzymes and pathways, redirecting carbon toward the product, importance of five regions in the genetic constructs and optimization of the cultivation system. Also included are recent contributions addressing butanol tolerance, recovery of the produced photosynthetic butanol, life cycle assessment on environmental impacts, energy demand of photosynthetic butanol production and public acceptance of genetically engineered algae/cyanobacteria for biofuel production.

2.1 Goal

Cyanobacteria, prokaryotic microorganisms with basically the same oxygenic photosynthetic system as higher plants, are excellent green cell factories for a sustainable generation of renewable chemicals and fuels from solar energy and carbon dioxide. The goal of the work described in this chapter is to develop cyanobacteria as biocatalysts for the truly carbon dioxide neutral production of photosynthetic butanol.

2.2 Basic background

Cyanobacteria, prokaryotic microorganisms with oxygenic photosynthesis, evolved on the Earth billions of years ago. They convert solar energy, CO_2 and water into chemical energy while releasing O_2 into the atmosphere. In addition, some strains are able to convert N_2 into ammonia. Compared to other oxygenic photosynthetic

Acknowledgments: The authors acknowledge the financial support by the European Union Horizon 2020 Framework Program under the grant agreement number 640720 (*photofuel*), the Kamprad Family Foundation for Entrepreneurship, Research and Charity (*photosynthetic butanol*), the NordForsk NCoE program "NordAqua" (project number 82845) and the Swedish Energy Agency (*cyanofuels*, project number P46607-1).

https://doi.org/10.1515/9783110716979-002

organisms, cyanobacteria possess the highest solar energy–capturing efficiency with corresponding adequate CO_2 concentrating mechanisms and CO_2 fixation. With their modest nutrient requirements and ability to grow on non-arable land and thereby not compete with, for example, food production, cyanobacteria have emerged as potential green cell factories for sustainable generation of renewable chemicals and fuels. The rapid progress in synthetic biology and genetic engineering have made it possible to custom design microorganisms including cyanobacteria. This may include deletion of native and addition of new capacities as well as modified and optimized metabolic flow toward desired product(s). Until today, cyanobacteria have been engineered, as proof of concept, to synthesize numerous non-native products. In parallel, fast-growing cyanobacteria are not only being discovered but also genetically engineered, which may lead to higher yields and rates – for instance, the strain *Synechococcus elongatus* UTEX 2973 [1]. This chapter presents attempts and progress to develop cyanobacteria as biocatalysts for truly carbon dioxide neutral production of photosynthetic butanol.

Butanol is a four-carbon alcohol (C_4H_9OH) occurring in four structural isoforms: 1-butanol, 2-butanol, isobutanol and *tert*-butanol. It is mainly used as a solvent, as an intermediate in chemical synthesis and as a fuel. The global market for this important bulk chemical and excellent blend-in fuel is very large, projected to reach USD 5.6 billion by 2022 [2]. Butanol is mainly produced from fossil resources. In addition, there are biological routes for renewable butanol production, mainly to produce 1-butanol and isobutanol. Existing bio-based 1-butanol is produced from starch, sugar or cellulose such as wheat, beet, corn and wood. Products of these fermentation processes additionally include acetone and ethanol. It is the understanding that the existing, and further developed, bio-butanol market will grow significantly in the near future. Photosynthetic butanol directly from solar energy and carbon dioxide will be the most sustainable and carbon dioxide neutral production of this important bulk chemical and fuel.

2.2.1 Heterotrophic butanol production

Microbial production of isobutanol has been studied in food fermentations and alcoholic beverages since the 1970s [3]. More recently, bio-produced isobutanol was examined as a biofuel. It was observed that only small amounts of isobutanol can be produced naturally in *Saccharomyces cerevisiae* through the degradation of the amino acid valine [4, 5]. However, naturally there are no bacteria with a fermentative capacity to produce isobutanol. Instead, there is an artificial isobutanol synthesis pathway: the 2-keto acid pathway, which was established by a metabolic engineering approach based on the native valine synthesis pathway. Since valine synthesis exists widely in microorganisms, the 2-keto acid pathway has been employed for isobutanol production in various microorganisms, including *Escherichia coli* [6, 7] and *Corynebacterium glutamicum* [8]. Recently, it was found that an isobutanol synthesis pathway exists in *Klebsiella pneumoniae* [9]. However, this pathway is normally dormant in wild-type cells. Considerable isobutanol may be produced in α-acetolactate decarboxylase knock-out strain at microaerobic and neutral pH conditions.

The 2-keto acid pathway shares the precursor 2-ketoisovalerate with the valine synthetic pathway. 2-Ketoisovalerate is first decarboxylated to isobutyraldehyde by Kivd, a thiamine diphosphate-dependent 2-keto acid decarboxylase from *Lactococcus lactis*. Generated isobutyraldehyde is reduced by ADH, an alcohol dehydrogenase, to form isobutanol. Apart from the last two reaction steps for isobutanol synthesis, there are three additional reaction steps involved in the 2-keto acid pathway, starting from the central metabolite pyruvate: AlsS (α-acetolactate synthase from *Bacillus subtilis*) converts two molecules of pyruvate to 2-acetolactate. The 2-acetolactate is then reduced to 2,3-dihydroxyisovalerate by IlvC (acetohydroxy acid isomeroreductase from *E. coli*), which is further converted to 2-ketoisovalerate by IlvD (dihydroxyacid dehydratase from *E. coli*).

In bacterial metabolism, 1-butanol can be natively produced by the members of the genus *Clostridium*, particularly *Clostridium acetobutylicum*, during the acetone–butanol–ethanol fermentation [10–12]. This microbial 1-butanol biosynthetic pathway was determined and named clostridial pathway, which has already been used to develop a direct 1-butanol-producing process in *E. coli* and other microorganisms. Subsequently, three distinct artificial 1-butanol biosynthetic pathways were constructed in *E. coli*, consisting of the 2-keto acid pathway [6], the reversed β-oxidation pathway [13] and the acyl carrier protein (ACP)-dependent pathway [14].

In the clostridial pathway, the biosynthesis of 1-butanol is accomplished by six reactions of reductions and dehydrations, using acetyl-CoA as the substrate and involving multiple acyl-CoA thioester intermediates [10, 11]. First, two acetyl-CoA initiate the pathway, forming acetoacetyl-CoA catalyzed by acetyl-CoA acetyltransferase. In the second step of the pathway, acetoacetyl-CoA is reduced to the corresponding 3-hydroxybutyryl-CoA by 3-hydroxybutyryl-CoA dehydrogenase. Then, crotonase catalyzes the dehydration of 3-hydroxybutyryl-CoA to crotonyl-CoA in the third step. In the fourth step, crotonyl-CoA hydrogenation is catalyzed by butyryl-CoA dehydrogenase to synthesize butyryl-CoA. The reduction from butyryl-CoA to butyraldehyde is then catalyzed by aldehyde dehydrogenase in the fifth step. ADH works as the final step by reducing butyraldehyde into 1-butanol. In *C. acetobutylicum*, the first four steps are catalyzed by Thl, Hbd, Crt and Bcd-EtfAB, respectively, while the last two steps were sequentially catalyzed by the bifunctional enzyme AdhE2. The clostridial pathway was first implemented in *E. coli*, resulting in 1.2 g L^{-1} of 1-butanol production [10]. To facilitate the targeted 1-butanol biosynthesis in *E. coli*, Thl was replaced with a homologous enzyme AtoB from *E. coli*, leading to a marked improvement in 1-butanol productivity [11]. More importantly, a *trans*-enoyl-CoA reductase (Ter) was demonstrated as an ideal enzyme to supply crotonyl-CoA reduction, which is a bottleneck of clostridial pathway [12]. Bcd-EtfAB catalyzes a reversible reaction in this bottleneck step, utilizing both NADH and reduced ferredoxin as sources of reducing power, while Ter can catalyze an irreversible step and utilize NADH as the direct reducing equivalent, acting as driving forces to 1-butanol formation. Indeed, when Bcd-EtfAB was substituted with Ter from *Treponema denticola*, a breakthrough in product titer to 30 g L^{-1} 1-butanol was achieved. This work demonstrates the significant potential of clostridial pathway for economically feasible production of 1-butanol. The optimized clostridial pathway including the enzyme combination AtoB-Hbd-Crt-Ter-AdhE2 is widely used for 1-butanol production in heterotrophic microorganisms, for example, *S. cerevisiae*, *R. eutropha*, *Methylobacterium extorquens* and even *Clostridium* species.

Besides the clostridial pathway, the artificial biosynthetic pathways have also attracted much interest in developing heterotrophic microorganisms for 1-butanol production. Atsumi et al. demonstrated that both isobutanol and 1-butanol can be produced via implementation of the 2-keto acid pathway in *E. coli* [6]. The Kivd in the 2-keto acid pathway is not limited to decarboxylate 2-ketoisovalerate for isobutanol formation but also can decarboxylate 2-ketovalerate to butyraldehyde, following with an ADH to reduce butyraldehyde to 1-butanol. Although 2-ketoisovalerate and 2-ketovalerate are both native intermediates in amino acid metabolism, 1-butanol production was usually lower than isobutanol production in the 2-keto acid pathway. In total, 1-butanol is formed through ten steps from two native substrates, acetyl-CoA and pyruvate, in the 2-keto acid pathway [15]. In this case, 1-butanol production was also much lower than the highest 1-butanol titer (30 g L^{-1}) reported from the clostridial pathway in *E. coli*. Even so, the 2-keto acid pathway was successfully applied to *S. cerevisiae* for 1-butanol biosynthesis.

The second artificial pathway for 1-butanol biosynthesis in *E. coli* is called reversed β-oxidation pathway [13]. Although both CoA-dependent pathways, the clostridial pathway and first cycle of reversed β-oxidation, are basically identical, they differ in the conversion of acetoacetyl-CoA to 3-hydroxybutyryl-CoA, then to form crotonyl-CoA. While two individual enzymes are required for this conversion in the clostridial pathway, FadB (hydroxyacyl-CoA dehydrogenase/enoyl-CoA hydratase from *E. coli*) catalyzes the sequential two reactions in the first cycle of reversed β-oxidation. Engineering of the reversed β-oxidation pathway resulted in 1-butanol at a titer of 14.5 g L^{-1}, approaching those achieved in the clostridial pathway. Furthermore, some 1-butanol-forming enzymes in the reversed β-oxidation pathway can act on longer acyl-CoA intermediates, making the pathway extended for biosynthesis of other higher alcohols or fatty acids in two or more cycles of reversed β-oxidation. The reversed β-oxidation pathway explored in *E. coli* also guided the biosynthesis of 1-butanol or other acyl-CoA-dependent biochemicals in *S. cerevisiae*.

Besides the two CoA-dependent pathways mentioned above, an ACP-dependent pathway was artificially constructed for 1-butanol biosynthesis [14]. In this pathway, acyl-ACP thioesterase from *Bacteroides fragilis* is applied to release butyryl-ACP from the first cycle of native fatty acid biosynthesis. The generated butyrate is reduced to butyraldehyde by a carboxylic acid reductase from *Mycobacterium marinum*, followed by an ADH to convert butyraldehyde to 1-butanol. The ACP-dependent pathway requires in total nine steps to form 1-butanol from acetyl-CoA which is known as the precursor of native fatty acid biosynthesis. The 1-butanol titer achieved in the engineered ACP-dependent pathway was 300 mg L^{-1}. However, the ACP-dependent 1-butanol biosynthetic pathway has only been studied in *E. coli*.

2.2.2 Butanol production by cyanobacteria

Cyanobacteria do not produce butanol naturally. However, a combination of synthetic biology and system biology strategies has been employed to enable photosynthetic butanol production in various cyanobacterial strains (Tab. 1).

Atsumi et al. pioneered the engineering of a unicellular cyanobacterium, *Synechococcus elongatus* PCC 7942 (*Synechococcus*), to synthesize isobutanol by overexpressing the five enzymes from 2-keto acid pathway, involving AlsS, IlvC, IlvD, Kivd and YqhD (an ADH from *E. coli*) [16]. This was the first proof-of-concept study to develop a cyanobacterium as a biocatalyst to produce isobutanol directly from solar energy and CO_2. In the same study, a higher productivity of isobutanol was observed when overexpressing ribulose 1,5-bisphosphate carboxylase/oxygenase (RuBisCO), the primary enzyme for carbon fixation in photosynthetic organisms. By knocking out *glgC*, encoding a glucose-1-phosphate adenylyltransferase to initiate glycogen synthesis in *Synechococcus*, 2.5 times more isobutanol was produced through the well-established 2-keto acid pathway in *Synechococcus* [17]. Additionally, Kivd and AdhA (ADH from *L. lactis*) from the 2-keto acid pathway were introduced into *Synechocystis* PCC 6803 (*Synechocystis*), another widely studied model cyanobacterial strain for both basic and applied research. This resulted in an engineered strain that is able to accumulate 298 mg L^{-1} isobutanol titer under mixotrophic conditions [18]. Based on further isotopomer analysis, a significantly decreased utilization of glucose was observed in engineered cells compared to wild-type cells.

For 1-butanol production, the four biosynthetic pathways discussed above for heterotrophic production opened the way for engineering cyanobacteria as hosts to produce photosynthetic 1-butanol. The clostridial pathway is most widely used in various microorganisms for 1-butanol production, with the highest 1-butanol titer and among the fewest catalytic steps from acetyl-CoA. Therefore, the clostridial pathway may be the optimal candidate for 1-butanol production in cyanobacteria, which explains why all published cyanobacterial 1-butanol biosynthesis research is using the clostridial pathway (Tab. 1).

In the first report, the enzyme combination AtoB-Hbd-Crt-Ter-AdhE2 in the optimized clostridial pathway was transferred into *Synechococcus* [23]. As a result, 14.5 mg L^{-1} 1-butanol was produced under dark and anaerobic conditions, while only trace amounts of 1-butanol were produced under aerobic condition. In cyanobacteria

Tab. 1: Overview of cyanobacteria engineered to produce butanol.

Cyanobacteria strain	Medium/carbon source	Cultivation mode/cultivation time[a]	Titer	Pathway	Reference
Isobutanol					
Synechococcus elongatus PCC 7942	BG11 with 10 mg L^{-1} thiamine addition/ NaHCO$_3$	Shake flask, constant lighting, fed-batch, photoautotrophic condition, 30 °C/6 days	450 mg L^{-1}	2-Keto acid pathway	[16]
Synechococcus elongatus PCC 7942	BG11/NaHCO$_3$	Screw-cap flask, constant lighting, fed-batch, photoautotrophic condition, 30 °C/12 days	297 mg L^{-1}	2-Keto acid pathway	[19]
Synechocystis PCC 6803	BG11/NaHCO$_3$	Shake rubber-cap flask with in situ solvent trap, constant lighting, fed-batch, photo-autotrophic condition, 30 °C/21 days	240 mg L^{-1}	2-Keto acid pathway	[18]
Synechococcus elongatus PCC 7942	BG11 with 10 mg/L thiamine addition/ NaHCO$_3$	Pyrex shaker flasks, constant lighting, fed-batch, photoautotrophic condition, 30 °C/8 days	550 mg L^{-1}	2-Keto acid pathway	[17]
Synechocystis PCC 6803	BG11/NaHCO$_3$	Shake plug-sealed flask, constant lighting, fed-batch, photoautotrophic condition, 30 °C/6 days	3 mg L^{-1} OD$_{750}$$^{-1 b}$	2-Keto acid pathway	[20]
Synechocystis PCC 6803	BG11/NaHCO$_3$	Shake plug-sealed flask, constant lighting, fed-batch, photoautotrophic condition, 30 °C/8 days	59.6 mg L^{-1}	2-Keto acid pathway	[21]
Synechocystis PCC 6803	BG11/NaHCO$_3$	Shake plug-sealed flask, constant lighting, fed-batch, photoautotrophic condition, 30 °C/8 days	911 mg L^{-1}	2-Keto acid pathway	[22]

1-Butanol

Organism	Medium	Conditions	Titer	Pathway	Ref.
Synechococcus elongatus PCC 7942	BG11/NaHCO$_3$	Dark test tube, batch, anaerobic condition, 30 °C/7 days	14.5 mg L^{-1}	Modified clostridial pathway	[23]
Synechococcus elongatus PCC 7942	BG11/NaHCO$_3$	Shake screw-cap flask, constant lighting, fed-batch, photoautotrophic condition, 30 °C/18 days	30 mg L^{-1}	Modified clostridial pathway	[24]
Synechococcus elongatus PCC 7942	BG11/NaHCO$_3$	Shake screw-cap flask, constant lighting, fed-batch, photoautotrophic condition, 30 °C/12 days	404 mg L^{-1}	Modified clostridial pathway	[25]
Synechocystis PCC 6803	BG11/NaHCO$_3$	Parafilm-sealed 24-well plates, constant lighting, batch, photoautotrophic condition, 28 °C/8 days	36 mg L^{-1}	Modified clostridial pathway	[26]
Synechocystis PCC 6803	BG11/CO$_2$	Multicultivator (Photon Systems Instruments), constant lighting, fed-batch, photoautotrophic condition, 30 °C/143 h (5.96 days)	120 mg L^{-1}	Modified clostridial pathway	[27]
Synechococcus elongatus PCC 7942	BG11/NaHCO$_3$	Shake screw-cap flask, constant lighting, fed-batch, photoautotrophic condition, 30 °C/6 days	418 mg L^{-1}	Modified clostridial pathway	[28]
Anabaena PCC 7120	BG11$_0$ (nitrogen-free medium)/CO$_2$	Culture under aeration, constant lighting, fed-batch, photoautotrophic condition, 30 °C/10 days	2.5 mg L^{-1}	Modified clostridial pathway	[29]
Synechocystis PCC 6803	BG11/NaHCO$_3$	Shake plug-sealed flask, constant lighting, fed-batch, photoautotrophic condition, 30 °C/28 days	4.8 g L^{-1}	Modified clostridial pathway	[30]

[a]The cultivation time refers to the time when the highest butanol production was reached.
[b]Titer (in mg L^{-1}) not reported.

under photosynthetic conditions, due to the prevalence of intracellular NADPH, ATP and oxygen, which are directly evolved through photosynthesis (Fig. 1), 1-butanol biosynthesis could be facilitated by substituting previous enzymes to the enzymes with cofactor preference to NADPH, ATP as a driving force and/or oxygen insensitivity [24]. Based on these efficient strategies, multiple modifications were done. First, AtoB, catalyzing the reversible condensation of two acetyl-CoA molecules, was replaced with NphT7 (acetoacetyl-CoA synthase from *Streptomyces* CL190). The NphT7-mediated irreversible condensation of acetyl-CoA with malonyl-CoA could couple with the irreversible acetyl-CoA carboxylation driven by an ATP force, establishing several driving forces for 1-butanol biosynthesis (Fig. 1). Then the two steps catalyzed by Hbd and Crt were substituted by the route catalyzed by PhaB (acetoacetyl-CoA reductase from *R. eutropha*) and PhaJ (enoyl-CoA hydratase from *Aeromonas caviae*), changing the cofactor requirements from NADH to NADPH. Meanwhile, the NADH-dependent bifunctional AdhE2 was replaced with Bldh (butyraldehyde dehydrogenase from *Clostridium saccharoperbutylacetonicum* NI-4) and YqhD, both using NADPH as a reducing power. These modifications enabled a functional 1-butanol biosynthesis under photoautotrophic conditions, as evidenced by an observed 1-butanol accumulation of 30 mg L^{-1} in 18 days. Later, the same research group substituted the oxygen-sensitive Bldh with an oxygen-tolerant PduP (CoA-acylating aldehyde dehydrogenase from *Salmonella enterica*), making an innovative and quantitatively important step in improving the 1-butanol production to an in-flask 1-butanol titer of 317 mg L^{-1} with a cumulative 1-butanol titer of 404 mg L^{-1} after 12 days in a fed-batch cultivation of the final *Synechococcus* strain [25]. The further optimized 1-butanol biosynthetic pathway with enzyme combination NphT7-PhaB-PhaJ-Ter-PduP-YqhD in cyanobacteria was completely changed from the original clostridial pathway in *C. acetobutylicum*.

Building on the optimized 1-butanol biosynthetic pathway in *Synechococcus*, the ribosome-binding site (RBS) of the PduP was optimized and acetyl-CoA carboxylase from *Yarrowia lipolytica* overexpressed [28], resulting in an improved 1-butanol titer to 418 mg L^{-1} in 6 days. Additional studies using engineered *Synechocystis* and *Anabaena* PCC 7120 showed various 1-butanol titers [26, 27, 29] (Tab. 1). Moreover, introduction of a phosphoketolase (PK) has the potential to facilitate 1-butanol biosynthesis with an increased titer in *Synechocystis* [26].

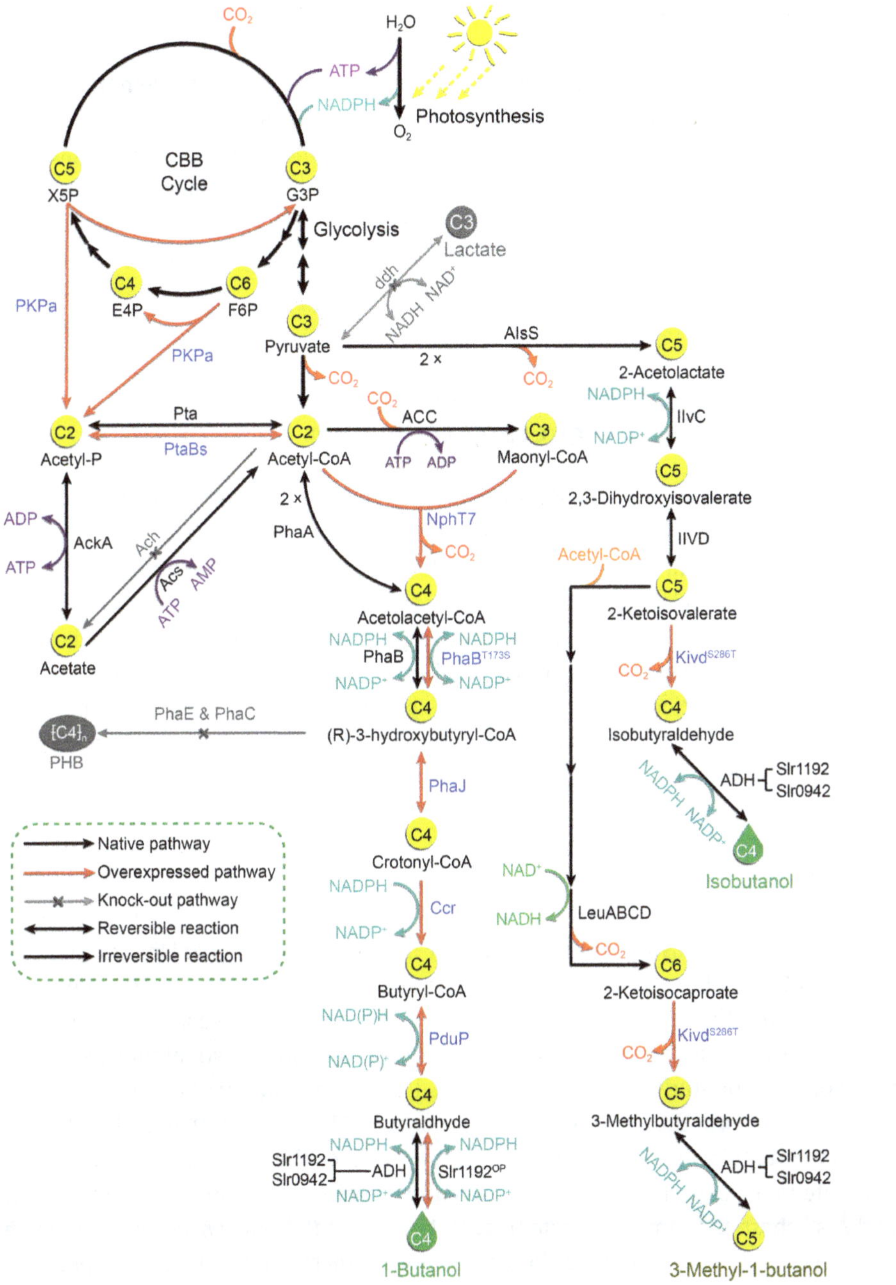

Fig. 1: Schematic diagram of optimized isobutanol and 1-butanol biosynthetic pathway from CO_2 in *Synechocystis* PCC 6803. Heterologous, endogenous and inactivated endogenous enzymes are shown in blue, black and gray font, respectively. Metabolite abbreviations: G3P, glyceraldehyde-3-phosphate; F6P, fructose-6-phosphate; E4P, erythrose-4-phosphate; X5P, xylulose-5-phosphate; Acetyl-P, acetyl-phosphate; PHB, poly-3-hydroxybutyrate. Modified from [30].

2.3 Methods involved

Compared to the systematic metabolic engineering developed in well-characterized heterotrophic model microorganisms, for example, *E. coli* and *S. cerevisiae*, there is a lack of comprehensive engineering strategies for cyanobacteria. Here, we present and discuss recent studies with the highest photosynthetic isobutanol and 1-butanol production in cyanobacteria. They are combined with the introduction of techniques and methods which are typically used when engineering cyanobacteria for specific purposes.

2.3.1 Engineering the introduced enzymes and pathways, combined with redirecting carbon flux

In view of the studies in the field of metabolic engineering, a typical step-wise path is very effectively used to steer the engineering of hosts for producing biochemicals. This path involves three different strategies, which are introducing and optimizing product-forming enzymes, deleting competing pathways and improving precursor supply. The first strategy aims to construct a biosynthetic pathway and enhance product genera-tion, while the other two strategies can redirect more carbon flux to the biosynthetic pathway of desired biochemicals. Thus, this step-wise path is the most straightforward and efficient way for targeted isobutanol and 1-butanol production in a selected cyano-bacterium, for example, *Synechocystis*.

2.3.1.1 Photosynthetic isobutanol

The 2-keto acid pathway has been applied in various microorganisms to produce isobutanol as well as other branched-chain higher alcohols. After introducing the single enzyme Kivd from *L. lactis* either on a self-replicating plasmid or in the ge-nome, *Synechocystis* cells attained the capacity to produce isobutanol [20]. This proof-of-concept study clearly demonstrated that *Synechocystis* has the native ca-pacities to perform the same reactions as enzymes AlsS, IlvC, IlvD and ADH (Fig. 1). Furthermore, D-lactate dehydrogenase (Ddh) catalyzes the formation of lactate from pyruvate in *Synechocystis*, a competing pathway with isobutanol synthesis, as Ddh and AlsS share the same substrate (Fig. 1). Deleting *ddh* from *Synechocystis* genome will direct more carbon toward the isobutanol synthesis pathway. After expressing KivD, *Synechocystis* cells will not only produce isobutanol but also biosynthesize 3-methyl-1-butanol [21]. The reason is that Kivd utilizes two different metabolites as substrates: a smaller substrate, 2-ketoisovalerate (KIV), results in isobutanol produc-tion; and a larger substrate, 2-ketoisocaproate (KIC), results in 3-methyl-1-butanol pro-duction (Fig. 1). Due to the difference in size of these two substrates, using protein engineering approach to decrease the size of the substrate-binding pocket of Kivd could be a strategy to achieve a desired substrate preferential shift for enhanced isobutanol production.

Protein engineering is an efficient approach for generating a desired enzyme which has improved activity, increased stability and preferable substrate specificity [31]. Two methods are generally used: directed evolution and rational design. For directed evolution, a large mutant library needs to be generated via random mutagenesis, followed by a high-throughput screening based on different requirements on the final product. It is an optimal approach for engineering a protein both with and without available crystal structure. It also expands the possibility of protein engineering since computational prediction is not capable to include all the effects caused by amino acid changes. For rational design, prior information on protein 3D structure is an essential requirement. More available knowledge about biophysical and biochemical characteristics of the protein provides more precise prediction on the design. Based on the information in hand, specific amino acid substitution, insertion or deletion can be done via an inexpensive and efficient method: Site-directed mutagenesis. However, the main drawback of rational design is the uncertainty on prediction accuracy since it is generally difficult to predict the solubility and folding motifs of proteins. Directed evolution and rational design are not mutually exclusive but rather complementary. Therefore, it is common to employ both methods to have more comprehensive engineering design. Protein engineering has become an important tool in metabolic engineering studies [32–35]. An engineered enzyme may have a higher catalytic efficiency and/or substrate specificity for the targeted metabolic pathway, resulting in enhanced desired products from living cells. Kivd engineering had been demonstrated for non-natural long-chain alcohol production in *E. coli* via enlarging the substrate-binding pocket [35]. Comparatively speaking, it is more challenging to decrease the size of substrate-binding pocket since using larger amino acid substitution at the active site may cause serious conformational collapses.

The rational design methodology was employed to engineer Kivd [21]. In this study, the 3D prediction on substrate-binding site using Discovery Studio Visualizer was performed and 11 different Kivd variants were generated using site-directed mutagenesis. Interestingly, even when only one amino acid was changed in each variant, protein expression level, total activity and substrate preference were affected strongly and differently (Fig. 2). Kivd variant S286T showed approximately double the amount of soluble protein expression compared to variant V461I, but total products titer from both variants were the same. In the case of Kivd variant S286Y, there was a more significant decoupling of expression level and products titer, where more protein did not result in more products compared to Kivd variants V461I and S286T. A combination of the two best mutants was also generated with an expectation of accumulative effect. Unfortunately, the soluble expression level of this Kivd variant, S286T&V461I, was extremely poor, resulted in a low isobutanol production titer, and even the substrate preference shifted successfully toward the smaller substrate KIV. Either this

Kivd could not be expressed in a high level or the solubility of the protein was affected by the combined amino acid substitutions. Even so, the *Synechocystis* strain overexpressing Kivd[S286T] (Fig. 3A) showed the highest isobutanol production. Several further investigations and improvements can be carried out in future studies. For example, different Kivd variants can be purified and analyzed in vitro to understand the correlation between their expression level and activities. The best variant even can be crystallized for a better knowledge on how the amino substitution affects the total activity and substrate preference. Finally, directed evolution can be employed for future optimization of the enzyme.

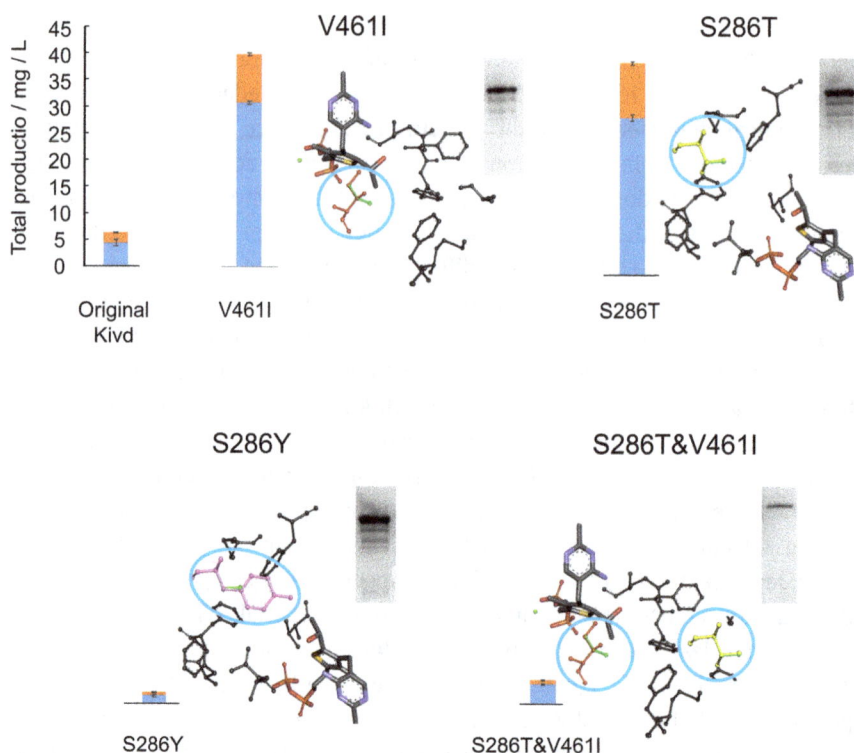

Fig. 2: Examples of predicted substrate-binding pocket of different engineered Kivd variants. Changes highlighted in blue circles, with original residues of each change indicated in green, covered by the substitutes in red, yellow and pink, respectively. Bar graph shows total titer of isobutanol (blue) and 3-methyl-1-butanol (orange) from *Synechocystis* strains expressing different Kivd variants. V461I: valine at position 461 replaced with isoleucine; S286T: serine at position 286 replaced with threonine; S286Y: serine at position 286 replaced with tyrosine; SV: S286T&V461I, combination of S286T and V461I. Western immunoblots show the expression level of each Kivd variant in *Synechocystis*. Modified from [21].

2.3.1.2 Photosynthetic 1-butanol

A clostridial pathway has been employed for 1-butanol production in *Synechocystis* [30]. After screening of various 1-butanol-forming enzymes, the best performing enzymes were selected and introduced to form an optimized clostridial pathway (Fig. 1). Both YqhD and a native ADH (Slr1192), catalyzing the last step of clostridial pathway, use NADPH as a cofactor which may be preferred by cyanobacteria. Despite the previously demonstrated important role of YqhD for clostridial pathway in cyanobacteria [24], overexpressing Slr1192 in *Synechocystis* showed a higher 1-butanol titer compared to when overexpressing YqhD. For the step from crotonyl-CoA to butyryl-CoA, previous reports of the driving forces to 1-butanol biosynthesis by irreversible activity of Ter made the Ter a widely used enzyme [12]. A Ccr from *Streptomyces collinus* can also irreversibly catalyze the operation of crotonyl-CoA reduction. Considering that Ccr utilize NADPH, Ccr may be favored over Ter which prefer NADH. As a result, substitution of Ter to Ccr led to an improvement of 1-butanol titer. For the step reducing acetoacetyl-CoA to (*R*)-3-hydroxybutyryl-CoA, three site-directed mutant PhaBs based on previous protein engineering research were examined. PhaB^{T173S}, one of the PhaB variants, has shown improved recognition of acetoacetyl-CoA and enhanced activity compared with wild-type enzyme in kinetic analysis study. Consequently, the mutant strains carrying PhaB^{T173S} did exhibit improved 1-butanol accumulation, compared to strains carrying original PhaB, making PhaB^{T173S} an ideal enzyme to use. After choosing the enzymes with better performance, the in-flask 1-butanol titer reached 572 mg L^{-1} in 8 days [30]. The further upgraded enzyme combination NphT7-PhaB^{T173S}-PhaJ-Ccr-PduP-Slr1192 in *Synechocystis* was completely changed from the commonly used enzyme combination AtoB-Hbd-Crt-Ter-AdhE2 in heterotrophic microorganisms. Half of the enzymes of the six-step 1-butanol biosynthetic pathway were modified compared to the optimized enzyme combination NphT7-PhaB-PhaJ-Ter-PduP-YqhD in *Synechococcus*. The final optimized 1-butanol biosynthetic pathway requires four NADPH per butanol produced (Fig. 1), as PduP proved to use either NADPH or NADH as a reducing power [36, 37].

After optimizing the 1-butanol biosynthetic pathway from acetyl-CoA to 1-butanol, deleting the competing pathways was performed to redirect the carbon flux to the 1-butanol pathway [30]. PHB accumulation, resulting from the irreversible conversion of intermediate (*R*)-3-hydroxybutyryl-CoA catalyzed by PHB synthase (PhaE and PhaC) in *Synechocystis*, directly competes with 1-butanol biosynthesis (Fig. 1). Hence, *phaEC* is a good choice of knock-out/integration site. We then focused on the acetate metabolism pathways. In the "triangular" acetate metabolism in *Synechocystis* (Fig. 1), the first pathway generates acetyl-P as an intermediate, catalyzed by phosphotransacetylase (Pta) and acetate kinase (AckA) successively in two reversible steps with ATP formed in the second step. The second pathway is irreversibly catalyzed by acetyl-CoA hydrolase (Ach) to form acetate directly from acetyl-CoA. The reverse reaction is catalyzed by acetyl-CoA synthetase (Acs), which produces acetyl-CoA directly from acetate, while consuming ATP. Therefore, the two potential competing pathways, catalyzed by the Ach and Pta, were intended to be deleted. As presented in our study, single knockout of *ach* or *pta* resulted in higher 1-butanol production [30]. Since previous results in *Synechocystis* suggested a PK pathway to increase the overall carbon flux from the Calvin cycle to acetyl-CoA [26], PK was installed to provide more precursor for 1-butanol production [30] (Fig. 1). Due to the apparent difference between the PK homologs, we systematically evaluated nine PK candidates from different sources where PKs were experimentally verified. A PK from *Pseudomonas aeruginosa* (PKPa) resulted in the highest 1-butanol production and may be relevant for biochemical production requiring acetyl-CoA as a substrate. Meanwhile, AckA- and Acs-catalyzed route (AckA–Acs route) and Pta-catalyzed route (Pta route) may further increase the carbon flux from acetyl-phosphate to acetyl-CoA when overexpressing PK (Fig. 1). Therefore, a Pta from *B. subtilis* (PtaBs) was overexpressed for precursor boosting. Meanwhile, a neutral site *slr0168* and the two competing pathway genes *phaEC* and *ach* were selected as three integration sites for overexpressing the complete 1-butanol pathway genes (Fig. 3B). There was a substantial improvement of in-flask 1-butanol titer to 836 mg L^{-1} in 10 days by the resulted final strain BOH78 [30]. The combination of PKPa-PtaBs and PKPa-AckA-Acs routes identified in this study can be employed to tune the central carbon metabolism for production of acetyl-CoA-derived products.

2.3.2 Importance of 5′-regions in genetic constructs

In synthetic biology, an expressing unit normally refers to a genetic construct, including 5′-region, gene(s) and terminator. The 5′-region comprises promoter, RBS and some other genetic sequence upstream of the gene(s) of interest. The gene expression levels, which will change enzyme levels, usually depend on the effectiveness of the 5′-region. However, in previous engineering attempts of cyanobacteria for butanol production, only one gene (*pduP*) expression was regulated through modifying the RBS [28], without systematically tuning expression of all the components. Therefore, the efficacy of the established isobutanol and 1-butanol pathway in *Synechocystis* still has potential to be increased considerably by systematic screening of the 5′-region in genetic constructs, especially as more promoters and other genetic elements are continuously developed for use in cyanobacteria.

PpsbA2, a native promoter in *Synechocystis* [38, 39], is often used when expressing enzymes in cyanobacteria. Additionally, the strong native and constitutive promoter Pcpc$_{560}$ [40] and the strong heterologous promoter Ptrc$_{20}$ [41] were included to identify the optimal promoter for 1-butanol production. Ptrc is a heterologous promoter family commonly used as strong inducible promoters but show constitutive expression in *Synechocystis* due to an inefficient regulation. Among them, Ptrc$_{20}$, a modified version of Ptrc with two *lac* operators, showed the highest strength in different Ptrc versions tested in *Synechocystis*. Even though the introduced gene of interest is transcribed under control of a promoter, the generated transcript may not always be translated into a corresponding protein. BCD, a bicistronic design, and RiboJ, a self-cleaving ribozyme, are strong systems to initiate translation and provide more reliable expression from previous reports [42–44]. Thus, they were translational coupling with the $Ptrc_{core}$, a $Ptrc$ promoter with no operator sites based on the construction strategy in BCD study, to obtain two artificial 5′-region $Ptrc_{core}$BCD and $Ptrc_{core}$RiboJ, respectively. The importance of the sequences in the 5′-region was exemplified in a recent study, in which codon optimized *hydA1* encoding a [FeFe] hydrogenase was expressed in *Synechocystis* [45]. Using the well-functional $Ptrc_{core}$, a transcript could easily be detected but not a corresponding protein. However, when the 5′-region was modified to contain a BCD design, the hydrogenase protein was readily synthesized.

Isobutanol production was observed in *Synechocystis* when the heterologous enzyme Kivd is expressed. By changing the P*psbA2* to $Ptrc_{core}$BCD to drive the expression of Kivd, both transcription and translation levels of *kivd* increased, leading to improved isobutanol production [20]. Since it has been reported that higher expression can be achieved by expressing genes on a self-replicating vector instead of in the genome [46], the $kivd^{S286T}$ coupled with $Ptrc_{core}$BCD was placed on a self-replicating vector (Fig. 3A), resulting in the highest isobutanol production.

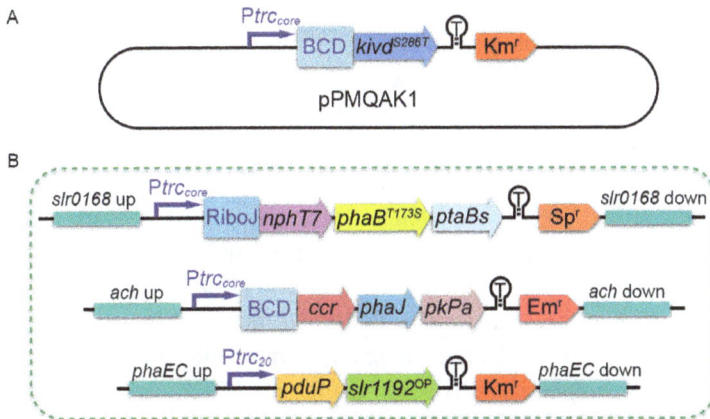

Fig. 3: Simplified schematics of genetic architecture for optimized 5′-regions and butanol biosynthetic genes in *Synechocystis* PCC 6803. (A) Illustration of optimized expression plasmid harboring kivd^{S286T} for the isobutanol biosynthetic pathways. (B) Illustration of three optimized expression units for integration of the 1-butanol biosynthetic genes into the genome. Modified from [21, 30].

Since the 1-butanol-forming genes are organized in three expressing units (Fig. 3B), the 5′-region of each component should be optimized [30]. Different sequences of the genes may additionally result in variations in efficiencies when using identical promoter or 5'-region sequences. Based on this understanding, the 5'-region in each expressing unit was systematically examined by using PpsbA2, $Ptrc_{20}$, $Ptrc_{core}BCD$, $Ptrc_{core}RiboJ$ and $Pcpc_{560}$. With numerous genetic modifications introduced into *Synechocystis*, the process to evaluate a large number of combinations of 5'-region and genes was performed by measuring the 1-butanol production followed throughout. Meanwhile, enzyme expression levels were visualized by Western immunoblots. The best-performing 5'-region combined with the optimized overall 1-butanol biosynthetic genes, including *pkPa* and *ptaBs*, resulted in the final strain BOH78 harboring $Ptrc_{20}$-*pduP-slr1192*^{OP}, $Ptrc_{core}BCD$-*ccr-phaJ-pkPa* and $Ptrc_{core}RiboJ$-*nphT7-phaB*^{T173S}-*ptaBs* units [30] (Fig. 3B).

The discussed work involves combinatorial engineering of the transcription and translation level of all the components by introducing different promoters, BCD and RiboJ [30]. This may give a good paradigm to integrate promoters and translational enhancers into the metabolic design for synergistically facilitating targeted products' biosynthesis in microorganisms.

2.3.3 Optimization of cultivation system

Isobutanol and 1-butanol are water-soluble alcohols with relatively high evaporation. Therefore, in order to correctly measure production levels and rates of butanol, a closed cultivation system using plug-sealed culture tissue flasks was selected [20] (Fig. 4A). In the developed system, 10% of the culture's volume was regularly removed and replaced with fresh BG11 media enriched with a carbon dioxide source ($NaHCO_3$). The bicarbonate enters the cell and is transported to the carboxysome, a microcompartment consisting of polyhedral protein shells filled with the enzymes RuBisCO and carbonic anhydrase (CA). CA generates increased levels of carbon dioxide from the bicarbonate close to RuBisCO. During this reaction, hydroxide is produced which increases the pH. Therefore, the pH of the cell cultures was adjusted to levels between 7.0 and 7.6 at the same time as the replacements of the media [22]. These were necessary steps for establishing a long-term closed cultivation system.

Growing the engineered *Synechocystis* strain expressing Kivd^{S286T} [21] in plug-sealed culture tissue flasks with a 10% fresh media replacement every second day resulted in a cumulative isobutanol production of 911 mg L^{-1} after 46 days with a maximal rate of 43.6 mg L^{-1} day^{-1} [22]. For 1-butanol production, using the strain *Synechocystis* BOH78, a long-term cultivation system with a replacement of the growth media every second day resulted in a cumulative 1-butanol level of 3.0 g L^{-1}, in-flask 1.86 g L^{-1} [30]. However, shortening the intervals between the replacements of growth media to once a day increased the cumulative 1-butanol production to 4.8 g L^{-1}, in-flask 2.13 g L^{-1}, with a maximal rate of 302 mg L^{-1} day^{-1} [30]. This clearly shows that the frequency of product removal, in this case 1-butanol, and fresh media replenishment, significantly affects the production dynamics of the strain. The cell cultures can

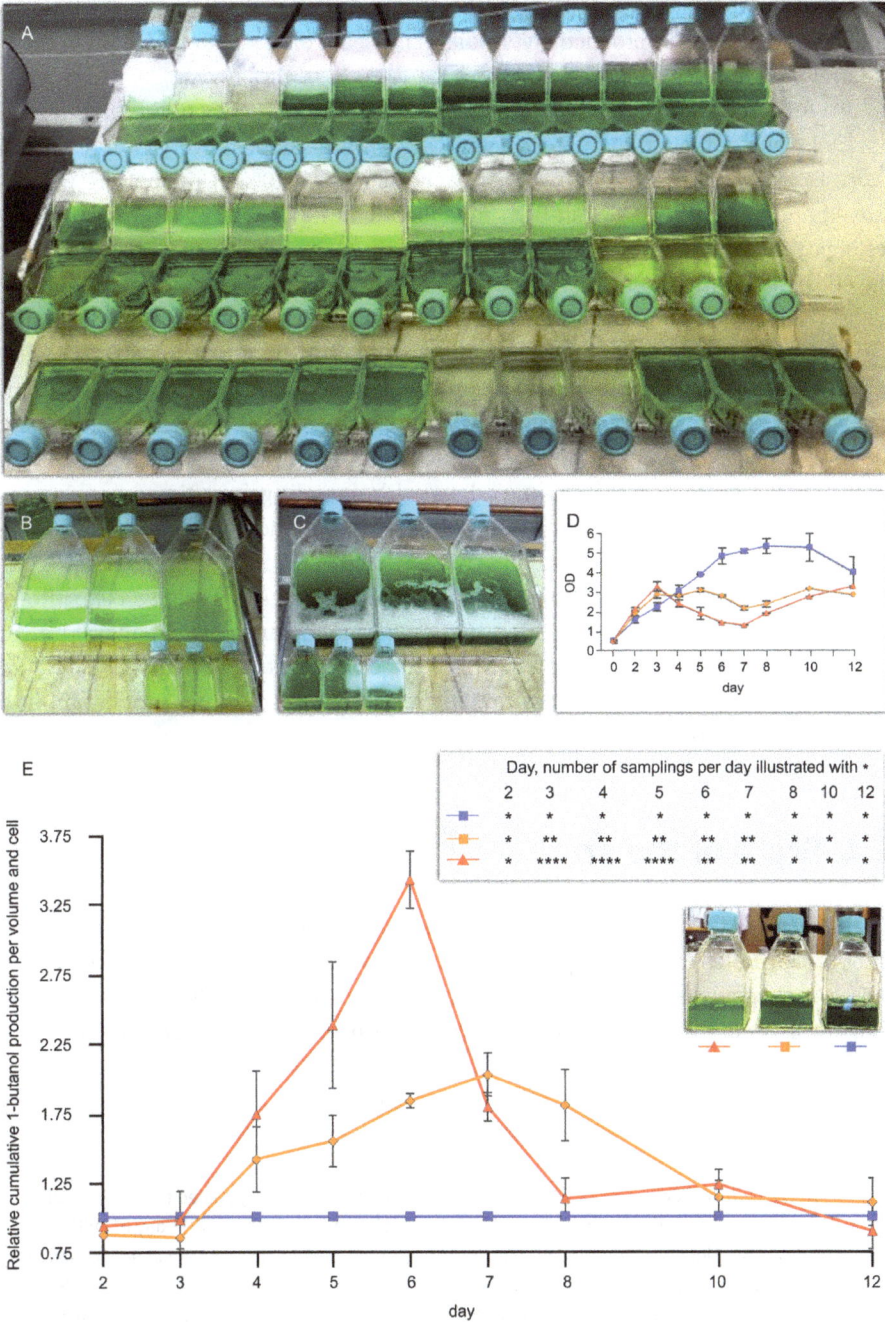

Fig. 4: Experimentally closed butanol production system using smaller (25 cm^2 culture area) and larger (175 cm^2 culture area) plug-sealed culture tissue flasks. Increased butanol production when applying increased withdrawal frequencies, and replacement of 10% of the culture volume of

be scaled up at least eight times from the smaller to the larger tissue flasks (Fig. 4B and 4C) without loss in productivity. Additionally, note the formation of foam in the 1-butanol-producing cell cultures.

To further observe the effect of increased removal frequency of the produced 1-butanol, cells of the strain *Synechocystis* BOH78 were sampled every 24 h (control), every 12 h and every 4 h (daytime). At each time point, 10% of the culture was removed and replaced with fresh BG11 media containing the corresponding antibiotics, every second day the media was additionally enriched with $NaHCO_3$. Growth of the cells and relative cumulative 1-butanol production are shown in Fig. 4D and 4E, respectively. The relative cumulative production is the total production, 1-butanol present in the cell culture plus what has been withdrawn, over a given time period in relation to the production in a cell culture sampled once a day. Increased sampling resulted in a dilution of the cell cultures (Fig. 4D). However, increased sampling, from once a day to several times a day, resulted in significantly more cumulative 1-butanol production, 3 days with sampling four times per day resulted in up to 3.4 times more 1-butanol per volume and cell compared to once per day (Fig. 4E). These results clearly show that multiple sampling and withdrawal of the cell cultures lead to higher productivity. It also gives clear directions for further experiments to identify the optimal balance between withdrawal of cell culture, growth and maintenance of the cell culture, and maximal 1-butanol production – in the end a fully automatic growth, and 1-butanol production and harvesting/collection system.

2.4 State of the art

Engineered cells of the cyanobacterium *Synechocystis* with a biocatalytic production of 600 mg photosynthetic 1-butanol per liter and day and a carbon partitioning of 60% have been developed [30, 47]. For an efficient complete production system, further key aspects are butanol tolerance and the recovery of the produced butanol.

Fig. 4 (continued)
cyanobacterial cultures. Cells cultivated in BG11 medium in small-scale plug-sealed flasks (A); scaling up by using larger plug-sealed flasks, 20 and 160 mL cyanobacterial cultures at first day of cultivation (B) and at day 4 (C). Note the formation of foam after a few days of cultivation (C). Relative cumulative butanol production using the strain *Synechocystis* PCC 6803 BOH78 [30] was grown in smaller tissue flask with increased sampling (D, E). Once per day (days 2 to 8, 10 and 12 – ▬▬, control), twice per day (once day 2, twice days 3 to 7, once days 8, 10, 12 – ◆) and four times per day (once day 2, four times days 3 to 5, twice days 6, 7, once days 8, 10, 12 – ▲). Growth (OD_{750}) and relative cumulative 1-butanol production are shown in graphs (D) and (E), respectively. Means +/– sd ($n = 3$). Insert photos (E) show representative cell cultures on day 7.

2.4.1 Butanol tolerance of cyanobacteria

Solvent-like products usually inhibit their producers, which is one of the bottlenecks for bioproduction. As reported, 1-butanol and isobutanol also exhibit inhibitory effects on cyanobacteria, with the cell growth of cyanobacteria severally inhibited when butanol accumulates [48]. 1-Butanol decreases the growth rate of *Synechocystis* when approximately 1.85 g L^{-1} (0.15 % v/v) was extracellularly added [49, 50]. Several studies have provided a very straightforward way to increase this tolerance considerably by expressing 1-butanol tolerance genes [48, 51–53]. While these studies are relevant for evaluating a strain for high-level productivity, it is likely that there will be different results with respect to tolerance when adding 1-butanol externally to wild-type cells, compared to using engineered cells internally producing the same level of 1-butanol. Interestingly, the engineered *Synechocystis* cells showed a tolerance of up to 2.13 g L^{-1} in-flask titer of 1-butanol [30]. Therefore, when the cells were engineered to produce 1-butanol and the amount of 1-butanol gradually increased in the cell's environment, a higher level of tolerance may be reached in a short period of time. Furthermore, reduced growth is not necessarily only negative, as it may reflect a shift in allocation of the carbon and energy from growth toward product formation.

2.4.2 Recovery of produced photosynthetic butanol

Produced photosynthetic butanol needs to be separated and recovered into pure butanol. In a recent study, the most promising butanol separation technologies (distillation, gas stripping, pervaporation and ionic liquid extraction) were evaluated to calculate the minimum butanol culture concentration required to render an energy-positive process [54]. With a break-even concentration of only 3.7 g butanol L^{-1}, ionic liquid extraction proved much more efficient than the distillation base-case scenario (9.3 g L^{-1}), while neither pervaporation (10.3 g L^{-1}) nor gas stripping (16.9 g L^{-1}) could compete on an energy basis with distillation. Despite this, due to the high costs of the ionic liquid solvent, the lowest capital costs were obtained for distillation (pilot plant scale, butanol culture concentrations of 10 g L^{-1}), while pervaporation carries the lowest utility costs, as a result of its low electrical energy demand. In addition, since the mixture of 1-butanol and water forms a heterogeneous azeotrope it is possible to use a two-column distillation system capable of handling significant fluctuations in the butanol concentrations in the feed from the cyanobacterial cultures [55]. Inspirations may come from systems developed to recover fermentative biobutanol into pure butanol.

2.5 Outlook: future perspective and economic feasibility

The concept of photosynthetic butanol is new. It is a further progression from bio-butanol, fermentative production using native or modified microorganisms, toward a truly carbon-neutral and sustainable technology to produce this blend-in fuel and important bulk chemical directly from solar energy and the greenhouse gas CO_2. Existing productivities will be further increased by a combination of selecting the strains to use, modification of the native metabolism to enhance or delete capacities and functions, and additional genetic sequences and functions to be introduced and regulated. Unexplored but important are the metabolic responses as a result of the introduced modifications and subsequent generation of photosynthetic butanol in the cells. Inspiration may be taken from a recent study comparing metabolic fluxes and proteomics in cells overexpressing two different carbon flux control enzymes in the Calvin–Benson–Bassham cycle where very different cellular responses were observed [56]. Obtained results may be used to guide future metabolic engineering to improve performances and efficiencies of butanol production. Interestingly, external addition of 1-butanol to wild-type cells resulted in changed transcript levels [57], and differences were observed if the alcohol was supplied exogenously or being produced by the cell [58].

2.5.1 Life cycle assessment of photosynthetic butanol

Recently, a first life cycle assessment of the environmental impacts and cumulative energy demand (CED) of cyanobacteria-based photosynthetic 1-butanol was performed [59]. Based on a hypothetical plant producing 5–85 m^3 1-butanol per year in northern Sweden and different scenarios, the greenhouse gas emissions (GHGe) ranged from 16.9 to 58.6 gCO$_2$eq/MJBuOH and the CED from 3.8 to 13 MJ/MJBuOH. A nearby supply of industrial waste sources for heat and CO_2 resulted in at least 60% GHGe savings compared to when fossil resources were used. However, in order to replace the generation of 1-butanol from fossil resources with sustainable photosynthetic butanol, further metabolic improvements of the cyanobacterial cells in, for example, light utilization and carbon partitioning to above 90%, and stable high-yield cultivation and butanol extraction technologies at scale, are all needed [47].

2.5.2 Public acceptance of engineered cyanobacteria

There are no studies specifically examining the public acceptance of engineered cyanobacteria for butanol production. However, a recent study addressed the opinion of European experts and stakeholders on genetically engineered algae, including cyanobacteria, for biofuel production [60]. The results of the survey-based study indicated that the majority of the respondents believe that biofuels produced by engineered algae can provide strong benefits compared to other fuels. This would be based on open communications of both the benefits of the technology and of potential risks. Concomitantly with the further development of production systems for photosynthetic butanol by engineered cyanobacteria, conditions and requirements for achieving public acceptance are of fundamental importance for successful societal integration of this novel and innovative technology.

References

[1]   Lin PC, Zhang F, Pakrasi HB. Enhanced production of sucrose in the fast-growing cyanobacterium *Synechococcus elongatus* UTEX 2973. Sci Rep 2020, 10(1), 390.
[2]   n-Butanol Market worth $5.6 billion by 2022. Northbrook: MarketsandMarkets™ INC, 2018. (Accessed March 10, 2021, at https://www.marketsandmarkets.com/PressReleases/n-butanol.asp).
[3]   Cronk TC, Mattick LR, Steinkraus KH, Hackler LR. Production of higher alcohols during Indonesian Tapé Ketan fermentation. Appl Environ Microbiol 1979, 37(5), 892–6.
[4]   Hazelwood LA, Daran JM, Van Maris AJ, Pronk JT, Dickinson JR. The Ehrlich pathway for fusel alcohol production: A century of research on *Saccharomyces cerevisiae* metabolism. Appl Environ Microbiol 2008, 74(8), 2259–66.
[5]   Chen X, Nielsen KF, Borodina I, Kielland-Brandt MC, Karhumaa K. Increased isobutanol production in *Saccharomyces cerevisiae* by overexpression of genes in valine metabolism. Biotechnol Biofuels 2011, 4, 21.
[6]   Atsumi S, Hanai T, Liao JC. Non-fermentative pathways for synthesis of branched-chain higher alcohols as biofuels. Nature 2008, 451(7174), 86–9.
[7]   Black WB, King E, Wang Y, Jenic A, Rowley AT, Seki K, et al. Engineering a coenzyme A detour to expand the product scope and enhance the selectivity of the Ehrlich pathway. ACS Synth Biol 2018, 7(12), 2758–64.
[8]   Smith KM, Cho K-M, Liao JC. Engineering *Corynebacterium glutamicum* for isobutanol production. Appl Microbiol Biotechnol 2010, 87(3), 1045–55.
[9]   Gu J, Zhou J, Zhang Z, Kim CH, Jiang B, Shi J, et al. Isobutanol and 2-ketoisovalerate production by *Klebsiella pneumoniae* via a native pathway. Metab Eng 2017, 43, 71–84.
[10]   Inui M, Suda M, Kimura S, Yasuda K, Suzuki H, Toda H, et al. Expression of *Clostridium acetobutylicum* butanol synthetic genes in *Escherichia coli*. Appl Microbiol Biotechnol 2008, 77(6), 1305–16.
[11]   Atsumi S, Cann AF, Connor MR, Shen CR, Smith KM, Brynildsen MP, et al. Metabolic engineering of *Escherichia coli* for 1-butanol production. Metab Eng 2008, 10(6), 305–11.
[12]   Shen CR, Lan EI, Dekishima Y, Baez A, Cho KM, Liao JC. Driving forces enable high-titer anaerobic 1-butanol synthesis in *Escherichia coli*. Appl Environ Microbiol 2011, 77(9), 2905–15.
[13]   Dellomonaco C, Clomburg JM, Miller EN, Gonzalez R. Engineered reversal of the β-oxidation cycle for the synthesis of fuels and chemicals. Nature 2011, 476(7360), 355–9.
[14]   Pasztor A, Kallio P, Malatinszky D, Akhtar MK, Jones PR. A synthetic O_2-tolerant butanol pathway exploiting native fatty acid biosynthesis in *Escherichia coli*. Biotechnol Bioeng 2015, 112(1), 120–8.

[15] Atsumi S, Liao JC. Directed evolution of *Methanococcus jannaschii* citramalate synthase for biosynthesis of 1-propanol and 1-butanol by *Escherichia coli*. Appl Environ Microbiol 2008, 74(24), 7802–8.

[16] Atsumi S, Higashide W, Liao JC. Direct photosynthetic recycling of carbon dioxide to isobutyraldehyde. Nat Biotechnol 2009, 27(12), 1177–80.

[17] Li X, Shen CR, Liao JC. Isobutanol production as an alternative metabolic sink to rescue the growth deficiency of the glycogen mutant of *Synechococcus elongatus* PCC 7942. Photosynth Res 2014, 120(3), 301–10.

[18] Varman AM, Xiao Y, Pakrasi HB, Tang YJ. Metabolic engineering of *Synechocystis* sp. strain PCC 6803 for isobutanol production. Appl Environ Microbiol 2013, 79(3), 908–14.

[19] Shen CR, Liao JC. Photosynthetic production of 2-methyl-1-butanol from CO_2 in cyanobacterium *Synechococcus elongatus* PCC7942 and characterization of the native acetohydroxyacid synthase. Energy Environ Sci 2012, 5(11), 9574–83.

[20] Miao R, Liu X, Englund E, Lindberg P, Lindblad P. Isobutanol production in *Synechocystis* PCC 6803 using heterologous and endogenous alcohol dehydrogenases. Metab Eng Commun 2017, 5, 45–53.

[21] Miao R, Xie H, Ho FM, Lindblad P. Protein engineering of alpha-ketoisovalerate decarboxylase for improved isobutanol production in *Synechocystis* PCC 6803. Metab Eng 2018, 47, 42–8.

[22] Miao R, Xie H, Lindblad P. Enhancement of photosynthetic isobutanol production in engineered cells of *Synechocystis* PCC 6803. Biotechnol Biofuels 2018, 11, 267.

[23] Lan EI, Liao JC. Metabolic engineering of cyanobacteria for 1-butanol production from carbon dioxide. Metab Eng 2011, 13(4), 353–63.

[24] Lan EI, Liao JC. ATP drives direct photosynthetic production of 1-butanol in cyanobacteria. Proc Natl Acad Sci USA 2012, 109(16), 6018–23.

[25] Lan EI, Ro SY, Liao JC. Oxygen-tolerant coenzyme A-acylating aldehyde dehydrogenase facilitates efficient photosynthetic n-butanol biosynthesis in cyanobacteria. Energy Environ Sci 2013, 6(9), 2672–81.

[26] Anfelt J, Kaczmarzyk D, Shabestary K, Renberg B, Rockberg J, Nielsen J, et al. Genetic and nutrient modulation of acetyl-CoA levels in *Synechocystis* for n-butanol production. Microb Cell Fact 2015, 14, 167.

[27] Shabestary K, Anfelt J, Ljungqvist E, Jahn M, Yao L, Hudson EP. Targeted repression of essential genes to arrest growth and increase carbon partitioning and biofuel titers in cyanobacteria. ACS Synth Biol 2018, 7(7), 1669–75.

[28] Fathima AM, Chuang D, Lavina WA, Liao J, Putri SP, Fukusaki E. Iterative cycle of widely targeted metabolic profiling for the improvement of 1-butanol titer and productivity in *Synechococcus elongatus*. Biotechnol Biofuels 2018, 11, 188.

[29] Higo A, Ehira S. Anaerobic butanol production driven by oxygen-evolving photosynthesis using the heterocyst-forming multicellular cyanobacterium *Anabaena* sp. PCC 7120. Appl Microbiol Biotechnol 2019, 103(5), 2441–7.

[30] Liu X, Miao R, Lindberg P, Lindblad P. Modular engineering for efficient photosynthetic biosynthesis of 1-butanol from CO_2 in cyanobacteria. Energy Environ Sci 2019, 12(9), 2765–77.

[31] Pleiss J. Protein design in metabolic engineering and synthetic biology. Curr Opin Biotechnol 2011, 22(5), 611–7.

[32] Tsuge T, Hisano T, Taguchi S, Doi Y. Alteration of chain length substrate specificity of *Aeromonas caviae* R-enantiomer-specific enoyl-coenzyme A hydratase through site-directed mutagenesis. Appl Environ Microbiol 2003, 69(8), 4830–6.

[33] Zha W, Shao Z, Frost JW, Zhao H. Rational pathway engineering of type I fatty acid synthase allows the biosynthesis of triacetic acid lactone from D-glucose in vivo. J Am Chem Soc 2004, 126(14), 4534–5.

[34] Nair NU, Zhao H. Evolution in reverse: Engineering a D-xylose-specific xylose reductase. Chembiochem 2008, 9(8), 1213–15.

[35] Zhang K, Sawaya MR, Eisenberg DS, Liao JC. Expanding metabolism for biosynthesis of nonnatural alcohols. Proc Natl Acad Sci USA 2008, 105(52), 20653–8.

[36] Luo LH, Seo JW, Baek JO, Oh BR, Heo SY, Hong WK, et al. Identification and characterization of the propanediol utilization protein PduP of *Lactobacillus reuteri* for 3-hydroxypropionic acid production from glycerol. Appl Microbiol Biotechnol 2011, 89(3), 697–703.

[37] Luo LH, Kim CH, Heo SY, Oh BR, Hong WK, Kim S, et al. Production of 3-hydroxypropionic acid through propionaldehyde dehydrogenase PduP mediated biosynthetic pathway in *Klebsiella pneumoniae*. Bioresour Technol 2012, 103(1), 1–6.

[38] Lindberg P, Park S, Melis A. Engineering a platform for photosynthetic isoprene production in cyanobacteria, using *Synechocystis* as the model organism. Metab Eng 2010, 12(1), 70–9.

[39] Englund E, Liang F, Lindberg P. Evaluation of promoters and ribosome binding sites for biotechnological applications in the unicellular cyanobacterium *Synechocystis* sp. PCC 6803. Sci Rep 2016, 6, 36640.

[40] Zhou J, Zhang H, Meng H, Zhu Y, Bao G, Zhang Y, et al. Discovery of a super-strong promoter enables efficient production of heterologous proteins in cyanobacteria. Sci Rep 2014, 4, 4500.

[41] Huang HH, Camsund D, Lindblad P, Heidorn T. Design and characterization of molecular tools for a Synthetic Biology approach towards developing cyanobacterial biotechnology. Nucleic Acids Res 2010, 38(8), 2577–93.

[42] Lou C, Stanton B, Chen YJ, Munsky B, Voigt CA. Ribozyme-based insulator parts buffer synthetic circuits from genetic context. Nat Biotechnol 2012, 30(11), 1137–42.

[43] Mutalik VK, Guimaraes JC, Cambray G, Lam C, Christoffersen MJ, Mai QA, et al. Precise and reliable gene expression via standard transcription and translation initiation elements. Nat Methods 2013, 10(4), 354–60.

[44] Clifton KP, Jones EM, Paudel S, Marken JP, Monette CE, Halleran AD, et al. The genetic insulator RiboJ increases expression of insulated genes. J Biol Eng 2018, 12(1), 23.

[45] Lindblad P, Fuente D, Borbe F, Cicchi B, Conejero JA, Couto N, et al. CyanoFactory, a European consortium to develop technologies needed to advance cyanobacteria as chassis for production of chemicals and fuels. Algal Res 2019, 41, 101510.

[46] Ng AH, Berla BM, Pakrasi HB. Fine-tuning of photoautotrophic protein production by combining promoters and neutral sites in the cyanobacterium *Synechocystis* sp. strain PCC 6803. Appl Environ Microbiol 2015, 81(19), 6857–63.

[47] Wichmann J, Lauersen KJ, Biondi N, Christensen M, Guerra T, Hellgardt K, et al. Engineering biocatalytic solar fuel production: The PHOTOFUEL consortium. Trends Biotechnol 2021, 39(4), 323–7.

[48] Anfelt J, Hallstrom B, Nielsen J, Uhlen M, Hudson EP. Using transcriptomics to improve butanol tolerance of *Synechocystis* sp. strain PCC 6803. Appl Environ Microbiol 2013, 79(23), 7419–27.

[49] Kamarainen J, Knoop H, Stanford NJ, Guerrero F, Akhtar MK, Aro EM, et al. Physiological tolerance and stoichiometric potential of cyanobacteria for hydrocarbon fuel production. J Biotechnol 2012, 162(1), 67–74.

[50] Tian X, Chen L, Wang J, Qiao J, Zhang W. Quantitative proteomics reveals dynamic responses of *Synechocystis* sp. PCC 6803 to next-generation biofuel butanol. J Proteomics 2013, 78, 326–45.

[51] Kaczmarzyk D, Anfelt J, Sarnegrim A, Hudson EP. Overexpression of sigma factor SigB improves temperature and butanol tolerance of *Synechocystis* sp. PCC6803. J Biotechnol 2014, 182–183, 54–60.

[52] Gao X, Sun T, Wu L, Chen L, Zhang W. Co-overexpression of response regulator genes *slr1037* and *sll0039* improves tolerance of *Synechocystis* sp. PCC 6803 to 1-butanol. Bioresour Technol 2017, 245, 1476–83.

[53] Bi Y, Pei G, Sun T, Chen Z, Chen L, Zhang W. Regulation mechanism mediated by *trans*-encoded sRNA Nc117 in short chain alcohols tolerance in *Synechocystis* sp. PCC 6803. Front Microbiol 2018, 9, 863.

[54] Wagner JL, Lee-Lane D, Monaghan M, Sharifzadeh M, Hellgardt K. Recovery of excreted *n*-butanol from genetically engineered cyanobacteria cultures: Process modelling to quantify energy and economic costs of different separation technologies. Algal Res 2019, 37, 92–102.

[55] Luyben WL. Control of the heterogeneous azeotropic n-butanol/water distillation system. Energy Fuels 2008, 22(6), 4249–58.

[56] Yu King Hing N, Liang F, Lindblad P, Morgan JA. Combining isotopically non-stationary metabolic flux analysis with proteomics to unravel the regulation of the Calvin-Benson-Bassham cycle in *Synechocystis* sp. PCC 6803. Metab Eng 2019, 56, 77–84.

[57] Matsusako T, Toya Y, Yoshikawa K, Shimizu H. Identification of alcohol stress tolerance genes of *Synechocystis* sp. PCC 6803 using adaptive laboratory evolution. Biotechnol Biofuels 2017, 10, 307.

[58] Wang Y, Chen L, Zhang W. Proteomic and metabolomic analyses reveal metabolic responses to 3-hydroxypropionic acid synthesized internally in cyanobacterium *Synechocystis* sp. PCC 6803. Biotechnol Biofuels 2016, 9, 209.

[59] Nilsson A, Shabestary K, Brandão M, Hudson EP. Environmental impacts and limitations of third-generation biobutanol: Life cycle assessment of n-butanol produced by genetically engineered cyanobacteria. J Ind Ecol 2020, 24(1), 205–16.

[60] Varela Villarreal J, Burgues C, Rosch C. Acceptability of genetically engineered algae biofuels in Europe: Opinions of experts and stakeholders. Biotechnol Biofuels 2020, 13, 92.

Hanna C. Grimm, Robert Kourist

3 Cyanobacteria as catalysts for light-driven biotransformations

Abstract: The capacity of photosynthetic microorganisms for light-driven oxidation of water can be utilized to provide the reduced cofactors NADPH and reduced ferredoxin (Fd_{red}) for enzymatic redox reactions. This makes the stoichiometric addition of organic cosubstrates obsolete, thereby, greatly improving the atom economy of enzymatic redox reactions. Furthermore, intracellularly produced oxygen can be used as cosubstrate for oxyfunctionalization reactions, which helps to overcome the low gas–liquid mass transfer rate. This chapter presents the application of natural and genetically modified photosynthetic microorganisms for the biocatalytic synthesis of valuable chemicals and introduces the strengths and current limitations of the approach.

3.1 Goal

High energy consumption, depletion of fossil resources and a growing awareness for the environmental consequences of our modern lifestyle are driving demand for sustainable alternatives. Conventional chemistry to produce fine chemicals often requires harsh reaction conditions and is accompanied by waste production and low selectivity. The 12 principles of green chemistry summarize the requirements for environmentally friendly processes such as minimized waste production, an optimal atom economy or the use of renewable resources [1]. In the past decades, biocatalysis has emerged as an environmentally friendly method for the production of a wide range of chemicals, ranging from active pharmaceutical ingredients to commodities. In particular, enzymatic redox reactions have found wide application due to the mild reaction conditions and the outstanding chemo-, regio- and enantioselectivities of many oxidoreductases. The application of oxidoreductases for reactive redox transformations is still hampered by the need for the addition of external cofactors such as NADPH. As the high price of NADPH prevents its addition in stoichiometric amounts, an appropriate system for the re-use of the cofactor is required. Well-established cofactor regeneration systems rely on the addition of sacrificial cosubstrates like glucose or isopropanol but are associated with a low atom economy. In this context, the highly optimized natural process of oxygenic photosynthesis constitutes an attractive, alternative electron source for NADPH regeneration.

Acknowledgments: This work was funded by the Austrian Science Funds FWF (project no. P31001-B29) and the German Federal Environmental Foundation (project no. AZ30818-32).

https://doi.org/10.1515/9783110716979-003

Cyanobacteria have been studied as whole-cell biocatalysts since 1986 [2] and were used to reduce numerous substrates such as aldehydes, ketones, enones and terpenes [3]. While first studies focused on non-modified wild-type strains and qualitatively evaluated conversions, the genetic modification of cyanobacteria for tailored processes gained attraction in recent years. The reaction rates and space-time yields obtained with cyanobacteria have not fulfilled yet the criteria for successful biotechnological processes [4]. Nevertheless, a rapid technological progress is expected to pave the way for industrial application in the near future. This chapter will provide a broad overview of the state of the art for light-driven biotransformations in wild-type and recombinant cyanobacteria. Moreover, the general applicability and possible limitations of the concept are discussed and evaluated.

3.2 Basic background

3.2.1 Whole-cell biotransformations in photoautotrophs

Biotechnological processes to produce chemicals can be divided into two categories: fermentations and biotransformations. Fermentations have played a crucial role for the development of the human civilization and are used for thousands of years [5, 6]. Multi-enzymatic pathways of the microbial metabolism are used to convert a carbon source into the final product. Accordingly, the starting material is directly fed into the central catabolism of the cell. Knowledge of the microbial background expanded the application for fermentations, for example, for the production of secondary metabolites such as antibiotics and vitamins, which are hardly accessible by chemical synthesis. Modern reactors allow exquisite control of cultivation conditions, including influent and effluent flows, addition of carbon, nitrogen and phosphate sources, and physical parameters such as pH, temperature or pressure, thereby heterotrophic microbial hosts can reach constant product titers in the order of 100 g L^{-1} surpassing productivities of 1 g L^{-1} h^{-1} [7].

In contrast to the utilization of a carbon source in fermentative processes, biotransformations are the enzymatic conversion of a substrate into a product. In industrial biotechnology, the term is more precisely specified: a specific substrate added to the reaction suspension is converted by one or more enzymatic steps while a supplemented carbon source is exclusively used for biomass production. For cyanobacterial biotransformations, this means that the fixation of CO_2 is used for the preparation of the whole-cell biocatalyst but does not necessarily play a role in the application as a catalyst. Accordingly, whole-cell biotransformation can be either performed using growing or resting cells. The latter option decouples the cultivation from the production phase and requires a harvesting step before the biotransformation is started by addition of substrate. This is beneficial if reaction conditions do not support cellular growth.

Since 1960, microalgae and cyanobacteria have been used for the fermentative production of food supplements such as β-carotene, astaxanthin, polyunsaturated fatty acids and polysaccharides [8]. The application in cyanobacterial whole-cell biotransformation is a rapidly growing field, especially, since genetic methods for the engineering of cyanobacteria became accessible. Since cell productivities of all reported fermentations and biotransformations are low, current research intensively concentrates on the increase of production rates [7], thereby fermentations are mainly limited by the photosynthetic carbon fixation efficiencies [7] while productivities of biotransformations are hindered by low intracellular enzyme concentrations and inefficient photosynthetic electron transport [9]. Thus, optimization strategies concentrate on (i) the development of genetic tools for engineering, (ii) increase of the photosynthetic efficiency and (iii) the development of suitable photobioreactors.

3.2.2 Cofactor recycling

Oxidoreductases catalyze electron transfer reactions that comprise the reduction of a compound (an oxidizing agent or an electron acceptor) with the concurrent oxidation of another compound (a reducing agent or an electron donor) [10]. A wide diversity of redox enzymes is involved in a multitude of processes for the production of fine chemicals and pharmaceutical ingredients. However, their industrial application is often hampered by the stoichiometric requirement for external cofactors. Over 50% of all registered oxidoreductase activities involve the transfer of a hydride (H^-) group from or to a nicotinamide adenine dinucleotide cofactor NAD(H) or NADP (H) [10]. Some representatives and their catalyzed reactions are summarized in Fig. 1. The high price of all nicotinamide cofactors makes their addition in stoichiometric amounts economically impossible. Therefore, an appropriate recycling system is indispensable for a sustainable process involving oxidoreductases.

Literature provides several strategies for cofactor recycling ranging from chemical, electrochemical, photochemical, microbial to enzymatic regeneration systems [11]. An optimal cofactor recycling system must meet high requirements. It should be inexpensive, easy to apply and sustainable. Catalysts suitable for cofactor recycling should be stable, highly selective and in the best-case reusable. Enzymatic systems are predominantly used for preparative applications which can be attributed to the ease of use and their inherent compatibility with enzymatic production systems. A second enzymatic reaction regenerates the cofactor with expense of a sacrificial cosubstrate (Fig. 2) [12–15]. Table 1 shows examples for frequently applied enzymes, cosubstrates and coproducts. An important parameter for the evaluation of a cofactor recycling system is the atom efficiency that compares how many atoms are introduced into the reaction and subsequently recovered in the desired product [1]. A low atom efficiency is accompanied with the production of side products that need to be separated from the reaction suspension [16]. For example, only two electrons

A) Reduction of aldehydes and ketones

D) Baeyer-Villiger oxidation reactions

B) Reduction of conjugated C=C double bonds

E) Hydroxylation of C-H bonds

C) Reduction of imines

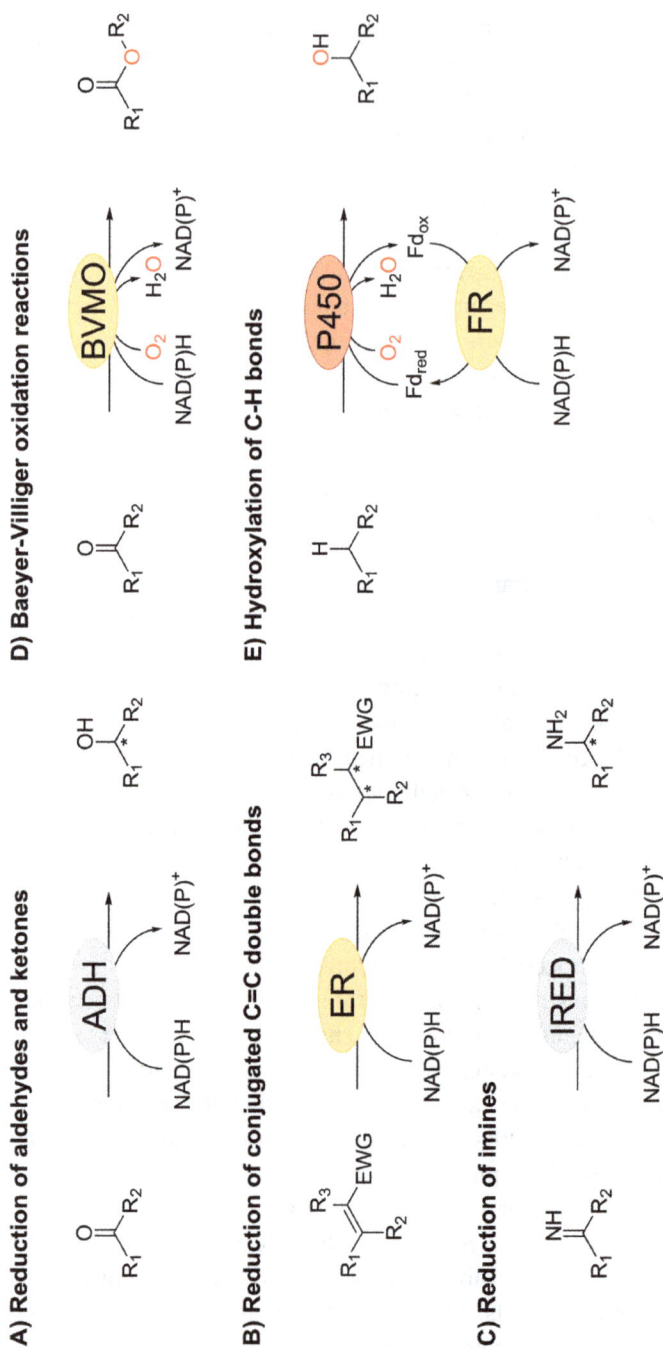

Fig. 1: Selection of representative, NAD(P)H-dependent oxidoreductases. Flavoenzymes harboring the flavin cofactor FMN or FAD are marked in yellow, heme-dependent enzymes are marked in red. ADH, alcohol dehydrogenase; ER, ene-reductase; IRED, imine reductase; BVMO, Baeyer–Villiger monooxygenase; P450, cytochrome P450 monooxygenase; FR, ferredoxin reductase.

are used for the oxidation of NADP$^+$ to NADPH in the GDH-catalyzed reduction of glucose to gluconic acid.

Enzymatic recycling of NAD(P)H

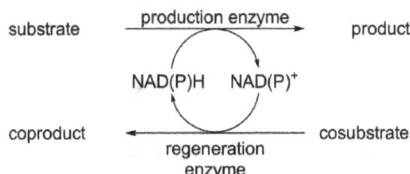

substrate ——production enzyme——▸ product

NAD(P)H   NAD(P)$^+$

coproduct ◂—————————— cosubstrate
regeneration
enzyme

NADPH recycling in heterotrophic cells

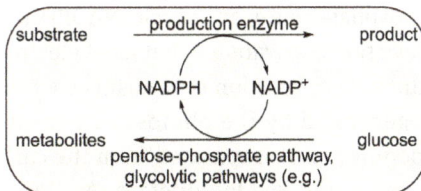

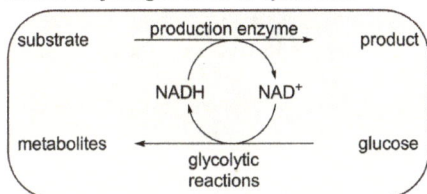

substrate ——production enzyme——▸ product

NADPH   NADP$^+$

metabolites ◂—————————— glucose
pentose-phosphate pathway,
glycolytic pathways (e.g.)

NADH-recycling in heterotrophic cells

substrate ——production enzyme——▸ product

NADH   NAD$^+$

metabolites ◂—————————— glucose
glycolytic
reactions

NADPH recycling in photoautotrophic cells

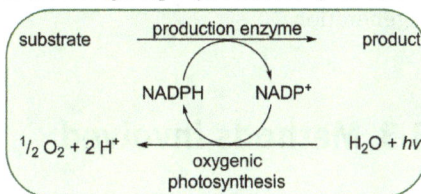

substrate ——production enzyme——▸ product

NADPH   NADP$^+$

$^{1}/_{2}$ O$_2$ + 2 H$^+$ ◂—————————— H$_2$O + hv
oxygenic
photosynthesis

Fig. 2: Examples for cofactor recycling strategies.

Tab. 1: Examples for enzymes, cosubstrates and coproducts used for cofactor recycling.

Regeneration system	Required cofactor	Cosubstrate	Coproduct
Formate dehydrogenase (FDH)	NAD(P)$^+$	Formate	Carbon dioxide
Glucose dehydrogenase (GDH)	NAD(P)$^+$	Glucose	Gluconic acid
Alcohol dehydrogenase (ADH)	NAD(P)$^+$	Isopropanol	Acetone
Phosphite dehydrogenase (PDH)	NAD(P)$^+$	Phosphorous acid	Phosphoric acid
Hydrogenase	NAD(P)$^+$	Hydrogen	–
NADH oxidase	NADH	Oxygen, protons	Water
Lactate dehydrogenase	NADH	Pyruvate	Lactate
Photosynthetic electron transport chain	NADPH, Fd	Water	–

Biotransformations in whole cells offer an attractive alternative to cell-free systems since the nicotinamide cofactor can be provided by the cellular metabolism [16]. Both, the NADH/NAD$^+$ and NADPH/NADP$^+$ ratio, constitute the reductive pool of the cell. The difference between both is a phosphate group at the 2′-OH group of the adenine ribose ring in NADPH/NADP$^+$. This modification does not affect the electron transfer capability but has a physiological function [10]. Enzymes with preference toward NADH/NAD$^+$ are mostly involved in catabolic reactions while reaction in anabolism is mainly catalyzed by NADPH/NADP$^+$-dependent oxidoreductases. Accordingly, the NADPH/NADP$^+$ pool is often more reduced to effectively provide reducing power for biomass production [10]. NADH regeneration in heterotrophic host cells mainly occurs by glycolytic reactions, while pathways like the pentose-phosphate pathway recycle NADPH (Fig. 2). Both require the use of glucose and have the disadvantage that most electrons are lost for biomass production. In photo-autotrophs, reaction equivalents such as NADPH and reduced ferredoxin (Fd$_{red}$) are regenerated by the photosynthetic electron transport chain (PETC). This energy module can be separated from the carbon assimilation and used to fuel reactions catalyzed by oxidoreductase. As this approach uses water as a sacrificial cosubstrate, the atom efficiency is much superior to other approaches for cofactor regeneration.

3.3 Methods involved

3.3.1 Genetic modification of cyanobacteria

Examples of recombinant cyanobacteria used for biotransformations include the unicellular genera *Synechocystis* and *Synechococcus*. Both genetic backgrounds are well investigated and all basic tools for genetic engineering are available [8]. Plasmids and regulatory elements for gene expression in cyanobacteria are also discussed in chapter 2. The genes of the oxidoreductases were either integrated into the chromosome [9, 17–21], an endogenous plasmid of the host cell [22], or maintained on self-replicating plasmids [4, 23–27]. Integration of recombinant genes is achieved via homologous recombination into a genomic locus which is not essential for cell survival such as the locus *slr0168* of *Synechocystis* sp. PCC 6803 [9, 17–19]. The multiple cloning site is flanked by two sequences, referred to as the upstream (us) and downstream (ds) sequences, which are homologous to the flanking gene segments of the integration locus. The expression cassette, which contains the regulatory elements for gene expression, the recombinant gene and a selection marker, is located between the us and ds sequences. After successful transformation, the expression cassette is integrated into the selected gene locus and the rest of the suicide vector is digested by the cell. Since cyanobacteria are polyploid [28, 29], integration of the expression cassette into every copy of the chromosome needs to be ensured. This so-called segregation can take up to 8 weeks which slows down the working process. Full segregation is controlled by PCR and band patterns are compared to a non-modified wild-type strain. In the example shown in Fig. 3, both strains S1 and S2 still show a strong band like the wild-type control indicating incomplete segregation.

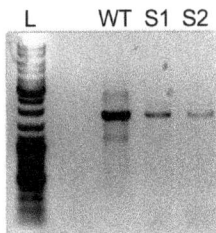

Fig. 3: Example result for segregation check PCR. Band pattern of recombinant strains are compared to the pattern of the wild-type strain. Strains are fully segregated if no wild-type signal is found. In this case, both strains S1 and S2 are not fully segregated. L, DNA-Ladder; WT, non-modified wild-type strain; S1 and S2, strains harboring the expression cassette for another heterologous gene. © Jelena Spasic, TU Graz.

Self-replicating plasmids are generally non-specific for cyanobacteria but based on broad-range plasmids with origin of replications such as RSF1010 [30]. They are easy to construct and quickly inserted into the cell. However, instability, loss of the plasmid in case of insufficient selection pressure and varying copy numbers might comprise the reliability and behavior of the system [30]. Since chromosome copy number also depends on the growth phase and nutrient availability [28], strict control of cultivation conditions is an essential prerequisite for reproducible experiments.

3.3.2 Biotransformations in wild-type and recombinant cells

For biotransformations, cells are cultivated to the mid-exponential growth phase, harvested and supplemented with fresh medium or buffer before reactions are initiated by the addition of the substrate (Fig. 4). In more recent studies, supplementation of carbon dioxide during cultivation and the influence of the light intensity on cellular growth gain importance but are not yet standardized for specific strains in terms of maximal cell activity. Reactions were often normalized to optical density (OD), chlorophyll a (chl a) content or cell dry/wet weight (CDW or CWW) for better reproducibility. Reference to the CDW allows comparison of the reaction parameters to heterotrophs. This parameter, however, is usually determined indirectly by measuring the optical density (OD_{750}). As in cyanobacteria, the ratio between OD and CDW might easily change with different mutants and depending on the cultivation conditions, the reference to the chlorophyll offers advantages in the practical use. It is important to note, that normalization is facilitated for unicellular strains from genera such as *Synechocystis* or *Synechococcus* but difficult for filamentous strains from genera such as *Anabaena* and *Nostoc*. In the latter case, an exact cultivation protocol is of major importance to allow reproducibility of biological replicates. Reaction times for whole-cell biotransformation vary from few hours to several days (Tabs. 3 and 4). For industrial application, processes with constant production rates over several days or catalyst reusability are desired. Simultaneous cultivation and biotransformation, biofilm entrapment [23] or catalyst immobilization [31, 32] can reduce production costs and facilitate reaction set up as well as ds processing.

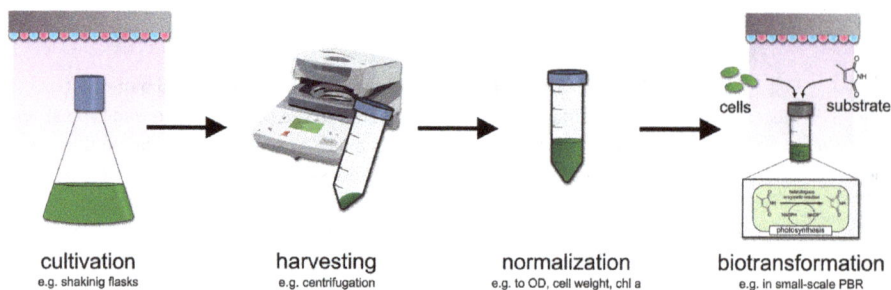

cultivation
e.g. shakinig flasks

harvesting
e.g. centrifugation

normalization
e.g. to OD, cell weight, chl a

biotransformation
e.g. in small-scale PBR

Fig. 4: Workflow for whole-cell biotransformations in recombinant cyanobacteria. OD, optical density; chl a, chlorophyll a; PBR, photobioreactor.

3.3.3 Analysis of biotransformations

Whole-cell biotransformations can be analyzed by general methods such as high-performance liquid chromatography, gas chromatography (GC) or NMR. Figure 5 summarizes the workflow for GC analyses with prior organic phase extraction of the respective sample. Substances migrate differently in dependence of the column material and can be detected, for example, by flame ionization. Obtained peak areas correlate with the concentration of the substances allowing quantification.

The time curve of the reaction designates the analysis of the reaction after different time points. It enables the calculation of important parameters (Tab. 2) such as the specific activity, the productivity or the space-time yield. These are usually associated with a known measurand like the concentration of the catalyst, the time of the reaction or the volume of the reactor. Specific chiral columns for chromatographic methods enable the separation of enantiomers. Here, the peak area is used to determine the enantiomeric excess (ee) and thus, the stereoselectivity of the reaction.

3.4 State of the art

3.4.1 Whole-cell biotransformation in non-modified cyanobacterial strains

Cyanobacteria provide a wide array of highly selective oxidoreductases [3, 33]. Since 1986 [2], cyanobacterial wild-type strains have been shown to catalyze a broad range of biotransformations summarized in Tab. 3.

One of the most investigated wild-type strains for light-driven biotransformations is the unicellular *Synechococcus* sp. PCC 7942 (Fig. 6). It was first shown to selectively reduce several aryl methyl ketones with different electron pulling moieties as

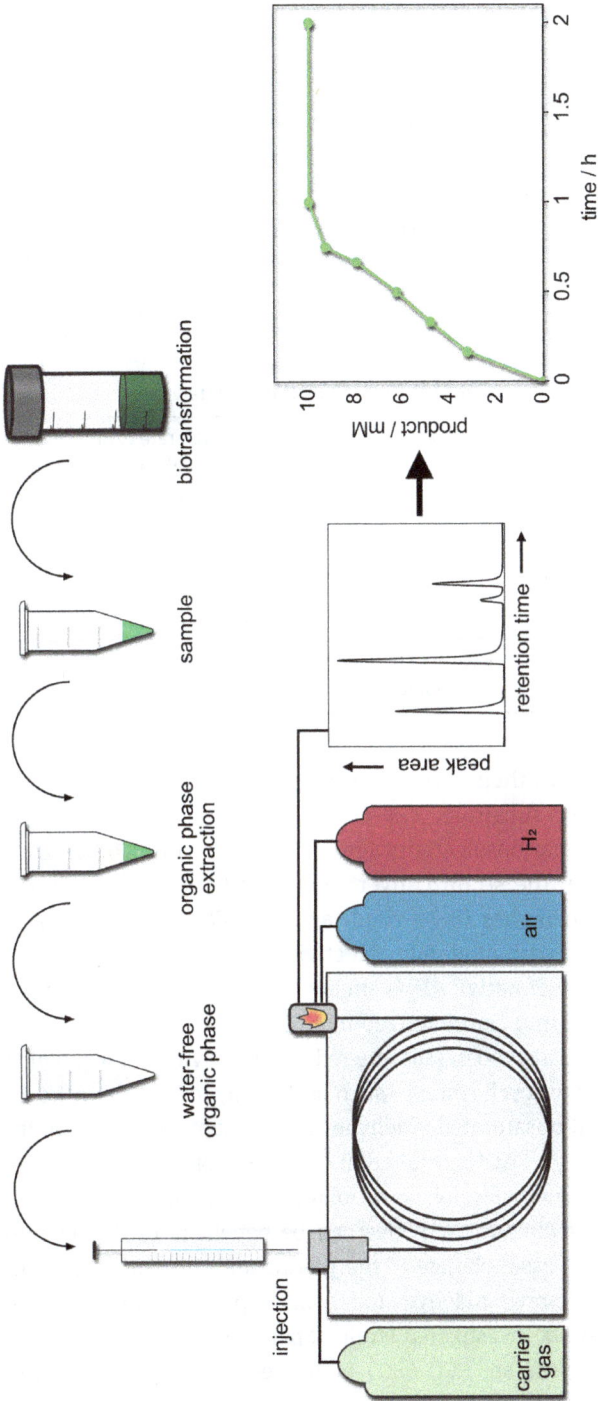

Fig. 5: Analysis of biotransformations by gas chromatography. Samples are extracted with organic solvent, dried to remove the residual water and injected into the device. Substances are separated due to different migration speed inside the column. Inert helium or nitrogen are commonly used carrier gases. Substances can be detected, for example, by a flame ionization detector (FID). Standard curves for each compound are required to translate peak area into substance concentration for quantification. Analysis of several time samples leads to the time curve (right) of the reaction that can be used to calculate required parameters.

Tab. 2: Important process parameters for biotransformation processes.

Parameter	Commonly used unit	Description
Conversion (C)	$[mol_{product}\ mol_{substrate}^{-1}]$	Calculated from the substrate depletion expresses how much of the initially provided substrate was transformed into product
Isolated yield (Y)	$[mol_{product}\ mol_{substrate}^{-1}]$	Detected product concentration in correlation to the provided amount of substrate or the total amount of isolated product
Initial reaction rate	$[g_{product}\ h^{-1}]$, $[mol_{product}\ h^{-1}\ L^{-1}]$	Velocity of the reaction in a time interval with less than 10% conversion. Reaction is not limited by the concentration of substrate or catalyst
Specific activity $[A_{spec.}]$	$[\mu mol_{product}\ g_{catalyst}^{-1}\ min^{-1}]$	Calculated from the initial reaction rate expresses how much product is converted by the given amount of catalysts in a specific time interval
Productivity	$[g_{product}\ g_{catalyst}^{-1}]$, $[mol_{product}\ g_{catalyst}^{-1}]$	Total product formed by a specific amount of catalyst
Space-time yield	$[g_{product}\ L^{-1}\ h^{-1}]$	Quantity of product per unit of volume and time
Enantiomeric excess (ee)	[%]	Excess of one enantiomer in a mixture of both enantiomers

halogens or methoxide groups to their corresponding (S)-alcohols [34–37]. About 4.1 mM of the fastest converted substrate, 2′-3′-4′-5′-6′-pentafluoroacetophenone, was reduced with excellent enantioselectivity (ee > 99%) and within 48 h (82% conversion) [36]. Furthermore, the strain converts (+)- and (–)-camphorquinone yielding (–)-(3S)-exo-hydroxycamphor (86% yield) and (+)-(3R)-exo-hydroxycamphor (56% yield), respectively, with moderate selectivity [38]. Alcohol production can be attributed to the activity of native ADHs such as the NAPDH-dependent ketoreductase identified, isolated and characterized by Hölsch et al. [39, 40].

Synechococcus sp. PCC 7942 also possesses ene-reductase activity [33, 41], which was proved by its ability to reduce cyclic enones such as 2-methyl-cyclopenten-1-one, (–)- and (+)-carvone [42] or α,β-unsaturated aldehydes like cinnamaldehyde [43]. Interplay between ene-reductases and ADHs resulted in overreduction of substrates like 2-methyl-cyclohexen-1-one to 2-methylcyclohexan-1-ol [42] and cinnamaldehyde to 3-phenylpropan-1-ol [43]. Such sequential activities can be beneficial or detrimental [18] to the process. In the latter case, change of the production host can be considered. For instance, reduction of cinnamaldehyde to the aimed product cinnamyl alcohol was catalyzed by all other six investigated strains apart from Synechococcus sp. PCC 7942 [43]. Here, Synechocystis sp. PCC 6803 performed best showing highest yield (Y > 98%) and strict chemoselectivity [43].

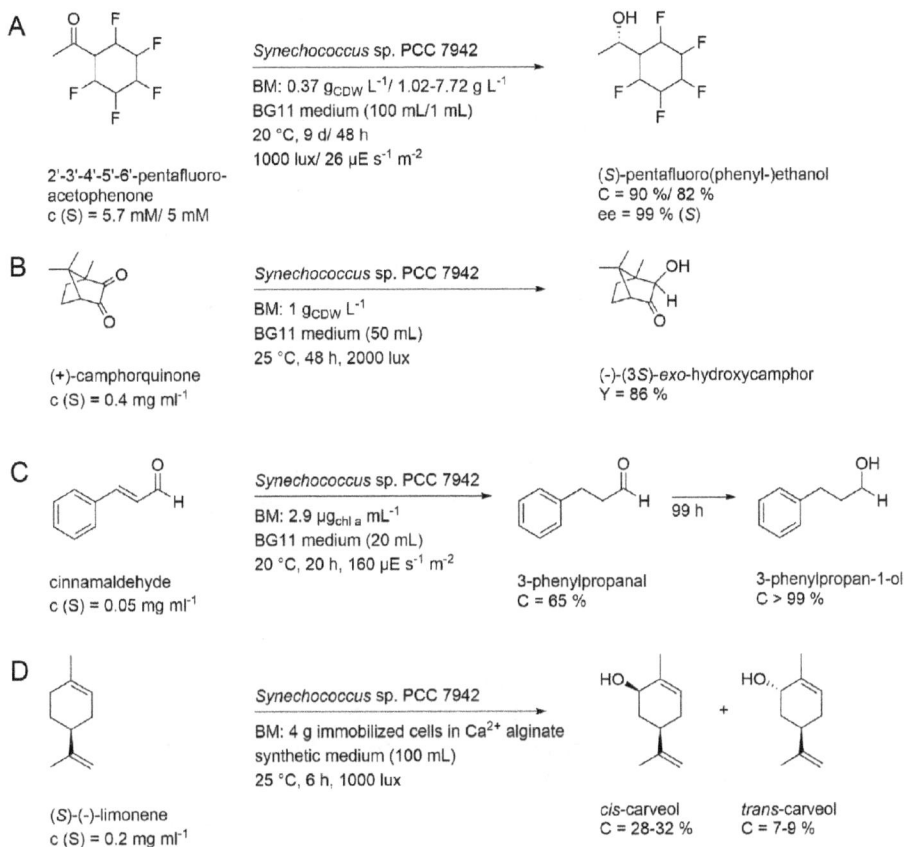

A

2'-3'-4'-5'-6'-pentafluoro-acetophenone
c (S) = 5.7 mM/ 5 mM

Synechococcus sp. PCC 7942

BM: 0.37 g_{CDW} L^{-1}/ 1.02-7.72 g L^{-1}
BG11 medium (100 mL/1 mL)
20 °C, 9 d/ 48 h
1000 lux/ 26 µE s^{-1} m^{-2}

(S)-pentafluoro(phenyl-)ethanol
C = 90 %/ 82 %
ee = 99 % (S)

B

(+)-camphorquinone
c (S) = 0.4 mg ml^{-1}

Synechococcus sp. PCC 7942

BM: 1 g_{CDW} L^{-1}
BG11 medium (50 mL)
25 °C, 48 h, 2000 lux

(-)-(3S)-exo-hydroxycamphor
Y = 86 %

C

cinnamaldehyde
c (S) = 0.05 mg ml^{-1}

Synechococcus sp. PCC 7942

BM: 2.9 $µg_{chl\ a}$ mL^{-1}
BG11 medium (20 mL)
20 °C, 20 h, 160 µE s^{-1} m^{-2}

3-phenylpropanal
C = 65 %

99 h

3-phenylpropan-1-ol
C > 99 %

D

(S)-(-)-limonene
c (S) = 0.2 mg ml^{-1}

Synechococcus sp. PCC 7942

BM: 4 g immobilized cells in Ca^{2+} alginate
synthetic medium (100 mL)
25 °C, 6 h, 1000 lux

cis-carveol
C = 28-32 %

+

trans-carveol
C = 7-9 %

Fig. 6: Selected biotransformations in *Synechococcus* sp. PCC 7942. Native alcohol dehydrogenases catalyze the reduction of (A) aryl methyl ketones [34, 36] and (B) monoterpenes [38]. (C) Cinnamaldehyde is first reduced to the corresponding saturated aldehyde by an ene-reductase and further to the alcohol by an alcohol dehydrogenase [43]. (D) Hydroxylation of (S)-(-)-limonene [31] suggests monooxygenase activity. c(S), initial substrate concentration; BM, biomass; C, conversion; ee, enantiomeric excess; Y, isolated yield.

The hydroxylation of (S)-(+)-limonene and (+)-limonene oxide by *Synechococcus* sp. PCC 7942 [31] is a rare example for a biocatalytic oxyfunctionalization in wild-type cyanobacteria (Fig. 6). Cells immobilized in Ca^{2+} alginate beads constantly converted (S)-(+)-limonene in seven subsequent batches yielding 25–32% *cis*-carveol and 7–9% *trans*-carveol within 6–7 h. Each batch was performed in 100 mL scale containing 20 mg substrate [31].

The bioreduction in *Synechococcus* sp. PCC 7942 depends on the light-driven regeneration of NADPH via photosynthesis [44]. Accordingly, reactions in darkness [43] and treatment with the photosynthetic inhibitor 3-(3,4-dichlorophenyl)-1,1-dimethylurea (DCMU), an *N*-phenyl urea derivative, reduced activity [35, 37]. It is worth mentioning that cyanobacteria can provide sufficient reaction equivalents from storage components to completely convert low substrate concentrations (<5–7 mM) in darkness [17, 19].

It was a striking observation that stereoselectivity of the asymmetric ketoreduction of several alcohols in *Synechococcus* sp. PCC 7942 and other cyanobacteria such as *Anabaena variabilis* and *Nostoc muscorum* was reduced in darkness [35, 36, 38]. While the light-driven bioreduction of 4-chloroacetophenone by *A. variabilis* exhibited excellent enantioselectivity, observed *ee*'s between 50% and 80% were considerably lower under dark conditions [36]. A possible explanation of this surprising finding is the involvement of different alcohol dehydrogenases with opposing selectivity. Furthermore, the present redox state of the cells can favor enzymatic activities in dependence of their cofactor preference. Tamoi et al. investigated the NADP(H) to NAD(H) ratio in *Synechococcus* sp. PCC 7942, the NADPH concentration under light is approximately 6.5-fold higher than the NADH concentration but only 3.8-fold higher in darkness [45]. Altering cultivation conditions can change the expression patterns of involved enzymes in an unpredictable manner which can have tremendous effects on the outcome of the biotransformation. Light intensity and wavelength composition of the light source can as well influence the reaction as both are directly connected to the photosynthetic efficiency.

Based on the studies in the beginning of the twenty-first century, Takemura et al. tried to further increase the enantioselectivity of light-driven biotransformation by knocking out oxidoreductases in the genome of *Synechocystis* sp. PCC 6803 [46]. While the mutants Δ*slr0942* and Δ*slr1825* exhibit increased (*R*)-selectivity for the reduction of *t*-butyl acetoacetate (ee = 53% and ee = 44%, respectively), the mutant Δ*slr0315* mainly produced the (*S*)-alcohol (ee = 91%). On the other side, no mutant showed excellent enantioselectivity (>ee = 95%) and furthermore, no effect was observed for the other investigated substrates [46]. The study reveals one of the main problems when working with wild-type strains: the enzymes responsible for the observed activity often remain unknown, which hinders process optimization by genetic engineering. Böhmer et al. first screened the genome of the eukaryotic microalgae *Chlamydomonas reinhardtii* for members of the Old Yellow Enzyme family [47] and isolated three of four identified, putative ene-reductases. Characterization of in vitro led to a knowledge about their activities, the substrate scope and the optimal reaction conditions [47]. Accordingly, the potential of *C. reinhardtii* as host for whole-cell biotransformations was evaluated resulting in promising initial rates of 0.53 mM h^{-1} for the reduction of *N*-methylmaleimide [47].

Filamentous cyanobacteria are less often investigated compared to unicellular strains but are known as a rich source of versatile secondary metabolites such as fatty acids, saccharides, lipopeptides and alkaloids [48]. Therefore, these strains are a potential source for enzymatic activity. *Nostoc muscorum* PTCC 1636 converted hydrocortisone and although the conversion was incomplete and not quantified, authors identified three androstane and pregnane derivatives after 5 days of biotransformations

[49]. Furthermore, Gorak et al. applied cyanobacteria with different morphology for wild-type biotransformations to produce chiral β-hydroxyalkylphosphonates, a class of organophosphorous xenobiotics with possible biological activity (Fig. 7) [48]. The cyanobacterium *Nodularia sphaerocarpa* completely hydroxylated the carbonyl moiety in direct vicinity to the aromatic ring of the substrate 2-oxo-2-phenylethylphosphonate. This was the first time that an enzymatic activity addressing this functionality was observed. Furthermore, *N. sphaerocarpa* exhibited a higher tolerance toward the substrate than previously described hosts for the reduction of oxophosphonates [48].

Lipok and coworkers demonstrated the potential of eight cyanobacterial strains for the reduction of chalcones and their substituted derivatives (Fig. 7) [50–52]. Chalcones belong to the flavonoid family and are naturally produced in plants. They are usually isolated from plant tissues, but because of the low concentrations in which they occur, this process is not efficient [50]. Authors showed the high potential of all strains to produce a large variety of reduced and hydroxylated chalcones in a preparative scale (250 mL reaction volume). In most cases, all strains completely converted 20 mg L^{-1} within 14 days, despite observed toxicity effects for hydroxylated and methoxylated chalcones [50–52].

A completely different approach how to use autotrophic organisms for biocatalysis was presented by Löwe et al. [53] who used the microalgae *Chlamydomonas reinhardtii* for cofactor regeneration of extracellular redox enzymes via an electron mediator. They exploited the ability of *C. reinhardtii* to produce formate from starch under darkness and anaerobic conditions. Excreted formate was reduced by an externally added formate dehydrogenase which regenerated the cofactor NADH required for the in situ formation of bulk amines catalyzed by an amine dehydrogenase [53]. About 1 mM of substrates such as *n*-butanal, *n*-hexanal, 2-butanone and cyclohexanone were successfully converted to the respective amine within 40 h [53].

In conclusion, most examples for whole-cell biotransformation in wild-type cyanobacteria investigated the reduction of substituted ketones by endogenous alcohol dehydrogenases [34–37, 42–44, 46]. In addition, their potential to convert terpenes [2, 31, 32, 38, 42, 54], hydrocortisones [49], chalcones [50–52] and xenobiotics like oxophosphonates [48] was evaluated. In nearly all cases, cyanobacterial strains exhibited high stereoselectivity and wide substrate specificity, thereby the addition of sacrificial cosubstrates such as glucose circumvented as cyanobacteria can fuel biotransformations by the pool of reduction equivalents or electron carriers regenerated via photosynthesis. The main disadvantage of the use of wild-type strains is perhaps the difficulty to control the concentration of the oxidoreductases, which in turn limits space-time yields to the lower millimolar range after several hours or days. Since the involved enzymes, their expression control mechanisms, cofactor preferences and selectivities often remain unknown, optimization of the reactions by genetic engineering is hindered or accompanied with the tedious search of the needle in the haystack. Biotransformations in recombinant cyanobacteria constitute an alternative since they allow the targeted production of desired oxidoreductases with known selectivities and substrate scopes.

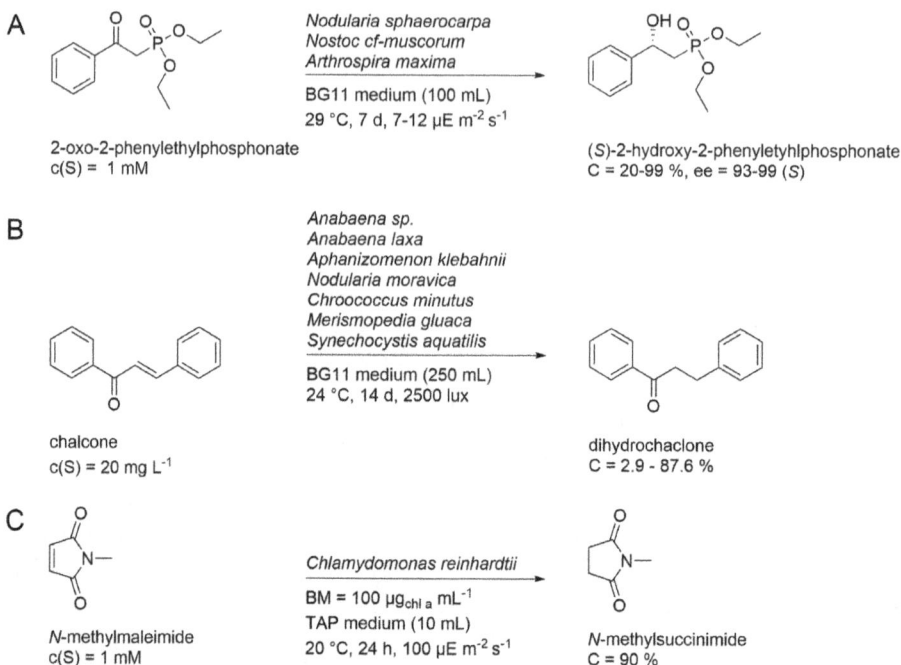

A

Nodularia sphaerocarpa
Nostoc cf-muscorum
Arthrospira maxima

BG11 medium (100 mL)
29 °C, 7 d, 7-12 µE m^{-2} s^{-1}

2-oxo-2-phenylethylphosphonate
c(S) = 1 mM

(S)-2-hydroxy-2-phenyletyhlphosphonate
C = 20-99 %, ee = 93-99 (S)

B

Anabaena sp.
Anabaena laxa
Aphanizomenon klebahnii
Nodularia moravica
Chroococcus minutus
Merismopedia gluaca
Synechocystis aquatilis

BG11 medium (250 mL)
24 °C, 14 d, 2500 lux

chalcone
c(S) = 20 mg L^{-1}

dihydrochaclone
C = 2.9 - 87.6 %

C

Chlamydomonas reinhardtii

BM = 100 µg$_{chl a}$ mL^{-1}
TAP medium (10 mL)
20 °C, 24 h, 100 µE m^{-2} s^{-1}

N-methylmaleimide
c(S) = 1 mM

N-methylsuccinimide
C = 90 %

Fig. 7: Selected examples for biotransformations in wild-type photoautotrophs. (A) Conversion of xenobiotic oxophosphonates was observed for three cyanobacterial strains [48]. (B) Seven cyanobacterial strains reduced chalcone [50]. (C) Ene-reduction was characterized in the microalgae *Chlamydomonas reinhardtii* [47].

3.4.2 Whole-cell biotransformation in genetically modified cyanobacteria

The use of recombinant bacteria as host for whole-cell biotransformations combines advantages of host organism and recombinant enzyme which can have a synergistic effect on a process. Most often, the enzyme possesses attractive characteristics for a biotechnological purpose such as high selectivity and/or stability. However, its source organism might not be suitable for biotechnological application, for example, because it requires specific cultivation conditions or is slowly growing. In the recombinant cell, genetic tools enable the targeted expression of a foreign gene in higher quantities compared to natural enzyme production. All examples are summarized in Tab. 4.

Tab. 3: Biotransformation performed in wild-type cyanobacterial strains. According to the different parameters mentioned in the respective studies, cell densities are either provided in cell dry weight (CDW), optical density (OD) or amount of chlorophyll a (chl a). For the outcome of the biotransformation, either isolated yields (Y), conversion (C) or detected product concentrations are given.

Organism	Substrate (S)	c (S)	LI (µE m^{-2} s^{-1})	Cell density	Product (P)	Outcome	Time	ee/de (%)	Ref.
Synechococcus sp. PCC 6911	Several aldehydes and ketones	1 µg/5 µg$_{CDW}$	n.a.	n.a.	Corresponding alcohol	n.a.	1–7 d	n.a.	[2]
Synechococcus sp. PCC 6716									
Synechococcus sp. PCC 6803									
Anabaena variabilis									
Anabaena oscillarioides									
Synechococcus sp. PCC 7942	2'-3'-4'-5'-6'-Pentafluoroacetophenone	5.7 mM	1000 lux	0.37 g$_{CDW}$ L^{-1}	(S)-Alcohol	Y > 90%	9 days	> 99 (S)	[34]
Synechococcus sp. PCC 7942	α, α-Difluoroacetophenone	0.5 mM	40	1 g$_{CDW}$ L^{-1}	(R)-Alcohol	C = 77%	24 h	66 (R)	[35]
Synechococcus sp. PCC 7942	(S)-Limonene Limonene oxide	0.1 mg mL^{-1}	1000 lux	1 g$_{CWW}$ L^{-1}	cis-Carveol trans-Carveol (1S,2S,4R)-Limonene-1,2-Diol (1S,4R)-Limonene-1-ol-2-one	C = 11% C = 9% C = 32% C = 16%	6 h	n.a.	[31]

(continued)

Tab. 3 (continued)

Organism	Substrate (S)	c (S)	LI (µE m⁻² s⁻¹)	Cell density	Product (P)	Outcome	Time	ee/de (%)	Ref.
Synechococcus sp. PCC 7942	2-Methyl-cyclopentenone	0.2 mg mL⁻¹	n.a.	2 g_cww	(S)-Ketone	C > 99%	1 day	98 (S)	[42]
	2-Methyl-cyclohexenone[a]					C = 86%	1 day	85 (S)	
	2-Ethyl-cyclohexenone					C = 17%	3 days	83 (S)	
	(−)-Carvone					C > 99%	3 days	80 (S, R)	
	(+)-Carvone					C > 99%	3 days	81 (S, S)	
	2,5,5-Trimethylcyclohex-2-en-1-one					C = 15%	3 days	86 (S)	
	2-Methylenecyclohexan-1-one[a]					C = 82%	1 day	71 (S)	
	2-Propylidenecyclohexanone					C = 7%	1 day	72 (S)	
Nostoc muscorum PTCC 1636	Hydrocortisone	0.5 mg mL⁻¹	40	n.a.	Steroid derivatives	n.a.	5 days	n.a.	[49]
Synechococcus sp. PCC 7942	2′-3′-4′-5′-6′-Pentafluoroacetophenone	5 mM	26	1.0–7.7 g L⁻¹	(S)-Alcohols	c(P) = 4.1 mM	48 h	>99 (S)	[36]
	Ethyl-4-chloroacetate	100 mM		1.7–8.8 g L⁻¹		c(P) = 8.8 mM		96.8 (S)	
Anabaena variabilis	2′-3′-4′-5′-6′-Pentafluoroacetophenone	5 mM		1.1–10.9 g L⁻¹		C > 99%		>99 (S)	
Nostoc muscorum	Ethylbenzoylacetate	100 mM				c(P) = 2.3 mM		>99 (S)	
	2′-3′-4′-5′-6′-Pentafluoroacetophenone	5 mM				c(P) = 3.8 mM		97 (S)	
	4-Chloroacetophenone	5 mM				c(P) = 2.1 mM		91 (S)	
Synechococcus sp. PCC 7942	(+)-Camphorquinone	0.4 mg mL⁻¹	2000 lux	1 g_CDW L⁻¹	Mixture of α-keto alcohols	Y = 92%	48 h		[38]
	(−)-Camphorquinone					Y = 91%	144 h		
Synechocystis sp. PCC 6803	(+)-Camphorquinone					Y = 92%	48 h		
	(−)-Camphorquinone					Y = 91%	144 h		

Organism	Substrate (S)	c (S)	LI (µE m⁻² s⁻¹)	Cell density	Product (P)	Outcome	Time	ee (%)	Ref.
Synechococcus sp. PCC 7942	α, α, α-Trifluoroacetophenone in the presence of iodoacetic acid	0.1 mg mL⁻¹	13.4	OD₇₂₀ = 1	(R)-Alcohol	C = 43% / C = 61%	12 h	85 (R)	[37]
Synechocystis sp. PCC 6803 / Δslr0942 / Δslr1825 / Δslr0315	t-Butyl acetoacetate	1.4 mM	49	OD₇₃₀ = 0.6	(R)-Alcohol / (R)-Alcohol / (R)-Alcohol / (S)-Alcohol	C = 32% / C = 30% / C = 37% / C = 6%	24 h	39 (R) / 53 (R) / 44 (R) / 91 (S)	[46]
Synechocystis sp. PCC 6803 / Synechocystis sp. PCC 6714 / Fischerella muscicola UTEX 1301 / Anabaena cylindrica IAM M1 / Plectonema boryanum IAM M101 / Anabaena sp. PCC 7120 / Synechococcus sp. PCC 7942	Cinnamyl aldehyde	0.05 mg mL⁻¹	160	3.8 µg_Chl mL⁻¹ / 3.4 µg_Chl mL⁻¹ / 2.0 µg_Chl mL⁻¹ / 1.8 µg_Chl mL⁻¹ / 1.2 µg_Chl mL⁻¹ / 3.0 µg_Chl mL⁻¹ / 2.9 µg_Chl mL⁻¹	Cinnamyl alcohol / 3-Phenylpropanol	C = 98% / C ≈ 47% / C ≈ 35% / C ≈ 28% / C ≈ 20% / C ≈ 42% / C > 99%	3 days / 44 h / 102 h / 24 h / 50 h / 45 h / 99 h	n.a.	[43]

(continued)

Tab. 3 (continued)

Organism	Substrate (S)	c (S)	LI (µE m^{-2} s^{-1})	Cell density	Product (P)	Outcome	Time	ee/de (%)	Ref.
Synechococcus sp. MCCS 034	Limonene	≈ 6.1 mM	n.a.	40 beads of cells immobilized in Ca^{2+} alginate	cis-Carveol	C = 38.1%	24 h	n.a.	[32]
	Carvone	≈ 6.3 mM			trans-Carveol	C = 1.3%			
Synechococcus sp. MCCS 035	Thymol	≈ 6.3 mM			carvone	C = 2.3%			
	carvone	≈ 6.3 mM			cis-Limonene oxide	C = 4.3%			
					trans-Limonone oxide	C = 2.3%			
					cis-Dihydro carvone	C = 3.4%			
					trans-Dihydro carvone	C = 12.7%			
					Dihydro carveol	C = 6.7%			
					Thymoquinone	C = 1.5%			
					trans-Dihydro carvone	C = 53.7%			
Nostoc cf-muscorum	2-Oxo-2-phenylethylphosphonate	1 mM	7–12	n.a.	(S)-Hydroxy product	C = 26%	7 days	>99 (S)	[48]
Nodularia sphaerocarpa	2-Oxo-2-phenylethylphosphonate		7–12	n.a.	(S)-Hydroxy product	C = 99%		93 (S)	
Arthrospira maxima	Diethyl 2-oxopropylphosphonate		5–9	n.a	(S)-Hydroxy product	C = 20%		>99 (S)	
Leptolyngbya foveolarum	Different oxophosphonates		7–12	n.a.	(S)-Hydroxy product	C < 5%		n.d.	
Spirulina platensis	Chalcone	20 mg L^{-1}	2500 lux	n.a.	Dihydrochalcone	Y = 3.1%	14 days	n.a.	[50]
Anabaena laxa						Y > 99%			
Aphanizomenon klebahnii						Y > 99%			
Nodularia moravica						Y > 99%			
Merismopedia glauca						Y = 91.8%			
Synechocystis aquatilis						Y > 99%			

Organism	Substrate (S)	c (S)	LI (µE m^{-2} s^{-1})	Cell density	Product (P)	Outcome	Time	ee (%)	Ref.
Chroococcus minutus Anabaena sp.	Chalcone	20 mg L^{-1}	2500 lux	n.a.	Dihydrochalcone Cinnamic acid Hydrocinnamic acid Dihydrochalcone Diphenylpropenol Diphenylpropanol	Y = 69.9% Y = 2.9 % Y = 9.8% Y = 87.6% Y = 6.7% Y = 5.7%	14 days	n.a.	[50]
Spirulina platensis Anabaena laxa Aphanizomenon klebahnii Nodularia moravica Merismopedia glauca Synechocystis aquatilis Chroococcus minutus Anabaena sp.	2'-Hydroxychalcone	20 mg L^{-1}	2500 lux	n.a.	2'-Hydroxy-dihydrochalcone	C = 4.3% C = 22.9% C = 5.4% C = 8.7% C = 6.7 and C = 15.8% C = 15.6% C = 16.3%	14 days		[51]
Spirulina platensis Anabaena laxa Aphanizomenon klebahnii Nodularia moravica Merismopedia glauca Synechocystis aquatilis Chroococcus minutus Anabaena sp.	2''-Hydroxychalcone	20 mg L^{-1}	2500 lux	n.a.	2''-Hydroxy-diphenylpropanone	C = 20.9% C = 98.7% C = 97.8% C = 97.5% C = 75.6% C = 98.6% C = 94% C = 97.5%	14 days		[51]

(continued)

Tab. 3 (continued)

Organism	Substrate (S)	c (S)	LI (µE m⁻² s⁻¹)	Cell density	Product (P)	Outcome	Time	ee/de (%)	Ref.
Spirulina platensis	4″-Hydroxychalcone	20 mg L^{-1}	2500 lux	n.a.	4″-Hydroxydihydro-chalcone (1), 4″-Hydroxydiphenyl-propanol (2) and 4″-dihydroxy-dihydrochalcone (3)	C = 99.3% (1)	14 days	n.a.	[51]
Anabaena laxa						C = 99.3% (1)			
Merismopedia glauca						C = 46.5% (1)			
Synechocystis aquatilis						C = 99.1% (1)			
Chroococcus minutus						C = 98.5% (1)			
Anabaena sp.						C = 31.2 % (1)			
Aphanizomenon klebahnii						C = 28.4% (2)			
						C = 16.2% (3)			
Nodularia moravica						C = 41.6% (1)			
						C = 20.0% (2)			
						C = 22.6% (3)			
						C = 82.2% (1)			
						C = 4.4% (2)			
						C = 6.8% (3)			

[a] Reduction of the saturated ketone to the alcohol was observed
[b] Diastereomeric excess.

LI, light intensity; c(S), initial substrate concentration; CDW, cell dry weight; chl, chlorophyll a; c(P), detected product concentration; Y, isolated yield; C, conversion; n.d., not detected; n.a., not analyzed.

3.4.2.1 Examples for reductive biotransformations

In 2016, Köninger et al. demonstrated the biocatalytic application of cyanobacterial cells overexpressing the genes of heterologous redox enzymes [17]. Authors integrated the gene of the ene-reductase YqjM from *Bacillus subtilis* [55, 56] into the genome locus *slr0168* of *Synechocystis* sp. PCC 6803. This flavoprotein catalyzes the *trans*-hydrogenation of C=C double bonds in enones with NADPH as a preferred electron donor (Fig. 8) [9, 17, 57, 58]. In *Synechocystis* sp. PCC 6803, the expression of the YqjM gene was controlled by the psbA2 promotor and seven enone substrates were applied for whole-cell biotransformations. Cell activities were considerably lower in darkness or in the presence of the photosynthesis inhibitor DCMU [17]. The results proved that the cofactor for the ene-reduction in *Synechocystis* sp. PCC 6803 is directly regenerated via photosynthesis, but to some part also by storage components. The best accepted substrate was 2-methylmaleimide. About 10 mM was successfully converted within 3 h achieving a specific cell activity of 123 U g_{CDW}^{-1} and a space-time yield of 1.1 g $L^{-1} h^{-1}$ [17].

Chiral amines are important chemical building blocks for a variety of pharmaceutical and agrochemical intermediates [59–61]. Estimations suggest that around 40% of all pharmaceuticals contain at least one chiral amine moiety [60]. Imine reductases (IREDs) selectively catalyze the reduction of imines to their corresponding amines and are therefore of biotechnological interest (Fig. 1). Since IREDs are strictly NAPDH dependent, they are excellent candidates for bioreductions in cyanobacteria. Büchsenschütz and Vidicme-Risteski et al. integrated the genes of three different IREDs into *Synechocystis* sp. PCC 6803 and evaluated the resulting whole-cell biocatalysts for their potential to reduce a variety of imine substrates (Fig. 8) [19]. The study revealed two main limiting factors of the concept. The high toxicity of some imine substrates limited the approach to the conversion in the lower millimolar range. In contrast, heterotrophic hosts like *E. coli* were shown to tolerate and convert up to 100 mM [61]. Toxicity of substrates and products is an often observed problem for whole-cell biotransformation, even though it can be remediated to some extent by using a two-phase system [4, 24] or by substrate feeding [9]. The expression of the heterologous genes proved to be crucial for the cellular activity. Exchange of the promoter P_{psbA2} to the stronger promoter P_{cpc} increased the activity of cells producing IRED-A from *Streptomyces* sp. GF3587 toward 6-methyl-2,3,4,5-tetrahydropyridine from 2.7 U g_{CDW}^{-1} to 21 U g_{CDW}^{-1} [19].

Sengupta et al. integrated the gene of the NADPH-dependent alcohol dehydrogenase from *Lactobacillus kefir* (LkADH) into *Synechococcus elongatus* PCC 7942 which grows faster and exhibits a higher tolerance toward high light conditions compared to *Synechocystis* sp. PCC 6803 (Fig. 8) [26]. The recombinant strain was evaluated for (i) its potential to reduce acetophenone and 1-phenylethanol, and (ii) the effect of light and CO_2 content on its productivity [26]. Furthermore, the activity of growing and resting cells was compared. The recombinant *Synechococcus elongatus*

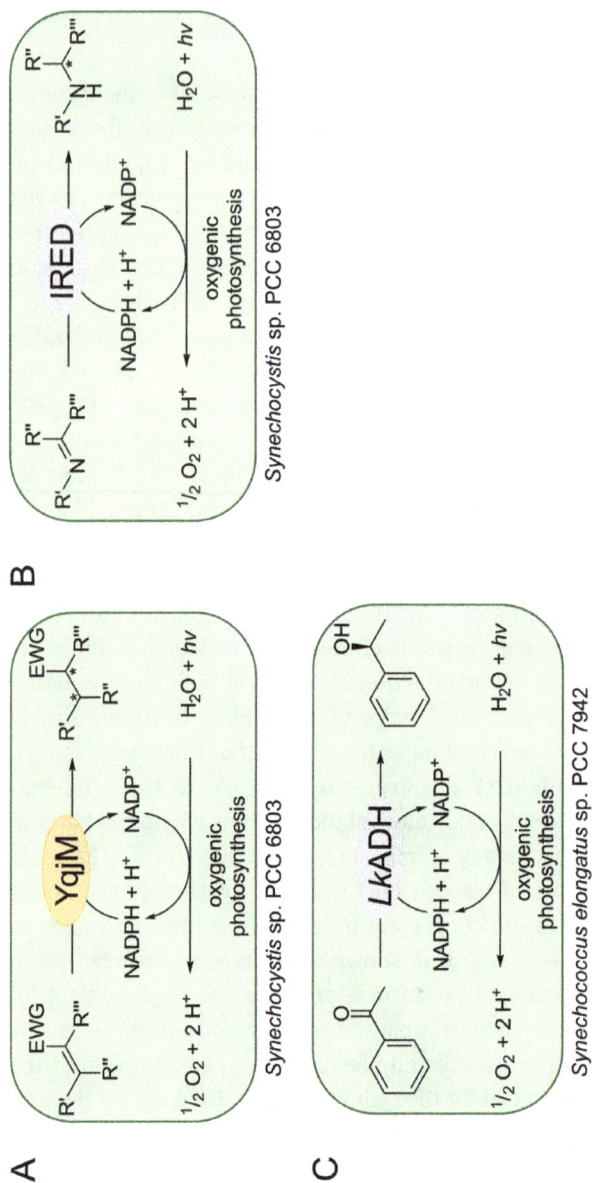

Fig. 8: Examples of reductive biotransformations in recombinant cyanobacteria with NADPH-recycling via photosynthesis. (A) Ene-reduction catalyzed by YqjM from *Bacillus subtilis* in *Synechocystis* sp. PCC 6803 [9, 17]. (B) Reduction of imines by heterologous imine reductases (IREDs) in *Synechocystis* sp. PCC 6803 [19]. (C) The alcohol dehydrogenase from *Lactobacillus kefir* (LkADH) catalyzes the conversion of acetophenone to (R)-1-phenylethanol in *Synechococcus elongatus* sp. PCC 7942 [26].

PCC 7942 strain showed the highest growth rate under high light (400 μE m^{-2} s^{-1}) and high CO_2 (1%, bubbling) without any signs of decreased cell viability. The highest productivity of the cells was achieved under moderate light intensity (150 μE m^{-2} s^{-1}) and 0.5% CO_2 supply. Cells successfully converted 20 mM acetophenone to (R)-1-phenylethanol (ee > 99%) within 6 h and with a rate of 3.1 mM h^{-1} [26]. Notably, the applied cell density of 0.66 g$_{CDW}$ L^{-1} [26] was rather low compared to the previous studies about light-driven ene-reduction (1.8 g$_{CDW}$ L^{-1}) [17] and imine reduction (3.6 g$_{CDW}$ L^{-1}) [19].

Yunus et al. investigated the production of alcohols with intermediate-chain length such as 1-octanol and 1-decanol from carbon dioxide in *Synechocystis* sp. PCC 6803 [27]. The approach differs from previous examples because no specific substrate was added to the cells. Instead, authors extended the native fatty acid biosynthesis pathway by integrating a three-step reaction starting from acyl-ACP as a precursor. Acyl-ACP is converted to the free fatty acids via a thioesterase (Tes3, *Anaerococcus tetradius*) and further to fatty aldehydes by a carboxylic acid reductase (CAR, *Mycobacterium marinum*). For activation of the CAR, a phosphopantetheinyl transferase (Sfp, *Bacillus subtilis*) was coproduced [27]. Finally, the aldehydes are reduced by the native aldehyde reductases Ah3 from *Synechocystis* sp. PCC 6803 to form the desired fatty alcohols. As *Synechocystis* sp. PCC 6803 is known to recycle fatty acids via an acyl-ACP synthetase (aas) [62], authors used a knock-out strain lacking this ability. The three recombinant enzymes were integrated into the cell via conjugation using the self-assembled plasmid pIY417 with RSF1010 origin for self-replication. Gene expression was investigated using different promoters and the cobalt-inducible promoter P$_{coa}$ proved to be most suitable for the approach. However, genetic instability of the used plasmid caused loss of cell activity upon several subculturing although an antibiotic pressure was applied [27]. Nevertheless, titers of more than 100 mg L^{-1} of 1-octanol and 1-decanol were achieved. For 1-octanol, the concentration of octanoic acid was the limiting factor and by feeding the acid, titers could be further increased to 905 mg L^{-1} over 8 h. Especially CARs are highly interesting enzymes for the application in cyanobacteria, since these oxidoreductases require both ATP and NADPH for their activity. The study proved that photosynthesis is sufficient to supply the CAR-catalyzed reaction with both molecules [27].

3.4.2.2 Examples for oxyfunctionalization

Molecular oxygen represents an ideal oxidant for oxidative catalyses as it is abundant, has a low molecular weight and is least expensive compared to other oxidants such as transition metals [63]. However, its use is accompanied with safety issues and harsh reaction conditions [63, 64]. Enzymatic oxidations are beneficial as catalysis occurs under mild reaction conditions and with high selectivity. Especially oxyfunctionalizations that involve the incorporation of one or both oxygen atoms into the final product are of major interest. Such reactions are catalyzed by oxygenases, enzymes that often require flavin adenine dinucleotide (FAD), NAD(P)H and/or ferredoxin (Fd). However, whole-cell biotransformations in aerobic, heterotrophic host are often limited by the oxygen concentrations. This has two major reasons: (i) the competition for oxygen between the reaction and the respiratory chain at increasing cell densities and (ii) the low gas–liquid mass transfer rate of oxygen in aqueous system under ambient conditions [25, 65, 66]. Cyanobacteria produce oxygen from water via photosynthesis directly within the cells and aqueous liquid phase. Therefore, the gas–liquid mass transfer limitation is mitigated which benefits introduced oxyfunctionalizations [4, 24, 25, 65].

One of the first oxygenases that was recombinantly produced in cyanobacteria is the mammalian cytochrome P450 monooxygenase CYP1A1 from *Rattus norvegicus* (Fig. 9) [20]. Cytochrome P450s are a versatile class of monooxygenases that catalyze multiple oxidation reactions including hydroxylations, dehydrogenations, epoxidations and O-dealkylations [20, 67, 68]. The splitting of molecular oxygen within the reaction mechanism requires the successive delivery of two electrons. These are often provided by Fd which is reduced in a NAD(P)H-dependent reaction catalyzed by a respective ferredoxin reductase (FnR; Fig. 1) [67]. Berepiki et al. integrated the gene of CYP1A1 into the *glpK* genomic neutral site of the cyanobacterium *Synechococcus* sp. PCC 7002. Gene expression was controlled by the promotor P_{cpc} deriving from *Synechocystis* sp. PCC 6803. The membrane bound, recombinantly produced enzyme was localized in the thylakoid membrane and thus in direct proximity to the photosynthetic machinery [20]. The activity of CYP1A1 was measured by a fluorescent assay monitoring the O-deethylation of 7-ethoxyresorufin by the enzyme. Notably, CYP1A1 was active without coproducing its native P450 oxidoreductase necessary for electron supply and activity decreased in darkness and presence of the photosynthesis inhibitor DCMU. These results showed that the reaction was fueled with electrons from photosynthesis – potentially from Fd directly [20]. Furthermore, the presence of CYP1A1 raised the optimal light intensity for photosynthetic electron transport. This indicated that CYP1A1 is an additional electron sink consuming electrons that are otherwise wasted or redirected [20].

The cyclohexanone monooxygenase (CHMO) from *Acinetobacter calcoaceticus* NCIMB 9871 is one of the best investigated Baeyer–Villiger monooxygenases because of its high activity [66]. Due to its low stability and its NADPH dependence, CHMO is often applied in whole-cell biotransformations [66, 69–71]. Böhmer et al. produced the CHMO in recombinant *Synechocystis* sp. PCC 6803 under control of the light-inducible promotor P_{psbA2} (Fig. 9) [18]. The obtained strain catalyzed the formation of lactones from several cyclic ketone substrates with activities in the range of 2–5 U g_{CDW}^{-1}. Thus, activity for the oxyfunctionalization was comparable with the activity for the imine reduction and lower than the activity of the ene-reduction in the same host [17–19]. Furthermore, the conversion was lowered by a concurrent side-reaction catalyzed by the native enzyme pool in the cyanobacterial cell. Alcohol

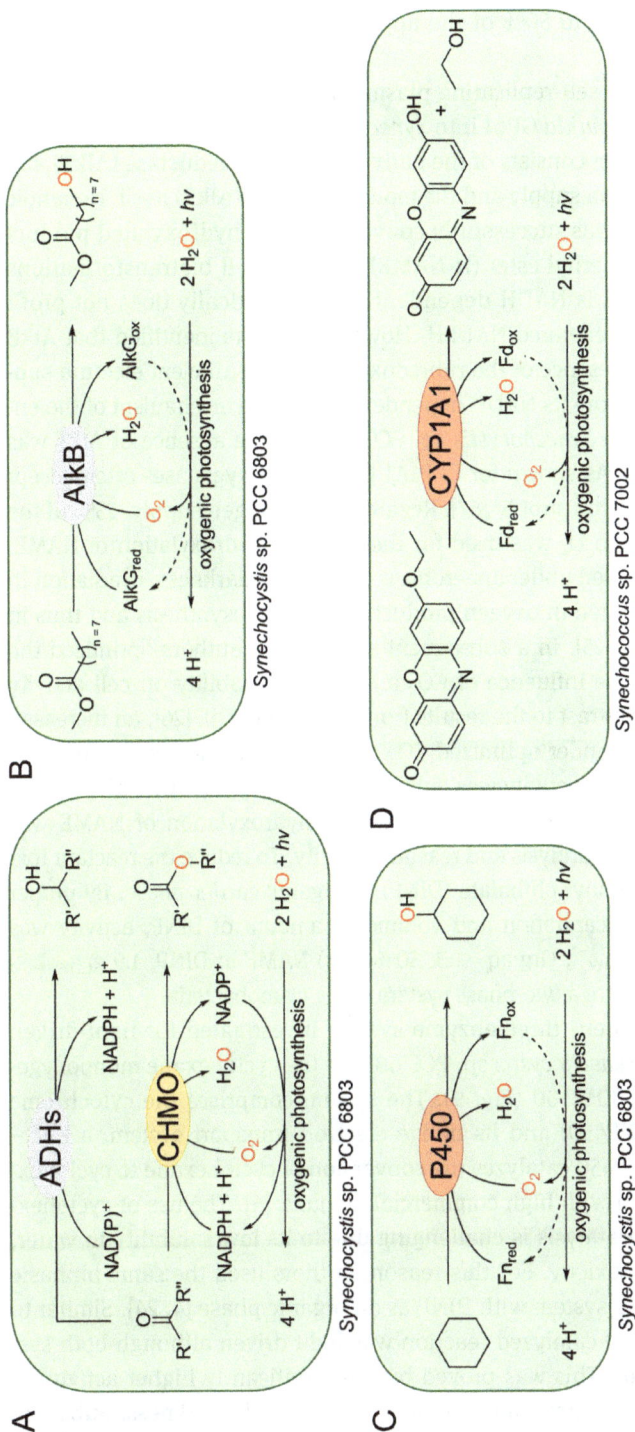

Fig. 9: Examples for oxyfunctionalizations in recombinant cyanobacteria. Dashed arrows indicate assumed electron flows. (A) Baeyer–Villiger oxidation is catalyzed by the NADPH-dependent cyclohexanone monooxygenase (CHMO) from *Acinetobacter calcoaceticus* NCIMB 9871. Since native alcohol dehydrogenases (ADHs) reduce the ketones to corresponding alcohols, the reaction is hindered [47]. (B) Nonanoic acid methyl ester (NAME) is hydroxylated by the monooxygenase AlkB from *Pseudomonas putida* GPo1 in recombinant *Synechocystis* sp. PCC 6803. Electrons are provided by its native rubredoxin AlkG and potentially derived from the photosynthetic electron transport [24, 25]. (C) Cyclohexane is converted to cyclohexanone by the P450 monooxygenase from *Acidovorax* sp. CHX100 which is supplied with electrons from the *Acidovorax* ferredoxin (Fn) [4, 23]. (D) O-Deethylation of 7-ethoxyresorufin by the P450 monooxygenase CYP151 from *Rattus norvegicus* and electrons is potentially derived from photosynthesis [20, 21].

dehydrogenases reduced up to 50% of the applied ketones to the corresponding alcohol [18].

Hoschek et al. used the self-replicating plasmid pAHO42 to transfer the AlkBGT system from *Pseudomonas putida* GPo1 into *Synechocystis* sp. PCC 6803 (Fig. 9) [25]. The three-component system consists of the native rubredoxin reductase (AlkT) and rubredoxin (AlkG) for electron supply and the monooxygenase (AlkB) itself. Nonanoic acid methyl ether (NAME) was successfully converted to the hydroxylated product ω-hydroxynonanoic acid methyl ester (H-NAME) in whole-cell biotransformations (1.5 U g_{CDW}^{-1}). The AlkBGT is NADH dependent and theoretically does not profit from photosynthetically regenerated NADPH. However, authors identified that AlkB activity only requires the presence of the rubredoxin AlkG for sufficient electron supply and is thus decoupled from its NADH dependence [24]. The mechanism of the endogenous AlkG reduction in *Synechocystis* sp. PCC 6803 in the absence of AlkT was not further elaborated [24]. As shown for CYP1A1 [20], monooxygenases often accept non-native systems for electron supply [67]. Regarding the oxygen supply, 25% of the photosynthetically generated O_2 was used for the terminal hydroxylation of NAME. While no activity was observed under anaerobic conditions in darkness, irradiation in the absence of oxygen resulted in oxygen production via photosynthesis and thus in an activity of 0.9 U g_{CDW}^{-1} [25]. In a subsequent study, same authors optimized the reaction by investigating the influence of CO_2 and light availability on cell activity [24]. Surprisingly and in contrast to the results from Sengupta et al. [26], an increased cyanobacterial growth rate under optimized CO_2 conditions was not associated with highest activity [24]. However, activity was enhanced to 3.0 U g_{CDW}^{-1} by NaHCO_3 supplementation and increase of the light intensity. The hydroxylation of NAME was mainly limited by substrate hydrolysis and reactant toxicity. To reduce the reactant toxicity, authors applied diisononyl phthalate (DINP) as organic carrier phase. By proper adjustment of substrate concentration and volumetric amount of DINP, activity was more than double (5.6 U g_{CDW}^{-1}, Org:aq = 1:3, 50% (v/v) NAME in DINP, 1.9 g_{CDW} L^{-1}) [24]. This is the first example of a two-phase system using cyanobacteria.

Another NADH-dependent three-enzyme system investigated for light-driven oxyfunctionalizations in *Synechocystis* sp. PCC 6803 is the cyclohexane monooxygenase from *Acidovorax* sp. CHX100 (Fig. 9). The system comprises the cytochrome P450 monooxygenase (CYP450) and its native electron transport system, a FAD–FnR and an Fd [4]. The CYP450 catalyzes the conversion of cyclohexane to cyclohexanol, a precursor for nylon with high commercial demand [4]. The use of cyclohexane as a substrate for biocatalysis is challenging due to its low solubility in water, its high volatility and its toxicity. For this reason, authors used the same biphasic system then for the AlkBGT system with DINP as an organic phase [4, 24]. Similar to AlkBGT system, the CYP450-catalyzed reaction was light driven although both systems are NADH dependent. This was proved by the significantly higher activity of 26.3 U g_{CDW}^{-1} under illumination compared to 6.2 U g_{CDW}^{-1} in darkness. Substrate feeding to the organic carrier phase further increased the activity to 39 U g_{CDW}^{-1} [4].

Upscaling of a reaction is usually accompanied with a loss of the space-time yields and, therefore, needs careful adjustment and optimization of cultivation and reaction conditions. Two different approaches for the upscaling of the light-driven production of cyclohexanol were reported. Biotransformation using the biphasic system in a 3 L stirred-tank photobioreactors resulted in the production of 2.6 g cyclohexanol in 52 h [4]. In a second approach, phototrophic *Synechocystis* sp. PCC 6803 was cocultivated with heterotrophic *Pseudomonas taiwanensis* VLB12. Both strains were genetically engineered and contained the expression cassette for the CYP450 system from *Acidovorax* sp. CHX100 for heterologous cyclohexanol production [4, 23, 72]. All previous examples have shown that oxygen production in phototrophs is highly advantageous for light-driven oxyfunctionalizations. It is also one of the main obstacles for the development of industrial-scaled photobioreactor. High-density cultivation of phototrophic organisms results in oxygen accumulation in closed systems like tubular photobioreactors which led to (i) photorespiration and (ii) toxicity effects [23, 73]. The ribulose-1,5-bisphosphate carboxylase/oxygenase (RuBisCo) is the core enzyme of the Calvin–Benson cycle and usually catalyzes the formation of two molecules of glycerate-3-phosphate from RuBisCo and CO_2. At high concentrations of dissolved oxygen, RuBisCo consumes O_2 and the resulting phosphoglycolate is recycled in an oxygen-dependent process called photorespiration [73]. Apart from that, a high oxygen concentration and the accompanied increase of reactive oxygen species (ROS) in the solution can inhibit growth and photosynthesis [23, 73]. Therefore, in situ O_2 removal is crucial for a stable cultivation of phototrophic organisms at high densities [23]. The direct utilization of O_2 for oxyfunctionalizations is a possible solution, however, due to low cell productivities it does not yet solve the problem. Cocultivation profits from the light-driven O_2 production of *Synechocystis* sp. PCC 6803 and the concurrent O_2 consumption by the oxyfunctionalizations and by respiration of *P. taiwanensis* [23]. Both strains formed a biofilm on the inside surface of a capillary reactor (3 mm diameter) constantly flushed with growth medium. Indeed, sole cultivation of *Synechocystis* sp. PCC 6803 led to low biofilm formation and clear growth inhibiting effects due to O_2 accumulation. Cocultivation of *P. taiwanensis*, the addition of citrate (0.4 g L^{-1}) and a flow without air segments optimized biofilm formation [23]. For biotransformations, both strains were cocultivated for 5 weeks without citrate supplementation but with segmented airflow before cyclohexane was added to the feed flow. Cyclohexanol was produced with a stable space-time yield of 0.2 g L^{-1} h^{-1} for a month with 50% of the produced oxygen being used for the biotransformation [23]. About 98.9% of the applied cyclohexane was converted and cyclohexanone (15.5%) was the only side-product. Authors named the complex medium and the accompanied material costs as well as the light distribution as main obstacles of their concept. Although biotransformations using only *P. taiwanensis* achieved 0.4 g L^{-1} h^{-1} [74] and the space-time yield was 125-fold lower to the existing chemical process [75], the results underline the potential of cocultivating phototrophs with heterotrophs with concurrent oxyfunctionalizations.

3.4.2.3 Engineering of the photosynthetic electron transport chain

A few years after Köninger et al. published the study of light-driven ene-reduction in *Synechocystis* sp. PCC 6803, Assil Companioni and Büchsenschütz et al. investigated the effect of different promotors onto the cell activity of the same reaction. The gene expression of YqjM from *Bacillus subtilis* was mediated by the three promoters P_{zia}, P_{psbA2} and P_{cpc}. As expected, the weak, Zn^{2+}-inducible promotor P_{zia} led to decreased activities while the strong P_{cpc} promotor improved cell activity by 1.3-fold [9]. Furthermore, authors used the NADPH assay for the detection of the YqjM activity in crude cell extracts. In the reaction setup, lysate was mixed with externally added NADPH and substrate. Thus, the activity was only restricted by the enzyme's concentration in the crude cell extract [9]. Surprisingly, the observed activity of the whole cells was not in line with the much higher production of YqjM by P_{cpc}-mediated gene expression compared with P_{psbA2}-based expression (1.7-fold improvement of the activity in crude cell extract). The light-driven ene-reduction still represents the fastest ever-reported reaction in recombinant cyanobacteria and this discrepancy observed by Assil Companioni and Büchsenschütz et al. raised the questions if the photosynthetic NADPH production is insufficient to meet the increased demand [9]. Indeed, pulse amplitude-modulated fluorescence spectroscopy [76] revealed a decrease of the intracellular NADPH concentration during biotransformation [9]. This indicates that the heterologous reaction constitutes a strong electron sink.

Reaching NADPH-limiting conditions, the space-time yield mainly depends on the maximum amount of reaction equivalents that can be provided by the phototrophic cell and, hence, from the efficiency of PETC. Increasing the catalyst concentration is a suitable strategy to improve space-time yields but because of self-shading, this is only possible to a certain extent for light-driven biotransformations. In dense cultures, cells conceal each other which diminishes light availability and photosynthetic efficiency [77, 78]. Depending on the diameter, light and dark zones within the photobioreactor results in cells with different physiological states [78]. Outside cells are directly exposed to light and if the produced pool of reductive equivalents exceeds the demand of the cell for metabolic processes, the generation of ROS and photoinhibition is the consequence [21]. In nature, cyanobacteria have developed various protective mechanisms to prevent cell damage under changing environmental conditions such as fluctuating light intensities [79]. Two of these electron sinks shall be discussed in more detail: the cyclic electron transport catalyzed by the NAD(P)H dehydrogenase-like complex 1 (NDH-1) and the water–water cycle catalyzed by flavodiiron proteins (FDPs). NDH-1 redirects electrons from reduced Fd back to plastoquinonewhich is accompanied by the pumping of protons to the thylakoid lumen driving ATP synthesis [21, 80]. This cyclic electron transport can be prevented by disrupting the functionality of the D2 subunit from the NDH 1 complex [21]. Instead, electrons can be redirected to a heterologous reaction, as exemplified by the *O*-deethylation of 7-ethoxyresorufin by the P450 monooxygenase CYP1A1 in *Synechococcus* sp. PCC 7002 [20, 21]. The knock-out strain*ΔndhD2* exhibited increased activity and higher photosynthetic rate compared to strain with intact NDH-1 complex [21].

FDPs [81–83] catalyze the transfer of excess electrons from reduced Fd to oxygen-producing water [84–86]. Thus, this reaction can be understood as a backreaction of the PSII-mediated water splitting. Under certain conditions, around 60% of the photosynthetically generated electrons might be recycled by FDPs and lost for potential biotransformations (assuming an NADPH photoproduction of 530–1070 µmol NADPH $mg_{chl\,a}^{-1}\,h^{-1}$) [9, 76]. To redirect electrons lost via the water–water cycle to the heterologous reaction, Assil Companioni and Büchsenschütz et al. introduced the gene for YqjM into the two FDP knock-out strains *Δflv1* and *Δflv3* [81, 87] which cannot form a functional Flv1/Flv3 heterooligomer. Biotransformations with both knock-out strains showed a significant increase in cell activity (up to 147 U g_{CDW}^{-1}) and product formation rate (up to 18 mM h^{-1}) [9].

The examples of light-driven *O*-deethylation and ene-reduction demonstrate that the reaction rate of redox biotransformations in cyanobacteria can be increased via engineering of photosynthetic electron transport [9, 21]. In both cases, NADPH availability for the heterologous redox reaction was significantly increased which is an important step for future acceleration of light-driven processes. This shows that metabolic engineering of the photosynthetic electron flow is a highly promising approach to achieve a further activity increase of cyanobacterial biocatalysts.

3.5 Outlook: future perspective and economic feasibility

Cyanobacteria have been investigated as host for whole-cell biotransformations since 1986 with the goal to replace heterotrophic production systems for industrial application. Reactions in wild-type strains are limited to conversion rate of few milligrams in several hours or days. Use of recombinant strains increased cell productivity by at least one order of magnitude and space-time yields between 0.04–1.1 g $L^{-1}\,h^{-1}$ have been reported [4, 17]. All studies revealed limitations of the concept such as substrate toxicity, a low intracellular enzyme concentration or insufficient NADPH photoproduction. Therefore, research in the field concentrates on the expansion of the genetic toolbox to design highly efficient photosynthetic cell factories and to solve above limitations. Furthermore, the cultivation of phototrophs is currently associated with high operating costs. Irradiation with artificial light sources, the complex composition of the minimal medium and the supply of carbon dioxide are the main cost factors that must be minimized to attract industrial interest [88]. Optimization of photobioreactors is therefore critical to exploit the main benefits associated with phototrophic microorganisms: the utilization of sunlight as energy source and the fixation of greenhouse

Tab. 4: Biotransformations performed in (A) recombinant Synechocystis sp. PCC 6803 and (B) recombinant Synechococcus elongatus sp. PCC7942.

	Recombinant enzyme, origin	Substrate (S)	c (S) (mM)	LI (µE m⁻² s⁻¹)	Cell density	C (%)	Time (h)	Spec. activity (U g_{CDW}^{-1})	Product formation	ee (%)	Ref.
A	YqjM, Bacillus subtilis	Cyclohexenone	15	150	OD₇₅₀ = 10, 1.8 g_{CDW} L⁻¹	70	4	39	4.1 mM h⁻¹	n.a.	[17]
		2-Methylcyclohexenone	15			42	24	21.1	2.1 mM h⁻¹	n.a.	
		Ketoisophorone	10			57	24	6.2	0.7 mM h⁻¹	n.a.	
		Cyclopentenone	15			>99	24	25.6	2.7 mM h⁻¹	n.a.	
		N-Methylmaleimide	15			94	24	53.3	5.7 mM h⁻¹	n.a.	
		2-Methylmaleimide	20			>99	3.3	90.9	9.6 mM h⁻¹		
		2-Methyl-N-methylmaleimide	10			>99	1	123	10 mM h⁻¹		
			20			>99	1	99.5	10.5 mM h⁻¹		
B	YqjM, Bacillus subtilis[a]	Cyclohexenone	10	150	OD₇₅₀ = 10, 2.4 g_{CDW} L⁻¹	n.a.	n.a	101	15.9 mM h⁻¹	n.a	[9]
		2-Methylcyclohexenone	10			n.a	n.a	15.8	2.3 mM h⁻¹	n.a	
		N-Methylmaleimide	60			n.a	n.a	107.1	15.5 mM h⁻¹	n.a	
		2-Methyl-N-methylmaleimide				n.a.	n.a	132.7	20.9 mM h⁻¹	n.a	
		2-Methylmaleimide				>99	1	146.9	18.3 mM h⁻¹	n.a.	
						>99	4	n.a.	n.a		
	CHMO, Acinetobacter calcoaceticus NCIMB 9871	Cyclohexanone	5	150	OD₇₅₀ = 10, 1.8 g_{CDW} L⁻¹	n.a	n.a	2.3	n.a.	n.a	[18]
		rac-2-Methylcyclohexanone				48	24	2.0		n.a	
		rac-3-Methylcyclohexanone				72	24	2.9		n.a.	
		4-Methylcyclohexanone				82	24	5.7			
		Cyclopentanone				>99	48	2.3			

Catalyst	Substrate				C				ee
IRED-A, Streptomyces sp. GF3587	2-Methylpyrroline	4	150	$OD_{750}=15$, 3.6 g_{CDW} L^{-1}	84	1	19.5	4.8 mM h^{-1}	> 94 (R)
	6-Methyltetrahydropyridine	8			65		21.8	6.3 mM h^{-1}	> 94 (R)
	7-Methyltetrahydroazepine	4			61		8.9	2.6 mM h^{-1}	> 94 (R)
	3,4-Dihydroisoquinoline	5			47		5.5	1.4 mM h^{-1}	> 99 (R)
	1-Methyldihydroisoquinoline	4			64		10.2	2.9 mM h^{-1}	–
	2,3,3-Trimethylindole	5			18		3.1	0.9 mM h^{-1}	> 71 (R)
		5			19		1.6	0.5 mM h^{-1}	> 88 (R)
AlkBGT, Pseudomonas putida GPo1 [24]	NAME Two-phase system[b]	10	30	1.8 g_{CDW} L^{-1}	n.a.	26	3.3	0.5 mmol g_{CDW}^{-1}	
	50% (v/v) in DINP			1.9 g_{CDW} L^{-1}			5.6	3.8 mmol g_{CDW}^{-1}	
CYP450, Acidovorax sp. CHX100 [4]	Cyclohexane Two-phase system[c]	5 mM	150	0.5 g_{CDW} L^{-1}	n.a.	2	26.3	n.a.	
	20% (v/v) in DINP					2	39.2	19.7 mmol g_{CDW}^{-1}	
	3 L photobioreactor[c]					52	34.9	4.9 g g_{CDW}^{-1}	
	5% (v/v) in DINP							g_{CDW}^{-1}	
B LkADH, Lactobacillus kefir [26]	Acetophenone	20	400	0.7 g_{CDW} L^{-1}	>99	6	n.a.	n.a.	>99 (R)

[a]In the knock-out mutant Syn::ΔFlv1,
[b]1:3 Org:Aq phase ratio with DINP,
[c]25% (v/v) (DINP).

DINP, diisononyl phthalate; NAME, nonanoic acid methyl ester; LI, light intensity; c(S), initial substrate concentration; OD, optical density; CDW, cell dry weight; C, conversion; ee, enantiomeric excess; n.a., not analyzed.

gas CO_2. In future, the question "how fast can we get?" will mainly depend on the amount of reaction equivalents that can be provided by the phototrophic cell. Since increasing the cell density is only possible to a certain extent due to self-shadowing, increasing the efficiency of PETC might be the way to go.

References

[1]   Anastas P, Eghbali N. Green chemistry: Principles and practice. Chem Soc Rev 2010, 39, 301–12.
[2]   Jüttner F, Hans R. The reducing capacities of cyanobacteria for aldehydes and ketones. Appl Microbiol Biotechnol 1986, 25, 52–4.
[3]   Schmermund L, Jurkaš V, Özgen FF, Barone GD, Büchsenschütz HC, Winkler CK, Schmidt S, Kourist R, Kroutil W. Photobiocatalysis: Biotransformations in the presence of light. ACS Catal 2019, 9, 4115–44.
[4]   Hoschek A, Toepel J, Hochkeppel A, Karande R, Bühler B, Schmid A. Light-dependent and aeration-independent gram-scale hydroxylation of cyclohexane to cyclohexanol by CYP450 harboring *Synechocystis* sp. PCC 6803. Biotechnol J 2019, 14, e1800724.
[5]   McGovern PE, Zhang J, Tang J, Zhang Z, Hall GR, Moreau RA, Nuñez A, Butrym ED, Richards MP, Wang CS, Cheng G, Zhao Z, Wang C. Fermented beverages of pre- and proto-historic China. Proc Natl Acad Sci U S A 2004, 101, 17593–8.
[6]   Cavalieri D, McGovern PE, Hartl DL, Mortimer R, Polsinelli M. Evidence for *S. cerevisiae* fermentation in ancient wine. J Mol Evol 2003, 57, 226–32.
[7]   Luan G, Zhang S, Lu X. Engineering cyanobacteria chassis cells toward more efficient photosynthesis. Curr Opin Biotechnol 2020, 62, 1–6.
[8]   Wijffels RH, Kruse O, Hellingwerf KJ. Potential of industrial biotechnology with cyanobacteria and eukaryotic microalgae. Curr Opin Biotechnol 2013, 24, 405–13.
[9]   Assil-Companioni L, Büchsenschütz HC, Solymosi D, Dyczmons-Nowaczyk NG, Bauer KKF, Wallner S, Macheroux P, Allahverdiyeva Y, Nowaczyk MM, Kourist R. Engineering of NADPH supply boosts photosynthesis-driven biotransformations. ACS Catal 2020, 11864–77.
[10] Sellés Vidal L, Kelly CL, Mordaka PM, Heap JT. Review of NAD(P)H-dependent oxidoreductases: Properties, engineering and application. Biochim Biophys Acta Proteins 2018, 1866, 327–47.
[11] Liu W, Wang P. Cofactor regeneration for sustainable enzymatic biosynthesis. Biotechnol Adv 2007, 25, 369–84.
[12] Kruse W, Hummel W, Kragl U. Alcohol-dehydrogenase-catalyzed production of chiral hydrophobic alcohols. A new approach leading to a nearly waste-free process. Recl des Trav Chim des Pays-Bas-J R Netherlands 1996, 115, 239–43.
[13] Wong CH, Drueckhammer DG, Sweers HM. Enzymatic vs. fermentative synthesis: Thermostable glucose dehydrogenase catalyzed regeneration of NAD(P)H for use in enzymatic synthesis. J Am Chem Soc 1985, 107, 4028–31.
[14] Weckbecker A, Hummel W. Cloning, expression, and characterization of an (*R*)-specific alcohol dehydrogenase from *Lactobacillus kefir*. Biocatal Biotransformation 2006, 24, 380–9.
[15] Ni Y, Holtmann D, Hollmann F. How green is biocatalysis? To calculate is to know. ChemCatChem 2014, 6, 930–43.
[16] Hollmann F, Schmid A. Electrochemical regeneration of oxidoreductases for cell-free biocatalytic redox reactions. Biocatal Biotransformation 2004, 22, 63–88.

[17] Köninger K, Gómez Baraibar Á, Mügge C, Paul CE, Hollmann F, Nowaczyk MM, Kourist R. Recombinant cyanobacteria for the asymmetric reduction of C=C bonds fueled by the biocatalytic oxidation of water. Angew Chemie Int Ed 2016, 55, 5582–5.

[18] Böhmer S, Köninger K, Gómez-Baraibar Á, Bojarra S, Mügge C, Schmidt S, Nowaczyk M, Kourist R. Enzymatic oxyfunctionalization driven by photosynthetic water-splitting in the cyanobacterium *Synechocystis* sp. PCC 6803. Catalysts 2017, 7, 240.

[19] Büchsenschütz HC, Vidimce-Risteski V, Eggbauer B, Schmidt S, Winkler CK, Schrittwieser JH, Kroutil W, Kourist R. Stereoselective biotransformations of cyclic imines in recombinant cells of *Synechocystis* sp. PCC 6803. ChemCatChem 2020, 12, 726–30.

[20] Berepiki A, Hitchcock A, Moore CM, Bibby TS. Tapping the unused potential of photosynthesis with a heterologous electron sink. ACS Synth Biol 2016, 5, 1369–75.

[21] Berepiki A, Gittins JR, Moore CM, Bibby TS. Rational engineering of photosynthetic electron flux enhances light-powered cytochrome P450 activity. Synth Biol 2018, 3, ysy009.

[22] Xu Y, Alvey RM, Byrne PO, Graham JE, Shen G, Bryant DA. Expression of genes in cyanobacteria: Adaptation of endogenous plasmids as platforms for high-level gene expression in *Synechococcus* sp. PCC 7002. Photosynth Res Protoc 2011, 684, 273–93.

[23] Hoschek A, Heuschkel I, Schmid A, Bühler B, Karande R, Bühler K. Mixed-species biofilms for high-cell-density application of *Synechocystis* sp. PCC 6803 in capillary reactors for continuous cyclohexane oxidation to cyclohexanol. Bioresour Technol 2019, 282, 171–8.

[24] Hoschek A, Bühler B, Schmid A. Stabilization and scale-up of photosynthesis-driven ω-hydroxylation of nonanoic acid methyl ester by two-liquid phase whole-cell biocatalysis. Biotechnol Bioeng 2019, 116, 1887–900.

[25] Hoschek A, Bühler B, Schmid A. Overcoming the gas–liquid mass transfer of oxygen by coupling photosynthetic water oxidation with biocatalytic oxyfunctionalization. Angew Chemie - Int Ed 2017, 56, 15146–9.

[26] Sengupta A, Sunder AV, Sohoni SV, Wangikar PP. The effect of CO_2 in enhancing photosynthetic cofactor recycling for alcohol dehydrogenase mediated chiral synthesis in cyanobacteria. J Biotechnol 2018, 289, 1–6.

[27] Yunus IS, Jones PR. Photosynthesis-dependent biosynthesis of medium chain-length fatty acids and alcohols. Metab Eng 2018, 49, 59–68.

[28] Zerulla K, Ludt K, Soppa J. The ploidy level of *Synechocystis* sp. PCC 6803 is highly variable and is influenced by growth phase and by chemical and physical external parameters. Microbiol (United Kingdom) 2016, 162, 730–9.

[29] Griese M, Lange C, Soppa J. Ploidy in cyanobacteria. FEMS Microbiol Lett 2011, 323, 124–31.

[30] Pinto F, Pacheco CC, Oliveira P, Montagud A, Landels A, Couto N, Wright PC, Urchueguía JF, Tamagnini P. Improving a *Synechocystis*-based photoautotrophic chassis through systematic genome mapping and validation of neutral sites. DNA Res 2015, 22, 425–37.

[31] Hamada H, Kondo Y, Ishihara K, Nakajima N, Hamada H, Kurihara R, Hirata T. Stereoselective biotransformation of limonene and limonene oxide by cyanobacterium, *Synechococcus* sp. PCC 7942. J Biosci Bioeng 2003, 96, 581–4.

[32] Rasoul-Amini S, Fotooh-Abadi E, Ghasemi Y. Biotransformation of monoterpenes by immobilized microalgae. J Appl Phycol 2011, 23, 975–81.

[33] Fu Y, Castiglione K, Weuster-Botz D. Comparative characterization of novel ene-reductases from cyanobacteria. Biotechnol Bioeng 2013, 110, 1293–301.

[34] Nakamura K, Yamanaka R, Hamada H. Cyanobacterium-catalyzed asymmetric reduction of ketones. 2000, 41, 6799–802.

[35] Nakamura K, Yamanaka R. Light-mediated regulation of asymmetric reduction of ketones by a cyanobacterium. Tetrahedron Asymmetry 2002, 13, 2529–33.

[36] Havel J, Weuster-Botz D. Comparative study of cyanobacteria as biocatalysts for the asymmetric synthesis of chiral building blocks. Eng Life Sci 2006, 6, 175–9.

[37] Yamanaka R, Nakamura K, Murakami A. Reduction of exogenous ketones depends upon NADPH generated photosynthetically in cells of the cyanobacterium *Synechococcus* PCC 7942. AMB Express 2011, 1, 1–8.

[38] Utsukihara T, Chai W, Kato N, Nakamura K, Horiuchi CA. Reduction of (+)- and (–)-camphorquinones by cyanobacteria. J Mol Catal B Enzym 2004, 31, 19–24.

[39] Hölsch K, Weuster-Botz D. Enantioselective reduction of prochiral ketones by engineered bifunctional fusion proteins. Biotechnol Appl Biochem 2010, 56, 131–40.

[40] Hölsch K, Havel J, Haslbeck M, Weuster-Botz D. Identification, cloning, and characterization of a novel ketoreductase from the cyanobacterium *Synechococcus* sp. strain PCC 7942. Appl Environ Microbiol 2008, 74, 6697–702.

[41] Fu Y, Hoelsch K, Weuster-Botz D. A novel ene-reductase from *Synechococcus* sp. PCC 7942 for the asymmetric reduction of alkenes. Process Biochem 2012, 47, 1988–97.

[42] Shimoda K, Kubota N, Hamada H, Kaji M, Hirata T. Asymmetric reduction of enones with *Synechococcus* sp. PCC 7942. Tetrahedron Asymmetry 2004, 15, 1677–9.

[43] Yamanaka R, Nakamura K, Murakami M, Murakami A. Selective synthesis of cinnamyl alcohol by cyanobacterial photobiocatalysts. Tetrahedron Lett 2015, 56, 1089–91.

[44] Nakamura K, Yamanaka R. Light mediated cofactor recycling system in biocatalytic asymmetric reduction of ketone. Chem Commun 2002, 2, 1782–3.

[45] Tamoi M, Miyazaki T, Fukamizo T, Shigeoka S. The Calvin cycle in cyanobacteria is regulated by CP12 via the NAD(H)/NADP(H) ratio under light/dark conditions. Plant J 2005, 42, 504–13.

[46] Takemura T, Akiyama K, Umeno N, Tamai Y, Ohta H, Nakamura K. Asymmetric reduction of a ketone by knockout mutants of a cyanobacterium. J Mol Catal B Enzym 2009, 60, 93–5.

[47] Böhmer S, Marx C, Gómez-Baraibar Á, Nowaczyk MM, Tischler D, Hemschemeier A, Happe T. Evolutionary diverse *Chlamydomonas reinhardtii* Old Yellow Enzymes reveal distinctive catalytic properties and potential for whole-cell biotransformations. Algal Res 2020, 50, 101970.

[48] Górak M, Zymańczyk-Duda E. Application of cyanobacteria for chiral phosphonate synthesis. Green Chem 2015, 17, 4570–8.

[49] Yazdi MT, Arabi H, Faramarzi MA, Ghasemi Y, Amini M, Shokravi S, Mohseni FA. Biotransformation of hydrocortisone by a natural isolate of *Nostoc muscorum*. Phytochemistry 2004, 65, 2205–9.

[50] Zyszka B, Anioł M, Lipok J. Highly effective, regiospecific reduction of chalcone by cyanobacteria leads to the formation of dihydrochalcone: Two steps towards natural sweetness. Microb Cell Fact 2017, 16, 1–15.

[51] Żyszka-Haberecht B, Poliwoda A, Lipok J. Biocatalytic hydrogenation of the C=C bond in the enone unit of hydroxylated chalcones – process arising from cyanobacterial adaptations. Appl Microbiol Biotechnol 2018, 102, 7097–111.

[52] Żyszka-Haberecht B, Poliwoda A, Lipok J. Structural constraints in cyanobacteria-mediated whole-cell biotransformation of methoxylated and methylated derivatives of 2′-hydroxychalcone. J Biotechnol 2019, 293, 36–46.

[53] Löwe J, Siewert A, Scholpp AC, Wobbe L, Gröger H. Providing reducing power by microalgal photosynthesis: A novel perspective towards sustainable biocatalytic production of bulk chemicals exemplified for aliphatic amines. Sci Rep 2018, 8, 2–8.

[54] Balcerzak L, Lipok J, Strub D, Lochyński S. Biotransformations of monoterpenes by photoautotrophic micro-organisms. J Appl Microbiol 2014, 117, 1523–36.

[55] Fitzpatrick TB, Amrhein N, Macheroux P. Characterization of YqjM, an old yellow enzyme homolog from Bacillus subtilis involved in the oxidative stress response. J Biol Chem 2003, 278, 19891–7.

[56] Kitzing K, Fitzpatrick TB, Wilken C, Sawa J, Bourenkov GP, Macheroux P, Clausen T. The 1.3 Å crystal structure of the flavoprotein YqjM reveals a novel class of old yellow enzymes. J Biol Chem 2005, 280, 27904–13.

[57] Hall M, Stueckler C, Ehammer H, Pointner E, Oberdorfer G, Gruber K, Hauer B, Stuermer R, Kroutil W, Macheroux P, Faber K. Asymmetric bioreduction of C=C bonds using enoate reductases OPR1, OPR3 and YqjM: Enzyme-based stereocontrol. Adv Synth Catal 2008, 350, 411–8.

[58] Pesic M, Fernández-Fueyo E, Hollmann F. Characterization of the old yellow enzyme homolog from *Bacillus subtilis*. ChemistrySelect 2017, 2, 3866–71.

[59] Mitsukura K, Suzuki M, Tada K, Yoshida T, Nagasawa T. Asymmetric synthesis of chiral cyclic amine from cyclic imine by bacterial whole-cell catalyst of enantioselective imine reductase. Org Biomol Chem 2010, 8, 4533–5.

[60] Ghislieri D, Turner NJ. Biocatalytic approaches to the synthesis of enantiomerically pure chiral amines. Top Catal 2014, 57, 284–300.

[61] Velikogne S, Resch V, Dertnig C, Schrittwieser JH, Kroutil W. Sequence-based in-silico discovery, characterisation, and biocatalytic application of a set of imine reductases. ChemCatChem 2018, 10, 3236–46.

[62] Von Berlepsch S, Kunz HH, Brodesser S, Fink P, Marin K, Flügge UI, Gierth M. The acyl-acyl carrier protein synthetase from *Synechocystis* sp. PCC 6803 mediates fatty acid import. Plant Physiol 2012, 159, 606–17.

[63] Gavriilidis A, Constantinou A, Hellgardt K, Hii KKM, Hutchings GJ, Brett GL, Kuhn S, Marsden SP. Aerobic oxidations in flow: Opportunities for the fine chemicals and pharmaceuticals industries. React Chem Eng 2016, 1, 595–612.

[64] Osterberg PM, Niemeier JK, Welch CJ, Hawkins JM, Martinelli JR, Johnson TE, Root TW, Stahl SS. Experimental limiting oxygen concentrations for nine organic solvents at temperatures and pressures relevant to aerobic oxidations in the pharmaceutical industry. Org Process Res Dev 2015, 19, 1537–43.

[65] Hoschek A, Schmid A, Bühler B. In situ O_2 generation for biocatalytic oxyfunctionalization reactions. ChemCatChem 2018, 10, 5366–71.

[66] Law HEM, Baldwin CVF, Chen BH, Woodley JM. Process limitations in a whole-cell catalysed oxidation: Sensitivity analysis. Chem Eng Sci 2006, 61, 6646–52.

[67] Hannemann F, Bichet A, Ewen KM, Bernhardt R. Cytochrome P450 systems-biological variations of electron transport chains. Biochim Biophys Acta - Gen Subj 2007, 1770, 330–44.

[68] Reisky L, Büchsenschütz HC, Engel J, Song T, Schweder T, Hehemann JH, Bornscheuer UT. Oxidative demethylation of algal carbohydrates by cytochrome P450 monooxygenases brief-communication. Nat Chem Biol 2018, 14, 342–4.

[69] Schmidt S, Büchsenschütz HC, Scherkus C, Liese A, Gröger H, Bornscheuer UT. Biocatalytic access to chiral polyesters by an artificial enzyme cascade synthesis. ChemCatChem 2015, 7, 3951–5.

[70] Schmidt S, Scherkus C, Muschiol J, Menyes U, Winkler T, Hummel W, Gröger H, Liese A, Herz HG, Bornscheuer UT. An enzyme cascade synthesis of ε-caprolactone and its oligomers. Angew Chemie - Int Ed 2015, 54, 2784–7.

[71] Schmidt S, Genz M, Balke K, Bornscheuer UT. The effect of disulfide bond introduction and related Cys/Ser mutations on the stability of a cyclohexanone monooxygenase. J Biotechnol 2015, 214, 199–211.

[72] Karande R, Debor L, Salamanca D, Bogdahn F, Engesser KH, Buehler K, Schmid A. Continuous cyclohexane oxidation to cyclohexanol using a novel cytochrome P450 monooxygenase from *Acidovorax* sp. CHX100 in recombinant *P. taiwanensis* VLB120 biofilms. Biotechnol Bioeng 2016, 113, 52–61.

[73] Huang Q, Jiang F, Wang L, Yang C. Design of photobioreactors for mass cultivation of photosynthetic organisms. Engineering 2017, 3, 318–29.

[74] Heuschkel I, Hoschek A, Schmid A, Bühler B, Karande R, Bühler K. Data on mixed trophies biofilm for continuous cyclohexane oxidation to cyclohexanol using *Synechocystis* sp. PCC 6803. Data Br 2019, 25, 104059.

[75] Fischer J, Lange T, Boehling R, Rehfinger A, Klemm E. Uncatalyzed selective oxidation of liquid cyclohexane with air in a microcapillary reactor. Chem Eng Sci 2010, 65, 4866–72.

[76] Kauny J, Sétif P. NADPH fluorescence in the cyanobacterium *Synechocystis* sp. PCC 6803: A versatile probe for in vivo measurements of rates, yields and pools. Biochim Biophys Acta - Bioenerg 2014, 1837, 792–801.

[77] Heining M, Sutor A, Stute SC, Lindenberger CP, Buchholz R. Internal illumination of photobioreactors via wireless light emitters: A proof of concept. J Appl Phycol 2015, 27, 59–66.

[78] Chen CY, Yeh KL, Aisyah R, Lee DJ, Chang JS. Cultivation, photobioreactor design and harvesting of microalgae for biodiesel production: A critical review. Bioresour Technol 2011, 102, 71–81.

[79] Grund M, Jakob T, Wilhelm C, Bühler B, Schmid A. Electron balancing under different sink conditions reveals positive effects on photon efficiency and metabolic activity of *Synechocystis* sp. PCC 6803. Biotechnol Biofuels 2019, 12, 1–14.

[80] Schuller JM, Birrell JA, Tanaka H, Konuma T, Wulfhorst H, Cox N, Schuller SK, Thiemann J, Lubitz W, Sétif P, Ikegami T, Engel BD, Kurisu G, Nowaczyk MM. Structural adaptations of photosynthetic complex I enable ferredoxin-dependent electron transfer. Science 2019, 363, 257–60.

[81] Allahverdiyeva Y, Ermakova M, Eisenhut M, Zhang P, Richaud P, Hagemann M, Cournac L, Aro E-M. Interplay between flavodiiron proteins and photorespiration in *Synechocystis* sp. PCC 6803. J Biol Chem 2011, 286, 24007–14.

[82] Allahverdiyeva Y, Mustila H, Ermakova M, Bersanini L, Richaud P, Ajlani G, Battchikova N, Cournac L, Aro EM. Flavodiiron proteins Flv1 and Flv3 enable cyanobacterial growth and photosynthesis under fluctuating light. Proc Natl Acad Sci U S A 2013, 110, 4111–6.

[83] Allahverdiyeva Y, Isojärvi J, Zhang P, Aro E-M. Cyanobacterial oxygenic photosynthesis is protected by flavodiiron proteins. Life 2015, 5, 716–43.

[84] Nikkanen L, Santana Sánchez A, Ermakova M, Rögner M, Cournac L, Allahverdiyeva Y. Functional redundancy between flavodiiron proteins and NDH-1 in *Synechocystis* sp. PCC 6803. Plant J 2020, 103, 1460–76.

[85] Santana-Sanchez A, Solymosi D, Mustila H, Bersanini L, Aro E-M, Allahverdiyeva Y. Flavodiiron proteins 1-to-4 function in versatile combinations in O_2 photoreduction in cyanobacteria. Elife 2019, 8, e45766.

[86] Sétif P, Shimakawa G, Krieger-Liszkay A, Miyake C. Identification of the electron donor to flavodiiron proteins in *Synechocystis* sp. PCC 6803 by in vivo spectroscopy. Biochim Biophys Acta - Bioenerg 2020, 1861, 148256.

[87] Helman Y, Tchernov D, Reinhold L, Shibata M, Ogawa T, Schwarz R, Ohad I, Kaplan A. Genes encoding A-type flavoproteins are essential for photoreduction of O_2 in Cyanobacteria. Curr Biol 2003, 13, 230–5.

[88] Gupta PL, Lee SM, Choi HJ. A mini review: Photobioreactors for large scale algal cultivation. World J Microbiol Biotechnol 2015, 31, 1409–17.

Oliver Lampret, Claudia Brocks, Martin Winkler*

4 O_2 escape strategies for hydrogenases in application

Abstract: Nature provides us with the most suitable and efficient biocatalysts in the form of redox enzymes to perform a variety of catalytic reactions [1]. While some enzymes are capable of catalyzing complex, organic multistep syntheses [2, 3], gas-processing metalloenzymes like nitrogenases, CO dehydrogenases and hydrogenases that activate or produce small molecules like N_2, CO_2 or H_2 gain increasing attention [4, 5]. These redox enzymes overcome significant energy barriers to catalyze the corresponding redox reactions with minimal overpotential requirements [4]; however, they all share a notorious sensitivity toward molecular O_2 [6–9]. While for each of these enzymes the mechanism and degree by which the often irreversible, O_2-induced inactivation takes place differs, they all originate from an era when molecular O_2 was not present in the Earth's atmosphere [10]. Nowadays, these old "pre-aerobic tools" experience a sheer hype when it comes to bioinspired fuel technologies and have hence emerged as promising candidates to satisfy industrial and technological needs for reducing greenhouse gas emissions. In this context, "renewable" H_2 is considered to play a key role in the future energy economy, rendering it an attractive and renewable energy source to fuel future biotechnologies and bioinspired technologies. Hence, this chapter pays attention to hydrogenases and their potential applicability for alternative energy technologies. Besides the more robust but less active [NiFe]-hydrogenases, a particular focus is placed on [FeFe]-hydrogenases as they represent the most active hydrogenase class. We highlight the extreme level of O_2 sensitivity that may vary significantly among different [FeFe]-hydrogenase subclasses. Beyond a detailed overview on different spectroscopic and analytical techniques allowing us to monitor the process of O_2-induced inactivation, we introduce novel strategies to increase the O_2 resistance of [FeFe]-hydrogenases on the molecular level, and also *in vitro* strategies to overcome the problem of O_2 sensitivity in individual [FeFe]-hydrogenase enzymes. Concerning biotechnological applications of hydrogenases on an industrial-scale level, novel semisynthetic approaches with robust biocatalysts being integrated into state-of-the-art biofuel cells will be presented.

4.1 Basics

4.1.1 [NiFe]-hydrogenases meet biofuel cells

While heterogeneous systems with immobilized biocatalysts are in principle ideal candidates to realize high-standard chemical production routes with reduced greenhouse

https://doi.org/10.1515/9783110716979-004

gas emissions, the use of non-immobilized enzymes still predominates in the global bioeconomy [11, 12]. Hence, novel technologies are needed to pave the way for a decarbonized chemical industry relying on robust and highly efficient heterogeneous biocatalysts. High enzymatic turnover rates in conjunction with outstanding (stereo-) selectivity is best achieved if a redox enzyme is immobilized on an electrode surface to allow for rapid charge transfer between the electrode and the enzyme's active site [13, 14]. During the last decade, enzymatic biofuel cells (BFCs) have emerged as remarkable benchmarks for sustainable biocatalytic systems [14]. Their modulatory composition easily allows them to be either miniaturized for testing or scaled up for production purposes, respectively [15]. Another great advantage of BFCs compared to conventional chemical processes is their ability to operate under mild reaction conditions, that is, at room temperature, low (gas) pressure and neutral pH [16, 17]. Extensive research on both hydrogenase enzymes and nanomaterials has been done, and in spite of significant progress, identification of the best combinations has proven to be quite challenging. Several proof-of-concept studies have shown the general possibility of implementing hydrogenases into BFCs [18, 19]. However, so far only O_2-tolerant [NiFe]-hydrogenases such as those from *Ralstonia eutropha* have been shown to be applicable for BFCs [20, 21]. Besides the basic concept of a classic H_2/O_2 BFC, even ambitious projects like fueling H_2-driven enzyme cascades attract attention [20, 21].

[NiFe]-hydrogenases characteristically feature a heterodimeric catalytic core unit consisting of a large (60 kDa) and a small subunit (29 kDa). The former harbors the active [NiFe] cofactor while the small subunit carries the electron transport chain, usually consisting of two [4Fe4S] clusters and one [3Fe4S] cluster [22]. The nickel site in the bimetallic active site complex is anchored to the protein environment by the thiolate groups of four conserved cysteines, two of which bridging the iron and nickel sites of the bimetallic cofactor [23]. The iron site is further coordinated by one CO and two CN^- ligands [24]. Depending on the redox state of the catalytic center, a ligand such as CO, H_2 or O_2 may occupy the free coordination site on the Ni atom. H_2 evolution/oxidation occurs by cycling between three different catalytic states (Ni-SI$_a$, oxidized open ready) from which reduction and protonation leads to Ni-C (oxidized hydride species, Ni^{III}-(H^-)-Fe^{II}). Another reduction step produces Ni-R (reduced hydride species, Ni^{II}-(H^-)-Fe^{II})). Reversible H_2 release occurs upon the second protonation step during the transition from Ni-R to Ni-Si$_a$ [23, 25]. O_2 and H_2 compete for the open coordination site in the Ni-Si$_a$ state, located in a bridging position between Ni and Fe sites. Under O_2 exposure, two inactive doublet states can be verified via EPR called Ni-A and Ni-B featuring different reactivation rates. The ratio of both states depends on the number of electrons available to reductively react with O_2 [26]. Ni-B dominates if O_2 exposure occurs under reducing conditions (0 mV vs. the normal hydrogen electrode (NHE)) while exposure at higher potentials (200 mV vs. NHE) mainly yields the Ni-A state, also known as "unready state." As it reactivates very slowly (time range of hours), Ni-A is part of the resting cofactor states [26]. The presence of H_2 or CO supports the reactivation of Ni-A [26, 27]. Ni-A is predominant in O_2-sensitive

[NiFe]-hydrogenases after O_2 exposure which can be referred to insufficient electron supply. Ni-A is discussed to coordinate either a hydroperoxide ligand, a mono-oxo- or di-oxo-species in bridging position between Fe and Ni sites or might exhibit a sulfoxide (S=O) or sulfenic acid group (SOH) [23, 28–31]. While the exact nature of the bridging species could not be verified, the distance between both metals in the Ni-Fe center is slightly larger in Ni-A (2.8 Å) than in Ni-B (2.7 Å). Ni-B is referred to as "ready state" and reactivates in seconds under reducing H_2 atmosphere or by supplying electrons and protons. This state can only be reached if a fast electron supply permits three reduction and protonation events. Consequently, a single H_2O molecule is released and the bridging position of the [NiFe] cofactor is occupied by a hydroxo ligand (OH$^-$) that can be quickly removed as a second H_2O molecule, following another reduction and protonation step. As both Ni-A and Ni-B can be reactivated, O_2 can be regarded as a largely reversible inhibitor of [NiFe]-hydrogenases [7, 26, 32, 33]. Despite their robustness against molecular O_2, [NiFe]-hydrogenases lag well behind [FeFe]-hydrogenases in terms of catalytic activity [34]. This is further accompanied by the fact that [NiFe]-hydrogenases are able to process H_2 in the presence of O_2 only in a competitive way, further rendering their overall H_2 processing activities less attractive for industrial scopes [35]. Initial experiments in which the extremely O_2-sensitive [FeFe]-hydrogenase DdH from *Desulfovibrio desulfuricans* was combined with a redox-active polymer on a gas-breathing electrode demonstrated the significant performance of such a bioanode that clearly outperformed previous bioanode concepts relying on [NiFe]-hydrogenases [36].

4.1.2 [FeFe]-hydrogenases – an extreme case of O$_2$ sensitivity

The exquisite performance of [FeFe]-hydrogenases to catalyze the reversible H$^+$/H$_2$ interconversion (often with turnover frequencies exceeding 10,000 s^{-1}) is merely matched by the rare and expensive platinum metals [37, 38]. The impossibility of developing a global hydrogen economy based on limited metal resources has stimulated research on alternative catalysts, for instance, to replace platinum in fuel cells by hydrogenases [19, 38]. The active site of [FeFe]-hydrogenases, a unique [6Fe6S] cluster, known as the "H-cluster" is composed of an unusually coordinated binuclear [2Fe2S] complex (the 2Fe$_H$ site) that is covalently linked to a [4Fe4S] cluster (the 4Fe$_H$ subsite) through a shared cysteine S-atom [39, 40]. The two Fe atoms of the 2Fe$_H$ site are differentiated as proximal (Fe$_p$) and distal (Fe$_d$) according to their distance to the 4Fe$_H$ cluster. The 2Fe$_H$ site is further coordinated and stabilized by two CN$^-$ and three CO ligands and cross-bridged by an azadithiolate (adt) ligand, with its central N-bridgehead hanging over the substrate coordination site of Fe$_d$ [41, 42]. During catalytic turnover, the H-cluster cycles between different catalytic states. H_2 binding occurs in the oxidized active ready state (H$_{ox}$) and starts with the heterolysis of H_2 into a terminally bound hydride species, coordinated at Fe$_d$ and a first proton which leaves the active site via the adt ligand and the proton transfer pathway (PTP) [43–45]. Successive oxidation and deprotonation steps lead from the hydride state (H_{hyd}) to the super-reduced H_{sred} state, the reduced states $H_{red}H^+$ and H_{red} back to H_{ox} ($H_{ox} \Rightarrow (+H_2/-H^+) \Rightarrow H_{hyd} \Rightarrow H_{sred} \Rightarrow (-e^-) \Rightarrow H_{red}H^+ \Rightarrow (-H^+) \Rightarrow H_{red} \Rightarrow (-e^-) \Rightarrow H_{ox}$). The reverse cycle ($H_{ox} \Rightarrow (+e^-) \Rightarrow H_{red} \Rightarrow (+H^+) \Rightarrow H_{red}H^+ \Rightarrow (+e^-) \Rightarrow H_{sred} \Rightarrow H_{hyd} \Rightarrow (+H^+/-H_2) \Rightarrow H_{ox}$) enables H_2 evolution via proton reduction [46]. The contribution of H_{sred} and $H_{red}H^+$ to fast catalytic turnover and the number and features of H_{hyd} states are still under debate [46–49]. The unique catalytic $2Fe_H$ complex is synthesized *in vivo* by a complex maturation system consisting of the highly specific maturases HydE, -F and -G, but can further be chemically synthesized ($2Fe_H^{MIM}$) to spontaneously activate an apo-form of the enzyme, hence providing a platform to rationally design semisynthetic [FeFe]-hydrogenases [39, 40, 50]. However, this catalytic cofactor is irreversibly damaged by O_2 which diffuses through the enzyme via intramolecular gas channels before rapid binding at the open Fe_d coordination site takes place [51–53]. This event is followed by successive protonation and reduction steps, finally yielding a reactive oxygen species (ROS) that causes irreversible H-cluster degradation [54]. Recently, the discovery and characterization of the O_2-resistant [FeFe]-hydrogenase CbA5H of *Clostridium beijerinckii* has ushered a new era for [FeFe]-hydrogenases and their potential use in industrial applications [55, 56].

4.1.3 The process of O_2-induced H-cluster degradation

Upon reaching the active site of [FeFe]-hydrogenases, O_2 initiates an irreversible degradative process that inactivates the H-cluster. Depending on the O_2 concentration and exposure time, the H-cluster degradation process can occur on a timescale of minutes [54, 57–61]. Besides being determined by the O_2 concentration and the duration of O_2 exposure, the level of O_2 sensitivity is further influenced by the redox potential of the cofactor, the buffer pH and the H_2 pressure, as demonstrated by electrochemical experiments [54, 57, 58]. In the "unmaturated" apo-form which lacks the catalytically vital $2Fe_H$ site but harbors the $4Fe_H$ cluster and accessory [FeS] clusters (if applicable), those clusters are not affected by O_2 (see Fig. 1) [53]. In line with this observation, a direct attack of O_2 on the [4Fe4S] clusters of the enzyme has been discussed to be thermodynamically unlikely [57, 58]. Thus, the $2Fe_H$ subsite appears to be essential for activating O_2 and yielding ROS that initiate the cluster attack [59]. It could further be shown that the H-cluster is more effectively attacked in the oxidized ready H_{ox} state $[4Fe-4S]^{2+}-Fe_p(I)Fe_d(II))$ compared to the more reduced H-cluster states H_{red}/H_{sred} [62–64]. The target site of O_2 interaction has been localized by Goldet and coworkers, comparing the behavior of O_2 in the presence of CO [57]. [FeFe]-

hydrogenases are reversibly inactivated upon CO binding, leading to 80–200-fold higher inhibition rates compared to O$_2$. CO has been demonstrated to transiently occupy the open coordination site at Fe$_d$ in H$_{ox}$ or H$_{red}$, yielding either H$_{ox}$CO or H$_{red}$CO, respectively [65]. Besides CO, formaldehyde and sulfide have also been shown to act as competitive reversible inhibitors of the H-cluster [66–68]. The observation that CO-inhibited [FeFe]-hydrogenases are protected against O$_2$ attack implies that both CO and O$_2$ compete for the same binding site (Fe$_d$) at the H-cluster [57]. Corresponding O$_2$-protective features have been demonstrated for sulfide inhibition and just like CO, sulfide has recently been shown to bind to the open coordination site at Fe$_d$ [69].

Orain and coworkers investigated and compared the kinetics of O$_2$-induced H-cluster inhibition/degradation for the smaller algal M1-type enzyme CrHydA1 of *Chlamydomonas reinhardtii* and the larger bacterial M3-type hydrogenase Ca-HydA via protein film electrochemistry, following the model:

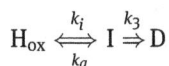

$$\text{H}_{ox} \underset{k_a}{\overset{k_i}{\rightleftarrows}} \text{I} \overset{k_3}{\Rightarrow} \text{D}$$

with the two constants k_i (rate constant for inhibition) and k_a (rate constant for reactivation) for the transition between the active ready state H$_{ox}$ and the inhibited first O$_2$ adduct (I) and a final rate constant k_3 for the irreversible transition from state I to H-cluster degradation (D) [70]. O$_2$ binding to Fe$_d$ has been described as an exothermic reaction [62, 71–73] followed by a series of redox reactions that lead to H-cluster loss. However, Orain and coworkers were able to demonstrate that generating the primary O$_2$ adduct (I) is a reversible step while only the transition to state D leads to irreversible damage. The significant difference in O$_2$ sensitivity determined between the two compared [FeFe]-hydrogenases can be largely explained by the 10-fold higher value for rate constant k_3 measured for the M1-type CrHydA1 enzyme, rendering it practically incapable of undergoing partial reactivation upon first O$_2$ binding, as observed for CaHydA. Two possible routes of reactivation from the first O$_2$ adduct (state I: Fe$_d$-O$_2$) exhibiting an apical superoxide (O$_2^-$) at Fe$_d$ have been proposed in a follow-up publication by Kubas and coworkers [54]. Here they differentiated between the extreme cases of high pH/high potential and low pH/low potential conditions. Under lack of proton pressure (high pH) and high potential, O$_2$ is less stabilized at Fe$_d$, partially leaving the active site upon detachment without causing any damage. However, under higher proton pressure and lower potential O$_2$ is quickly protonated and can undergo several transformations leading to ROS formation and cofactor attack, unless due to a sufficiently low potential, O$_2$ is four times reduced and protonated to yield two equivalents of H$_2$O. However, if not matched by a low enough potential, proton transfer to the primary O$_2$-ligand initiates the process of H-cluster degradation. Lyophilized samples of otherwise severely O$_2$-sensitive [FeFe]-hydrogenases, such as CpI and CrHydA1, can be stored under dehumidified air without harming the H-cluster. Due to their dry state, the enzymes are incapable of shuttling protons to the active site [64]. O$_2$ exposure to CrHydA1 variant C169A, featuring an H-cluster which is decoupled from proton transfer, traps the enzyme in the state of the first Fe_d-O_2 adduct, while significantly slowing down O_2-induced cluster degradation [74]. The lack of proton transfer activity, on the other hand, renders the enzyme catalytically inactive and reactivation is only possible for lyophilized wild-type enzyme, following a cumbersome procedure.

The site-specific degradative effects of oxygen exposure on the H-cluster architecture were recently elucidated by analyzing the development of the anomalous electron density at individual iron sites in numerous crystals of CpI from *Clostridium pasteurianum* which have been exposed to air for increasing time periods [53]. This study revealed a loss of specific Fe atoms from both the $2Fe_H$ and the $4Fe_H$ subclusters, exhibiting two parallel processes of degradation. For the $2Fe_H$ site, a dissociation of the distal Fe site and its diatomic ligands can be observed (see Fig. 1).

From the site-specific loss of electron density in the $2Fe_H$ cofactor, the authors concluded that proton supply and incomplete reduction may transform the ligand of the Fe_d-O_2 adduct into a hydroperoxo species (Fe_d-HO_2^-). Upon another protonation step, this intermediate may disproportionate to H_2O and an iron-oxo species ($Fe(IV)=O_2^-$) which would be capable to oxidatively attack the nearby diatomic CO_d and CN^-_d ligands, leading to their release as CO_2 and OCN^-. The increasingly unligated Fe_d would consequently undergo a solvation process and be released from the cofactor.

This may also destabilize the rest of the $2Fe_H$ complex as O_2-exposed inactive protein samples can be at least partially (approx. 50%) rematurated to active holoenzyme by supplying fresh $2Fe_H^{MIM}$ complex [53, 59].

The 50% loss of activity after rematuration is in line with the partial loss (by 40–50%) of two specific Fe sites in the $4Fe_H$ cluster architecture (C503 and C355 in CpI) leading to the conversion into a [2Fe2S] cluster. This may be referred to a double protonation and dissociation of the peroxo intermediate as H_2O_2. The ROS H_2O_2 is known to oxidatively attack the iron atoms of [FeS] clusters [75].

A crystal structure analysis of algal M1-type CrHydA1 suggests an oxidation of the thiol group of cysteine C169 to sulfenic acid [59]. C169 is located in H-bond distance to the $2Fe_H$ cluster and serves as an essential and strictly conserved constituent in the PTP of the enzyme as it shuttles protons as substrates or products of catalysis to and from the adt ligand [44, 45]. Cysteine or methionine oxidation in proximity of the $2Fe_H$ cluster could, however, not be confirmed for O_2-exposed crystals of CpI [53].

4.2 Methods, concepts and state of the art

4.2.1 Protein film electrochemistry

En route to the concept and application of [FeFe]-hydrogenases in BFCs, protein film voltammetry (PFE) has emerged as a powerful electrochemical tool more than 20 years ago [76]. Much of the above-described mechanisms that give insight into

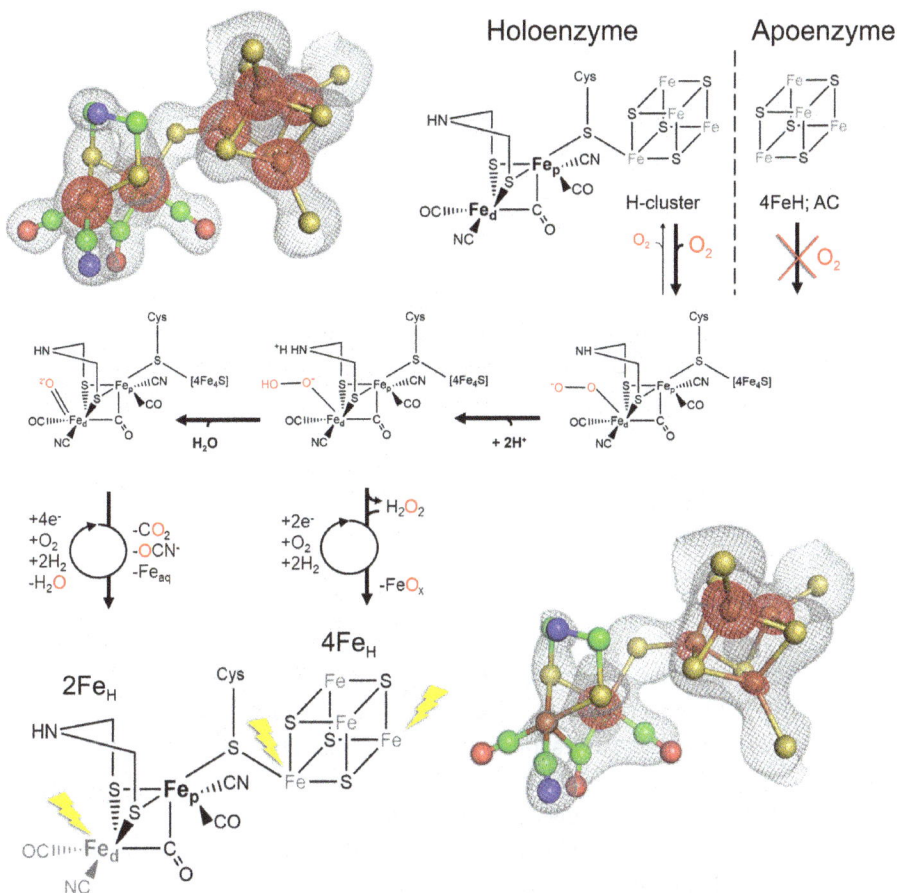

Fig. 1: O$_2$-induced H-cluster degradation in [FeFe]-hydrogenases documented via X-ray crystallography. O$_2$ enters and pervades the protein along accessible branches of the gas channel system. After reaching the 2FeH site, O$_2$ oxidatively coordinates to Fe$_d$ forming the first O$_2$ adduct (Fe$_d$-O$_2$). Protonation via the native proton transfer pathway yields a hydroperoxo intermediate which can dissociate and further protonate to H$_2$O$_2$ that upon release oxidatively attacks the 4FeH cluster. Alternatively, O$_2^-$ may oxidize the nearby diatomic ligands CO$_d$ and CN$_d$, causing their dissociation from the active site as CO$_2$ and OCN$^-$. The unligated Fe$_d$ is increasingly solvated and finally detached from the 2FeH cofactor. The selective degradation processes (red flashes) at the 4FeH site (primordial loss of 2 Fe-atoms) and the 2FeH site (selective loss of Fe$_d$, CO$_d$ and CN$_d$) have been documented via X-ray crystallography by analyzing O$_2$-exosed crystals of CpI holoenzyme and apoenzyme. The 4FeH cluster of apoenzyme is not affected by O$_2$.

structure–function relationships of [NiFe] and [FeFe]-hydrogenases have been derived from PFE experiments [77]. This technique relies on direct electron transfer from a submonolayer film of enzyme to the surface of a rotating disk electrode (RDE) which is used to minimize mass transport limitations of substrates and products [78]. The enzyme is adsorbed by drop-casting onto electrode materials like carbon. In most common cases, a pyrolytic graphite "edge" electrode is used [79], although several studies have demonstrated the advantages of using modified carbon electrodes where a "3D surface" is generated to increase the total enzyme coverage. These advantages are further discussed in Section 3.4. In the presence of substrates (H_2/H^+), electron transfer occurs directly to or from the electrode with the catalytic direction depending on the applied electrode potential, and continuously cycling through the above-described catalytic redox states of H_2 turnover [77]. While the resulting current is a proportional function of the enzyme's turnover frequency [80], PFE offers fast switches and control of various reaction parameters, including the electrode potential, temperature, gas composition and buffer pH. Especially the possibility to add influential compounds during a measurement, such as reversible and irreversible enzyme inhibitors like CO and O_2, respectively, allows for a detailed analysis of the kinetic parameters underlying inactivation [37]. Figure 2 shows a gas-tight electrochemical cell in a three-electrode configuration that is typically used to characterize hydrogenase enzymes. Here, the non-isothermal side-arm contains the reference electrode, typically a silver chloride electrode or saturated calomel electrode. A Luggin capillary connects the side arm to the main compartment which is water-jacketed to control the temperature. While a platinum wire is employed as a counter electrode, a rotating ring disk electrode (RDE) is used as a working electrode (WE). High rotation speed (>1000 rpm) of the WE ensures non-limiting substrate (H^+/H_2) supply conditions. The headspace is connected to a mass flow controller system to exchange gasses for substrate and/or inhibitor supply (H_2 and/or O_2/CO, respectively). Another syringe supply also allows for the injection of liquids like O_2-saturated buffer to add a defined amount of inhibitor to the enzyme.

A typical characterization experiment starts with a cyclic voltammogram (CV) of the immobilized hydrogenase enzyme (see Fig. 3A). In case of [FeFe]-hydrogenases, the potential may be reversibly cycled from −0.8 to +0.25 V (all potentials are given against the standard hydrogen potential (SHE)). The resulting CV informs about the catalytic H_2 reduction (red-colored potential window) and H_2 oxidation (green-colored potential window) activities. As [FeFe]-hydrogenases behave as highly reversible electrocatalysts (catalysis is performed with minimal overpotential), the current cuts sharply across the zero-current axis at the equilibrium potential of the $H_2/2H^+$ couple [81]. The ratio of the relative current magnitude in both catalytic directions at the same overpotential describes the catalytic bias which is a function of several parameters, including substrate-binding constants and product release at the active site (via gas channels), as well as the efficiencies of proton and electron relay systems [37, 81]. At potentials above 0 V versus SHE (all measurements at pH 7, 10 °C), H_2 oxidation is followed by high potential anaerobic inactivation (blue-colored window) resulting in a gradual loss of catalytic current, a mechanism that is typically observed for all [FeFe]-hydrogenase enzymes. However, this process is reversible when the potential is swept back to more reductive conditions, allowing reactivation of the enzyme film. Recently, Del Barrio et al. introduced a halide-dependent

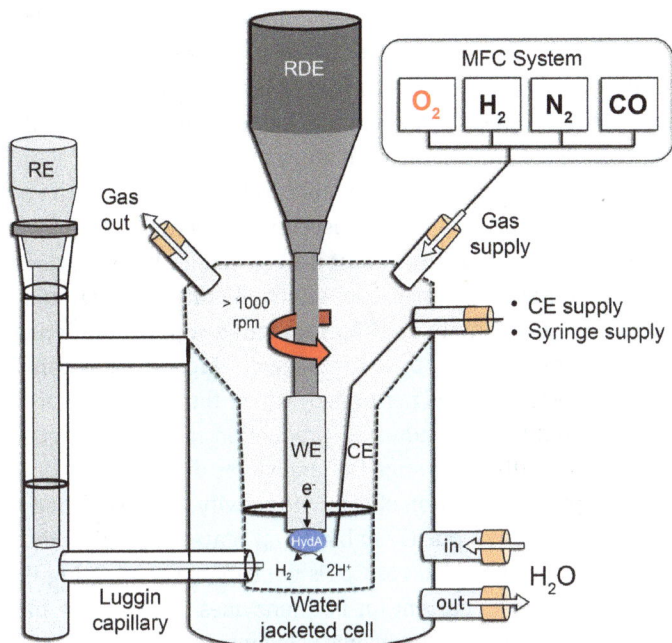

Fig. 2: Schematic representation of a gas-tight electrochemical cell in a three-electrode configuration (see main text for more details).

mechanism according to which the "standard" anaerobic, high potential inactivation might occur [82]. According to their mechanism, halides like Cl⁻ or Br⁻ ions interact with the amine group of the H-cluster where they act as uncompetitive inhibitors: They reversibly bind to catalytic intermediates of H$_2$ oxidation while leaving the binding site at the Fe$_d$ atom vacant. In contrast to CrHydA1 and CpI, the potential window of H$_2$ oxidation for CbA5H is significantly smaller (merely in a range of ~0.1 V compared to ~0.4 V for CrHydA1 and CpI) before CbA5H at −0.3 V undergoes reversible, high-potential-induced anaerobic inactivation. A recent mutagenesis study showed that for CbA5H, anaerobic inactivation follows a distinct mechanism which is further discussed in detail in Section 3.3 [56]. While the cyclic voltammetry experiments from Fig. 3A do not allow for a distinct comparison between CrHydA1, CpI and CbA5H in terms of O$_2$ sensitivity, chronoamperometry (CA) experiments (Fig. 3B) could prove a distinct level of O$_2$ sensitivity for each hydrogenase. In CA, the current is monitored as a function of time (t) at a fixed electrode potential E, rendering this method most suitable to probe the inactivation behavior and time-dependent kinetics of hydrogenases [83]. The left panel of Fig. 3B shows a comparison of the O$_2$ sensitivities of CrHydA1 (representing the most sensitive M1-type hydrogenase) and CpI (chosen as a representative of the slightly more robust M3-type hydrogenase). For CpI, the injection of 20 µM O$_2$ results in a loss of ~25% enzymatic activity whereas a more dramatic effect is observed for CrHydA1 where a total loss of ~75% activity takes place while a second injection further decreases the total activity by 50% and 95%, respectively. This clearly renders the M1 type to be the more sensitive subclass, most likely due to the lack of an additional internal [FeS]-relay system that could otherwise supply the H-cluster with sufficient electrons to prevent the formation of ROS.

When the more robust candidate CpI was compared to the O_2-stable hydrogenase CbA5H (Fig. 3B, middle panel) under five consecutive injections of O_2-saturated buffer (10 µM O_2 per injection), a loss in current is observed for both enzymes; however, CbA5H regains a significantly higher level of activity after each O_2 injection. As already mentioned, the O_2 stability of CbA5H underlies a potential-dependent protection mechanism that has further been investigated (Fig. 3B, right panel) [56]. Here, a potential jump experiment was performed, where the initial rate of H_2 production at −0.8 V is compared to the residual H_2 production rate after injection of 50 µM O_2 followed by a 5-fold buffer exchange (to remove residual O_2) at 0 V. As in Fig. 3B, CbA5H regains a significant fraction of catalytic activity, as the enzyme at 0 V reversibly switches to the inactive but O_2-stable H_{inact} state. For CpI, only a minor fraction of catalytic current is recovered as CpI is not capable of adopting the H_{inact} state. Note that the linear loss in current for both enzymes is caused by film loss which is typically observed during PFE experiments using unmodified electrodes.

4.2.2 ATR-FTIR spectroscopy

The strong signals originating from the vibrational stretching modes of the CO- and CN⁻ligands upon absorption of infrared (IR) radiation can directly be exploited in order to quantify the catalytically competent fractions of the $2Fe_H$ subsite [50, 84, 85]. In recent years, attenuated total reflection Fourier transform IR (ATR-FTIR) spectroscopy has been established as the preferred modus operandi aside from the classic transmission mode [5, 59, 65]. A major reason for this lies in the accessibility of the sample interface, offering the possibility to directly manipulate system parameters like gas composition, buffer pH and reactant/inhibitor concentration [50, 84, 85]. Figure 4A shows a schematic representation: In [FeFe]-hydrogenases (as well as in [NiFe]-hydrogenases), any changes in the electronic properties at the active site are reflected by specific shifts of the CO/CN⁻ vibrational signal pattern, thereby creating a unique fingerprint spectrum for each redox state [65, 86]. For this purpose, the protein sample is first dried under N_2 gas to accumulate the H_{ox} state. When monitoring the process of O_2-induced degradation, flushing the ATR cell headspace with defined levels of O_2 gas leads to a drastic decline in H-cluster signal intensities. Figure 4B (right panel) shows a decrease in H-cluster content for CpI by 70% after flushing the protein sample with 21% O_2 for 5 min, demonstrating the severe sensitivity of CpI toward molecular O_2.

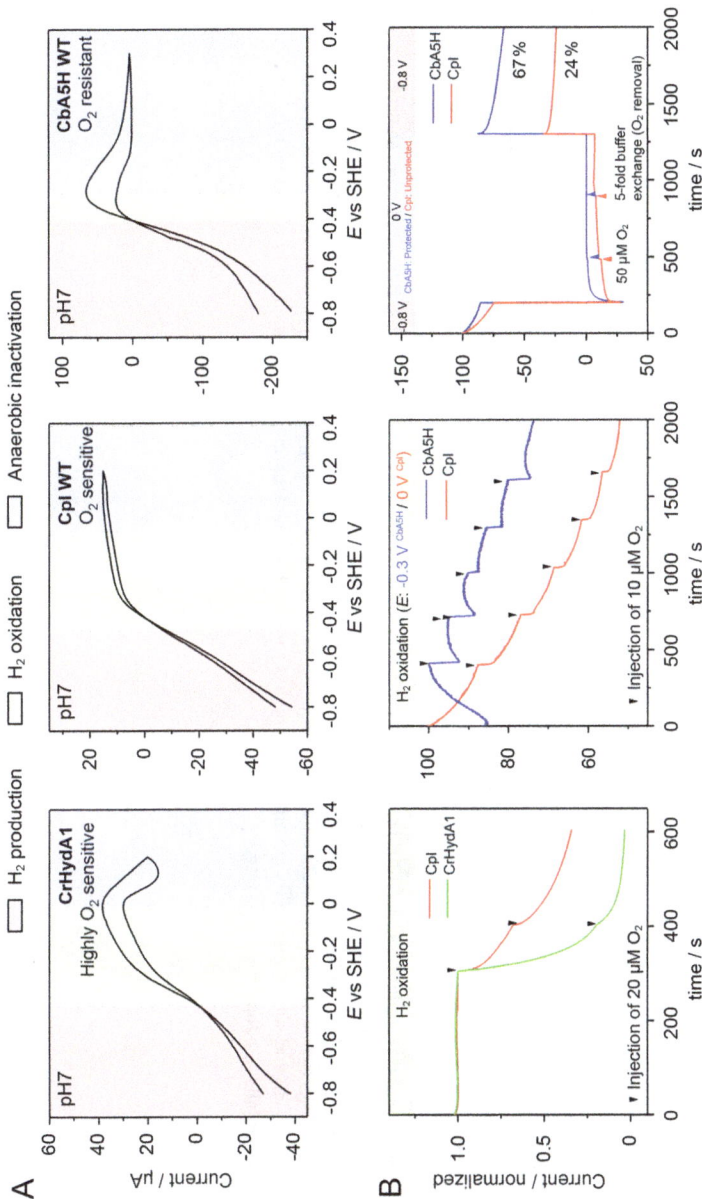

Fig. 3: Protein film electrochemistry (PFE) approaches used to compare the O₂ sensitivity levels of different [FeFe]-hydrogenases. (A) Typical cyclic voltammograms (CVs) obtained for wild-type (WT) enzymes of CrHydA1, CpI and CbA5H (experimental conditions: $T = 10\,°C$, 3000 rpm, 20 mV s⁻¹). (B) Left and middle panel: Chronoamperometric measurements depicting the consequence of repeated injections of air-saturated buffer on catalytic H₂ oxidation activity. Right panel: Chronoamperometric potential jump experiments where the initial catalytic H₂ production current at −0.8 V is compared to the residual H₂ production current after injection of 50 μM O₂ followed by a 5-fold buffer exchange (done to wash out residual O₂) at 0 V.

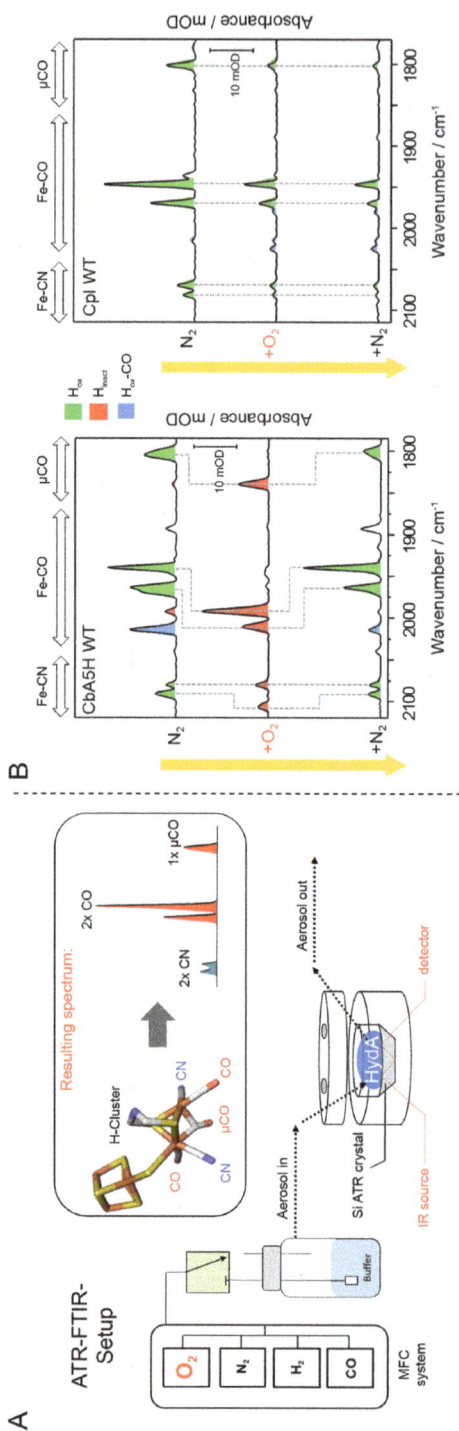

Fig. 4: Monitoring O_2-induced H-cluster degradation with ATR-FTIR spectroscopy. (A) Experimental ATR-FTIR setup where a defined enzyme sample is placed onto the ATR crystal surface and dried under a stream of N_2. A mass flow controller (MFC) system allows for adjustment of the gas flow composition to accumulate individual redox states. (B) Left panel: FTIR spectra of CbA5H after drying the sample under a N_2 stream (enriching H_{ox}) and subsequent gassing with air (21% O_2) resulting in a quantitative transition from H_{ox} to H_{inact}. A retransformation into H_{ox} is achieved after exposure to N_2 (5 min). Note that the CO and CN ligand signal intensities remain equally high throughout the measurement. In contrast, the right panel reflects the innate O_2 sensitivity of CpI which fails to adopt the H_{inact} state.

O$_2$-induced H-cluster degradation may alternatively lead to the accumulation of the H$_{ox}$CO-state due to a "self-cannibalization process" (CO molecules dissociate from damaged H-cluster parts and bind to the intact H-clusters of active enzymes nearby) [86, 87]. However, the continuous O$_2$ gas stream in Fig. 4B flushes away released CO molecules before they can reach the H-cluster. When the O$_2$-tolerant hydrogenase CbA5H is continuously flushed with O$_2$ for 5 min (Fig. 4B, left panel) the enzyme quantitatively adopts the H$_{inact}$ state (>99% conversion rate) [55, 56] where the signals of the CN$^-$ and CO ligands are blue-shifted to higher energies by ~20 to 50 cm^{-1}, adopting similar signal positions like the Fe$_d$-O$_2$ adduct of HydA1$_{Cr}$ variant C169A observed by Mebs et al. [74] (see Tab. 1). The H$_{ox}$ state for CbA5H is, however, quickly re-established after treatment with N$_2$ gas for 5 min, which demonstrates the ability of CbA5H to reversibly switch between H$_{ox}$ and H$_{inact}$, rendering this [FeFe]-hydrogenase a promising candidate for future applications. Thus, ATR-FTIR spectroscopy is a powerful tool to quantitatively monitor the process of O$_2$ degradation for O$_2$-labile hydrogenases and enables us to decipher the molecular properties that underlie the mechanism of O$_2$ resistance, a valuable indicator to identify potential hydrogenases for biofuel applications.

4.2.3 Mutagenesis approaches and screening systems

Enzyme design strategies are usually based on substitutions in the polypeptide chain. While most mutagenesis approaches such as a targeted substitution in the secondary ligand sphere of the H-cluster are based on rational design and simply require site-directed substitutions, concepts addressing less well-defined targets like strengthening the molecular sieve effect of dynamic gas channels are rather built on directed evolution strategies such as the generation of random or semirational site saturation mutagenesis libraries and suitable screening systems. Figure 5 shows a simple library screening procedure adapted from Lampret et al. [42] and extended to identify site saturation mutagenesis variants of M1-type [FeFe]-hydrogenase CrHydA1 with enhanced levels of O$_2$ resistance. Here expression construct libraries are generated by site saturation mutagenesis and CrHydA1 variants are heterologously expressed in *Escherichia coli* strain BL21 Δ*iscR*. In this host strain, the gene for the iron–sulfur cluster biosynthesis system repressor iscR has been knocked out, thus enabling a higher expression of [FeS] cluster proteins [88] like apo-CrHydA1. Coexpression of *hydA* with the genes of 2Fe$_H$ maturases HydE, -F and -G further enables complete H-cluster maturation *in vivo* within the host cell [89] and therefore allows holoenzyme expression in the colonies of *E. coli* transformants. A colony-based colorimetric assay already allows the pre-screening of site saturation mutagenesis libraries for transformants with properly folded and catalytically competent CrHydA1 variants, by sequentially applying cysteine (500 mM; O$_2$ removal and sulfur source), IPTG (40 mM; inductor of CrHydA1 expression) and after several hours of enzyme expression, methyl viologen (MV) solution (600 mM) right onto the colonies. Immediate incubation of the colony plates for 10 min in forming gas (90% N_2/10% H_2) enables to identify transformants expressing catalytically intact variants as their colonies adopt a bluish color due to H_2-dependent MV reduction. Precultures of active transformant colonies are collected and used to inoculate individual medium reservoirs of a deep well plate. After IPTG addition and several hours of expression time, cell lysates from transformant cultures are tested for *in vitro* H_2 oxidation activity prior (basic enzyme activity of variant) and after O_2 exposure (addition of air-saturated buffer) by adding MV and exposure to 10% H_2. H_2 conversion, indicative of the presence of active enzyme, is quantitatively determined spectrophotometrically by the accumulation of the blue-violet color of reduced MV in the sample and normalized in reference to the optical density (OD_{600}) of the individual culture sample (Fig. 5, reader). Residual activities are calculated as relative rates of MV reduction compared to the corresponding rates prior to O_2 addition and later compared to wild-type references (Fig. 5, green circles in microtiter plates). To identify substitutions leading to enhanced O_2 resistance, expression constructs of transformant cultures with enhanced residual activity compared to wild type are isolated and sequenced.

Bingham and coworkers established a random mutagenesis (RM) system employing error-prone PCR and a cell-free expression system to generate variants of CpI [90, 91]. Out of 3000 gene variants, just 20 (encoding for enzyme variants with 4–16 amino acid substitutions) still exhibited activity after *in vitro* expression. Enhanced residual activity after O_2 exposure was only determined for a single variant with 13 substitutions. Out of these, the major influence on O_2 resistance was later referred to a single exchange (I197V). Site saturation mutagenesis (SSM) at position 197 later demonstrated that by implementing Ser as a substitute of I197 the positive impact on O_2 resistance could even be significantly enhanced. Thus, a combinatorial approach of RM, used in a primary broad screening phase to identify positions with a basic influence, and SSM, used in a secondary phase to find the most impactful substituent for each of the selected positions, can be a very powerful strategy in enzyme design. Another important aspect for addressing the topic of O_2 sensitivity concerns the well-known phrase of directed evolution "you get what you screen for." Simple screening assays based on the comparison of residual activity after pre-incubation in O_2-saturated buffer favor the identification of variants with enhanced O_2 resistance. However, a more worthwhile aim of hydrogenase re-design would be true O_2 tolerance, that is, the capability to stay catalytically active in the presence of O_2 as known for O_2-tolerant [NiFe]-hydrogenases. Smith and coworkers established a test system that allows to verify the level of O_2 tolerance, which may become part of a larger screening procedure. It is based on the ferredoxin-$NADP^+$-oxidoreductase (FNR)-dependent reduction of ferredoxin which fuels the H_2 evolution activity of [FeFe]-hydrogenase independent of strong reductants like sodium dithionite (NaDT). Low potential reducing agents like NaDT used in standard *in vitro* tests to saturate the H_2 evolution activity of culture or enzyme samples may quickly diminish added O_2. The turnover of glucose-6-phosphate (G6P) via G6P dehydrogenase allows to maintain a stable level of NADPH to fuel FNR activity [91, 92]. This test enabled to determine an increased level of O_2 tolerance for CpI variant I197S [91] (see Tab. 1).

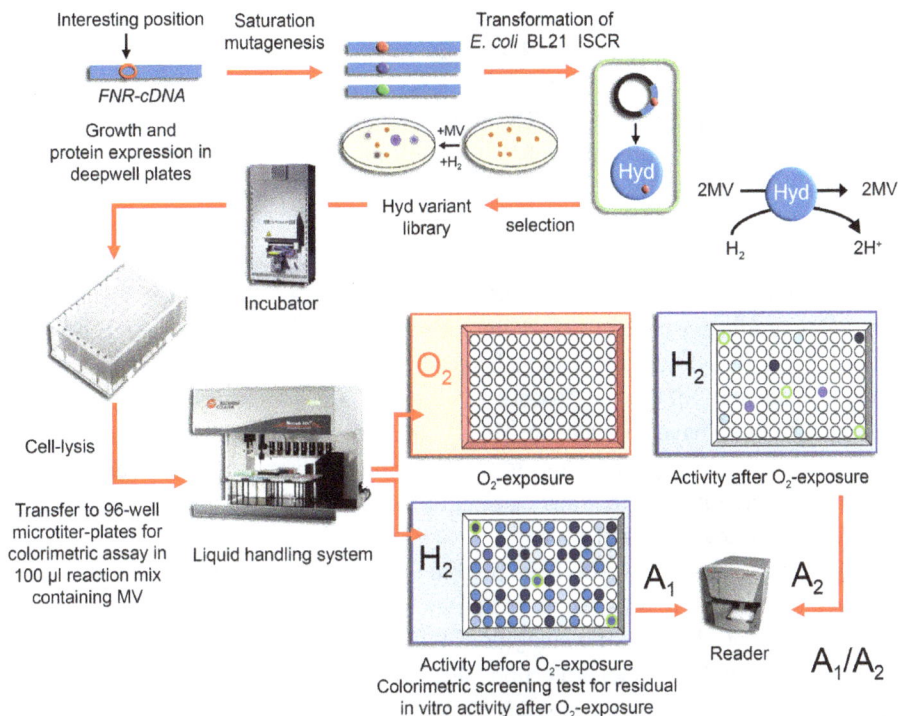

Fig. 5: Generation of saturation mutagenesis libraries and screening system for [FeFe]-hydrogenase variants with enhanced level of O$_2$ resistance.

4.3 Strategies to increase the O$_2$ resistance of [FeFe]-hydrogenases

Observing the process from O$_2$ entry up to oxidative H-cluster disintegration, we may derive potential design strategies for lowering the O$_2$ sensitivity of [FeFe]-hydrogenases:

1. Enhancing the molecular sieve effect within the gas channel system of the enzyme to prevent O$_2$ from reaching the active site
2. Increasing the electron supply to the H-cluster to enhance the reactivation rate (k_a) via reductive turnover of O$_2$ to H$_2$O
3. Modulating the protein environment of the active site cofactor to impact the steric accessibility of the open coordination site or the electrochemical features of the catalytic cofactor to decrease the binding affinity for O$_2$ (k_i) and/or downstream O$_2$ reactivity (k_3)

Following different versions of those strategies, [NiFe] and [FeFe]-hydrogenases have been analyzed both on a theoretical level and experimentally by investigating the effects of site-directed and RM targeting the gas channel system or amino acids in the secondary ligand sphere of the active site cofactor and accessory [FeS] clusters.

The high level of diversity of [NiFe]-hydrogenases [93, 94] is reflected by a broad spectrum of O_2 (in)stabilities. While there are many bidirectional representatives, [NiFe]-hydrogenases mostly favor H_2 oxidation over H_2 evolution [23, 95]. Yet, deepening our understanding of strategies to prevent O_2-dependent inactivation among [NiFe]-hydrogenases might inspire future engineering projects focusing on the highly productive and thus biotechnologically more promising – however severely O_2-sensitive – [FeFe]-hydrogenases [37]. Recently, new insights into the O_2-tolerance of some H_2-oxidizing [NiFe]-hydrogenases and into the higher O_2-resistance of other, bidirectional subtypes have been gained: This involves mutagenesis targets within their dynamic gas channel networks, close to the O_2-binding site or on native examples and design concepts that aim to enhance electron supply to the active site for a fast and complete O_2 turnover to water.

Similar strategies have also been tried to lower the O_2 sensitivity of [FeFe]-hydrogenases, unfortunately with rather limited success [54, 60, 70, 91, 96–98] (see Tab. 1). However, a novel H_2 evolving subtype of [FeFe]-hydrogenase has been discovered, which exhibits an in-built O_2 protection mechanism located in the immediate environment of the substrate-binding site [55, 56, 99]. This naturally evolved type 3 strategy enables a full O_2 resistance independent of further additives which may be introduced into the concepts of future biotechnologically applied [FeFe]-hydrogenase designs.

4.3.1 Enhancing the molecular sieve effect

The catalytic cofactors of [FeFe] and [NiFe]-hydrogenases are deeply buried in the protein core. Substrate (H_2) and inhibitors like CO and O_2 enter from the ambient solvent into a gas channel network which permeates the protein. It consists of hydrophobic caverns and intercavity passages that lead to a central cavity harboring the O_2-sensitive active site (see Fig. 6) [100, 101]. The difference in size between substrate H_2 (1.35 Å) and inhibitor O_2 (1.6 Å) [102, 103] implies the potential to exploit the molecular sieving effect of especially narrow passages that are less accessible for O_2 but still allow the smaller H_2 molecules to pass. A comparison of O_2-sensitive and O_2-tolerant [NiFe]-hydrogenases revealed a more complex and more accessible gas channel system in O_2-sensitive types, suggesting an impact of the gas channel accessibility on O_2 sensitivity [104]. To explore the effective channel accessibility for both gasses and the chances for enhancing the sieving effect at bottleneck locations of the channel network, molecular dynamics (MD) simulations of H_2 and O_2 diffusion based on volumetric solvent accessibility maps and temperature-controlled locally enhanced sampling have been done for the clostridial [FeFe]-hydrogenase CpI.

They confirmed the presence of two major pathways, one being a rather static channel (A) already visible in the available crystal structures of CpI, while the accessibility of the other (B) that is invisible in the rigid crystal structure largely depends on local packing defects that fluctuate in size and interconnections based on MD [103]. Analyzing crystals of *Desulfovibrio desulfuricans* hydrogenase DdH upon exposure to xenon suggested a single major Xe-binding site in the static gas channel midway on the trajectory to the H-cluster [101] surrounded by side chains of A427, F493 and L283 (numbering of CpI). To effectively explore the O_2-accessible internal volume of the enzyme, Mohammadi et al. used temperature-accelerated MD simulations to calculate the free energy landscape for O_2 diffusion throughout CpI, yielding a 3D potential of mean force map which reveals the most probable O_2 pathways according to their correspondingly low energy barriers (Fig. 6) [96].

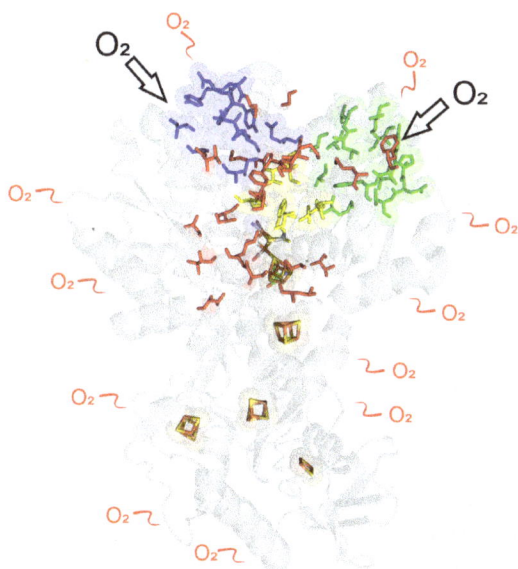

Fig. 6: Gas channel network and residues at the local free energy minima of the O_2 diffusion map (red sticks) in [FeFe]-hydrogenase CpI (PDB-ID:4XDC) shown in yellow (central cavity), green (static channel A) and blue (dynamic channel B) [103]. Indicated channel positions and color-coded iron–sulfur clusters (accessory clusters and H-cluster) shown as stick structures with half-transparent spheres. Overall structure presented as half-transparent gray cartoon. Besides the main channel entries to channels A and B (large black arrows), red wiggly lines indicate the numerous possible entrances for O_2 suggested for an overall dynamic gas channel network according to [96].

In line with the concept of dynamic gas channels, O_2 movement to the H-cluster is described as a step-wise hopping between dynamic protein cavities leading from one local minimum to another with the global minimum located very close to the H-cluster [54, 96, 103]. Diffusion to the active site merely takes 0.1–1 ms [54]. The O_2

diffusion time and number of possible cavities depends on the size and character-istics of the ligand [96, 103, 105]. Complementary to MD simulation approaches, a soak and freeze derivatization method applied on hydrogenase crystals can be used to follow O_2 diffusion experimentally, allowing the visualization of O_2-accessible cavities [100, 104].

Liebgott and coworkers employed PFE to determine and compare the overall rate constants for CO- and O_2-binding and -release of ten mutagenesis variants of *Desulfovibrio fructosovorans* [NiFe]-hydrogenase Df and two [FeFe]-hydrogenases (CaHydA of *Clostridium acetobutylicum* and DdH) [106]. The authors demonstrated that the rate of CO-induced enzyme inhibition is diffusion-limited, thus the CO inhi-bition rate could be used as a reliable indicator for effects on the molecular sieve effect. Both an increase in size and hydrophobicity of residues at target sites in the gas channel system lowered the gas diffusion rate in Df and a combination of both parameters led to additive effects. However, while CO and O_2 have similar diffusion rates, CO inhibition is by magnitudes faster showing that the rate of O_2-induced in-hibition is limited by the reactions of O_2 at the active site rather than by diffusion. An impact on O_2 sensitivity was only detectable when CO diffusion and thus the overall channel accessibility was strongly diminished.

Even if a successful channel obstruction may be deduced from the static X-ray crystallography data of an enzyme variant, side-chain dynamics may limit the im-pact of such rational design strategies under turnover conditions as shown by Ler-oux and coworkers [107]. This emphasizes the relevance of MD and the importance of diffusion rate measurements. In contrast to the [NiFe]-hydrogenase, CO- and O_2-induced inhibition of [FeFe]-hydrogenase CaHydA were quite comparable, suggest-ing here also a diffusion limitation of the O_2 inhibition [106].

In a different approach, the O_2 sensitivity of [FeFe]-hydrogenase CaHydA should be decreased by exchanging a selected number of smaller residues that line the channel walls of the pre-chamber and the Xe-binding site for larger substitutes, ex-ploiting the molecular sieve effect [60]. Out of eight single exchanges only A426L (A427L in CpI) had a promising impact on parameters that define molecular diffu-sion: It increased the Michaelis–Menten constant for H_2 binding and decreased the production of HD in H^+/D^+ exchange experiments as well as the rates for binding and release of O_2 (see Tab. 1). The same parameters, however, were entirely unaf-fected by the reversible inhibitor CO, indicating that O_2 progression via the ne-glected dynamic channel B might be more relevant for O_2 diffusion to the H-cluster. Here, protein cavities/packing defects fluctuate in size and interconnections depend on the motion state of the protein molecule. In contrast to a simple diffusion process along a static tunnel system, O_2 progression from surface to the active center may rather result from peristaltic motions due to random movements within the protein scaffold [103, 108].

Tab. 1: Site-directed and random mutagenesis variants of [NiFe] and [FeFe]-hydrogenases with increased or decreased O$_2$ resistance/tolerance.

Organism (enzyme)	Mutation	Analysis/method (reference)	Result	Reason
O$_2$-sensitive [NiFe]-hydrogenases (O$_2$ binds and destroys H-cluster)				
D. fructosovorans (Df)	V74M	Crystallography, spectroscopy [119]	Increased O$_2$ resistance	Methionine(s) in diffusion pathway
	L122M			
	V74M/ L122M			
	V47C	PFE, spectroscopy [120]	Increased O$_2$ resistance	Ni-B favored over -A
O$_2$-tolerant [NiFe]-hydrogenases				
E. coli (Hyd-1)	V78C	PFE, kinetics [121]	Increased O$_2$ tolerance	Slower inactivation faster reactivation
	D574N/ D118N	Crystallography, kinetics, PFE [124]	Decreased O$_2$ tolerance	Ni-B favored over -A
	D118N			
O$_2$-resistant [NiFeSe]-hydrogenases due to Se-Cys				
D. vulgaris	G491S	Crystallography, PFE, spectroscopy, kinetics [122]	Increased O$_2$ resistance	Blocked hydrophilic water channel
	G491A			
O$_2$-sensitive [FeFe]-hydrogenases (O$_2$ binds and destroys H-cluster)				
C. pasteurianum (CpI)	I197S	O$_2$ exposure of enzyme [91]	Increased O$_2$ resistance	Unknown
	T356V/ S357T	Tolerance test [97]	Increased O$_2$ tolerance	
	M387L	Tolerance test [98]		
C. acetobutylicum (CaHydA)	A426L	PFE; kinetics [60]	Increased O$_2$ resistance	Decreased intramolecular gas transport
C. reinhardtii (CrHydA1)	C169A	XAS, XES, NRVS [74]	Increased O$_2$ resistance/ inactive	O$_2$ adduct protonation affected

Tab. 1 (continued)

Organism (enzyme)	Mutation	Analysis/method (reference)	Result	Reason
O_2-resistant [FeFe]-hydrogenase (flexible loop enables safety cap mechanism)				
Clostridium beijerinckii (CbA5H)	L364F	PFE, kinetics, IR spectroscopy [56]	Decreased O_2 resistance	Decreases loop flexibility
	P386L			
	A561F			
	C267D			

IR, infrared; PFE, protein film electrochemistry; XAS, X-ray absorption spectroscopy; XES, X-ray emission spectroscopy; NRVS, nuclear resonance vibrational spectroscopy.

4.3.2 Increasing the electron supply to the H-cluster

A very effective strategy of hydrogenases to escape the persistent inhibition of the Ni-A state ([NiFe]-hydrogenases) or to prevent the harmful production of aggressive ROS which attack the active site and initiate its oxidative degradation ([FeFe]-hydrogenases) would be a fast and complete reduction of O_2 to H_2O. H_2O may leave the active site through water channels, existing in both hydrogenase types [109, 110].

In some [NiFe]-hydrogenases like the membrane-bound hydrogenase of *Ralstonia eutropha*, H_2-oxidation activity can be demonstrated even in the presence of O_2 [7]. Specific features in the electron relay of these O_2-tolerant hydrogenases enable them to overcome the electron bottleneck of O_2 detoxification to H_2O at the active site, as verified by the isotopic labeling of O_2 molecules [111]. Here O_2 tolerance also goes in line with a lack of the slowly reactivating Ni-A state [112–114] as a fast succession of protonation and reduction steps prevents the accumulation of incompletely reduced ROS like H_2O_2 [112]. To reliably accomplish the transition into the Ni-B state, three immediate reduction events are required. The first two electrons can be provided by the Ni site and the proximal accessory cluster. The third and fourth can be supplied from the medial and distal [FeS] clusters; however, the rate-limiting long-range transport of the third electron may lead to the persistent Ni-A intermediate [112]. In O_2-tolerant [NiFe]-hydrogenases, this is circumvented by an unusual proximal [4Fe3S] cluster coordinated by 4 + 2 cysteines. The two additional cysteines near the cluster are vital for reaching the Ni-B state, as their substitution by glycine renders the corresponding variants O_2-sensitive [112]. The ability to undergo a reversible cluster reconfiguration enables this unusual proximal [FeS] cluster to accomplish two instead of the usual single redox transitions comprising the oxidized $[4Fe3S]^{4+}$ and the superoxidized $[4Fe3S]^{5+}$ states (adopted in the Ni-B state) [115, 116]. It thus enhances the storage capacity of the electron relay for effective O_2

turnover [117, 118]. Apart from that, the unusual proximal and the distal [4Fe4S] cluster of O$_2$-tolerant [NiFe]-hydrogenases exhibit increased redox potentials (+100 mV) compared to sensitive conventional [NiFe]-hydrogenases. This stabilizes the reduced cluster states within the cluster relay and enables to re-route electrons previously gained from H$_2$ oxidation back to the active site for O$_2$ reduction to water in case of O$_2$ interaction.

As outlined in Section 4.1.3, a low-level water formation even occurs in [FeFe]-hydrogenases, with the rate of reactivation being higher in some cases (CaHydA and DdH) and hardly recognizable for others (M1-type enzyme CrHydA1) [54]. The strategy of O$_2$-tolerant [NiFe]-hydrogenases is difficult to transfer to [FeFe]-hydrogenases in rational re-engineering concepts. It would require a series of well-concerted steric and electrochemical modifications in a non-homologous protein scaffold. To explore the chances of recreating the unique cluster type of O$_2$-tolerant [NiFe]-hydrogenases in [FeFe]-hydrogenase CpI, Koo and coworkers followed a semirandom Cys-scanning mutagenesis approach. By systemically substituting the amino acids surrounding the proximal accessory [FeS] cluster to cysteine, they tested the chances to reconfigure the primary ligand sphere using different cluster scaffolds that exhibit surplus Cys ligands at various locations around the cubane cluster. Interestingly, some of the cysteine scanning variants indeed influenced the O$_2$ stability of the enzyme in one or the other direction. Whether this can be referred to a similar effect as described for O$_2$-tolerant [NiFe]-hydrogenases remains to be clarified [97] (Tab. 1).

4.3.3 Modulating the protein environment of the active site

The local environment of the catalytic cofactor and the substrate/inhibitor-binding site has an impact on the kinetic parameters of inhibitor interaction, the local steric accessibility and the electrochemical features of the substrate-binding site: affected are the rates of O$_2$-binding (k_i) and/or -release (k_a) as well as the initiation of potentially degradative O$_2$-turnover processes (k_3). Also, individual functional residues close to the active site may help to prevent O$_2$-binding or get rid of the bound O$_2$ species.

Substitutions of V47, located near the substrate-binding site of [NiFe]-hydrogenase Df *from Desulfovibrio fructosovorans*, show the influence of the secondary ligand sphere on O$_2$ sensitivity: a substitution to methionine reduces the rate of oxygen-induced inhibition by 10-fold, according to the X-ray data mainly due to the bulky Met residue which limits the accessibility for O$_2$ to the active site. Also, the methionine residue seems to facilitate the reactivation from the inactive Ni-A state by rearranging the hydroperoxo ligand for protonation by a nearby glutamate residue, resulting in a quickly detached H$_2$O$_2$ molecule [119] (Tab. 1). An exchange of V74 to cysteine attributes some features of O$_2$-tolerant hydrogenases [120] and increases the tolerance level of a natively O$_2$-tolerant [NiFe]-hydrogenase while limiting its catalytic activity [121] (Tab. 1).

The Se-Cys ligand of the active site cofactor of [NiFeSe]-hydrogenases plays a key role both for the comparatively high activity and for the oxygen tolerance of this class of [NiFe]-hydrogenases [122] (Tab. 1). While they are inactivated by high O_2 concentrations, the reversible reactivity of Se-Cys with O_2 supports their fast reductive reactivation [123].

In the O_2-tolerant [NiFe]-hydrogenase Hyd1 of *E. coli* the residues of two aspartates, one arginine and a proline, form a canopy-like cover over the [NiFe] cofactor. Replacement of the negatively charged residues D118 and D574 of the canopy by the neutral asparagine decreased the O_2 tolerance by stabilizing the resting Ni-B state (Ni^{III}–OH) [124].

In addition to the secondary ligand sphere, sulfide ions acting as competitors for the open coordination site can have a strong impact on the O_2 sensitivity of the H-cluster. Aerobically purified DdH from the sulfate-reducing bacterium *Desulfovibrio desulfuricans* was shown to be inactive but fully protected against O_2 prior to the first reductive activation [86]. However, once activated, DdH enzyme turned out to be even more O_2 sensitive than other [FeFe]-hydrogenases, such as CaHydA or CrHydA1 [95, 106]. The inactive but O_2-protected H_{inact} state of the isolated enzyme was only regained if external sulfide was added: This shows that the H_{inact} state DdH depends on external sulfide entering the enzyme and reversibly occupying the substrate-binding site as a competitive inhibitor of both H_2 (thus being inactive) and O_2 (being protected) (Fig. 7A) [125].

Confirming this, in a recently published crystal structure of DdH trapped in H_{inact}, a sulfur atom was doubtlessly identified as an apical ligand of the substrate-binding Fe_d site [126]. In contrast to other [FeFe]-hydrogenases, the recently isolated enzyme CbA5H of *Clostridium beijerinckii*was shown to reversibly shift between the O_2-resistant state H_{inact} and the catalytically active oxidized ready state H_{ox}, irrespective of the presence of sulfide [55, 56, 99]. Apart from O_2 exposure, H_{inact} formation also occurred upon treatment with oxidants (e.g., thionine or 2,6-dichlorophenolindophenol) while activity was quickly regained when the enzyme was exposed to reductants like sodium dithionite and H_2 [55] showing that adopting the protected H_{inact} state is a consequence of a high external potential rather than of O_2 interaction (Fig. 3B, CbA5H). Consequently, H_{inact} formation in CbA5H is visible in cyclovoltammetry by an early (low potential) onset of oxidative inactivation (Fig. 3A, CbA5H) [56, 99]. Recently the mechanism behind this unique intrinsic O_2 resistance was revealed by comparing crystal structures of air-exposed CbA5H (CbA5H^{air}, trapped in H_{inact}) with structures of CpI and DdH (H_{ox}). While matching with other parts of the catalytic H-domain, the peptide loop in the CbA5H^{air} structure with T365, S366 and C367 is relocated closer to the H-cluster (Fig. 7C). Consequently, the thiol residue of PTP position C367 (Fig. 7B) is shifted close enough to the substrate-binding site of Fe_d to block the access against incoming O_2 molecules; this mimicks the protective effect of sulfide binding (Fig. 7D). H_{inact} formation of CbA5H is enabled by the three "bystander" amino acids L364, P386 and A561 (see Tab. 1). Due to the smaller van der Waals radii of their

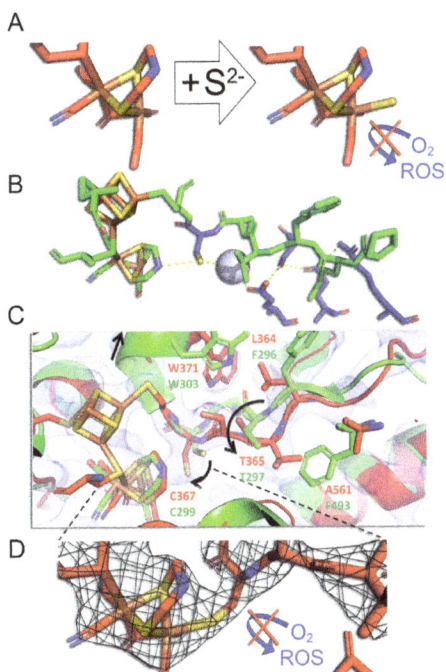

Fig. 7: External sulfide and a translocated Cys residue both protect the H-cluster of some [FeFe]-hydrogenases against O_2 attack. (A) Sulfide-binding at the open coordination site of [FeFe]-hydrogenases DdH and, for example, CrHydA1 yields the inactive but O_2-protected state H_{inact}. (B) In the active state of CpI and other [FeFe]-hydrogenases, the open coordination site of the 2FeH cluster is coupled to the proton transfer pathway (PTP), consisting of strictly conserved protonatable amino acid residues (blue stick structures) and a protein bound water molecule (sphere) aligned in an H-bond chain between H-cluster and protein surface. C299 in CpI is an essential position in the PTP which facilitates the H^+-exchange between the adt ligand of the $2Fe_H$ site and the remaining part of the PTP. (C) Structural alignment that highlights conformational differences at the active centers and H-domains of CbA5H^{air} (red) and CpI (green) including a shifted peptide loop. Depicted are amino acids within (T365, S366 and PTP position C367) and close to the TSC loop which influence anaerobic inactivation, O_2 resistance and H_{inact} formation visible as labeled stick structures. Under oxidative conditions (including O_2 exposure) the TSC-loop shift yields the catalytically inactive H_{inact} state, irrespective of the presence of sulfide by translocating C367 close enough to the substrate-binding site to block it against O_2 attack. (D) Removal of O_2 under reductive conditions restores the active ready state H_{ox}.

residues compared to the bulky counterparts in CpI (F296, L364 and F493), they provide the necessary rotational freedom in the H-domain of CbA5H to accomplish the reversible loop shift (see Fig. 7C). Comparing the kinetic constants of early oxidative inhibition in CbA5H wild-type and site-directed variants targeting positions C367 L364, P386 and A561 allowed to derive the potential and pH-dependent mechanism of reversible H_{inact} formation describing the transition between two active states (A1 and A2) and the inactive O_2-protected H_{inact} state (I) (Fig. 8).

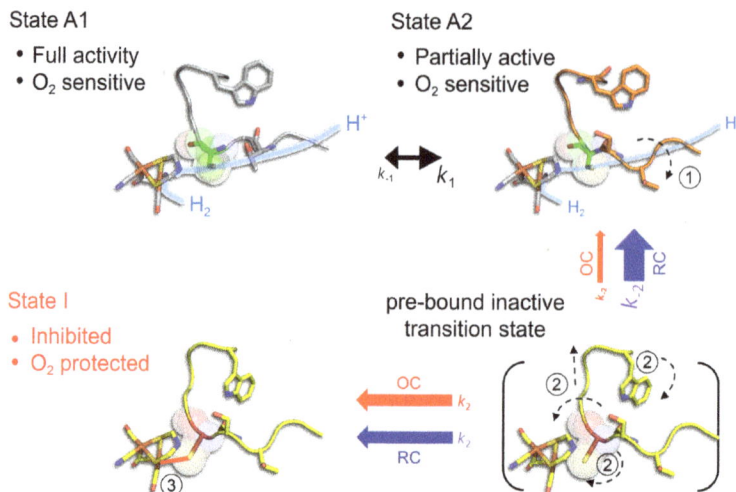

Fig. 8: Illustration of the AAI mechanism in CbA5H. Critical steps [1–3] in the transition from active ready H_{ox} state (A_1) (black-gray stick structure) via A_2 (orange) to H_{inact} (red-yellow) are marked by encircled numbers. 1: Partial shift of the TSC loop, including T365, initiating the $A_1 \rightarrow A_2$ transition. 2: W371-translocation and α-helix uplift, dragging along C367 closer to Fe_d (broken arrows). 3: Oxidation and deprotonation of C367 initiates its coordination to Fe_d, yielding the inactive but O_2-resistant H_{inact} state. Green/red coloration and different conformation of Cys367 indicates active (more distant from Fe_d)/inactive (closer to Fe_d) states. C367 binding to Fe_d (H_{inact}) is illustrated by a red line. Sizes of red and blue arrows reflect relative rates of k_1 and k_{-1} or k_{inact} and k_{react} define the dynamic equilibrium between A_1, A_2 and H_{inact} and have been derived from electrochemical experiments examining the kinetics of early oxidative inhibition in CbA5H. Under reductive conditions (RC, blue), the k_{inact}/k_{react} ratio favors A_2 as the main state while leading to H_{inact} accumulation under oxidative conditions (OC, red).

4.3.4 In vitro strategies to overcome O_2 sensitivity in applications

As already shown in Section 2.1, PFE can analyze the electrochemical behavior of hydrogenases on the molecular level. While [FeFe]-hydrogenases are in principle most suitable candidates for future biotechnological applications in BFCs in terms of catalytic efficiency (comparable to Pt nanoparticles in catalytic $2H^+/H_2$-turnover activity) [127], they suffer from high potential anaerobic inactivation and a severe sensitivity toward molecular O_2 [7, 128]. These two aspects represent major obstacles as hydrogen fuel cells operate under a defined H_2/O_2 atmosphere with a typical cell voltage of 0.6–0.7 V [19]. However, recently there is some progress in overcoming these limitations.

A great advantage of using hydrogenase-based BFCs compared to classic hydrogen fuel cells is the great selectivity of enzymes, making expensive membrane materials like Nafion and its derivatives unnecessary. Such materials are otherwise used to separate H_2 at the anode from O_2 at the cathode when electricity is generated from H$_2$ fuel [19]. A very first benchmark of an enzyme-based BFC was undertaken by the Armstrong lab in 2014 where an O$_2$-tolerant [NiFe]-hydrogenase (anode) was used in conjunction with a bilirubin oxidase (cathode). Such a BFC concept can be adapted by using an [FeFe]-hydrogenase instead (Fig. 9B, left panel). By operating the fuel cell in a parallel stack using an H$_2$/air mixture, Armstrong and coworkers could successfully power a "hydrogen house" with small electronic devices. In this study, the access of O$_2$ to the enzyme's active site was presumably restricted by using multi-walled carbon nanotubes (MWCNT) that generate a porous "3D surface" in order to increase the enzyme coverage at the electrode surface. While this usually leads to an order-of-magnitude enhancement in catalytic current [129], it may filter out O$_2$ relative to H$_2$, hence decreasing the effective O$_2$ diffusion rate. This is a feasible approach for O$_2$-tolerant hydrogenases and strongly resembles the molecular sieve effect discussed in Section 3.2, but it does not represent a practicable solution for O$_2$-sensitive enzymes like [FeFe]-hydrogenases as it only offers short-term protection. However, the introduction of hydrogenases into adequately designed redox polymers modified with low-potential viologen moieties yielded significant progress [130]. Further studies confirmed that viologen-modified hydrogel films do not only protect hydrogenases from high potential inactivation but also – most importantly – from O$_2$ damage [131–134]. These redox-hydrogel films also serve as an "electrical wire" to the electrode surface and thus as an effective immobilization matrix. In addition, redox mediators tethered to the polymer backbone are an electrical wire for the redox enzymes [135]. Electron transport between immobilized enzymes and the electrode surface is ensured via an electron-hopping mechanism across the polymer-bound redox mediators. This strategy has already proven to be suitable for bioelectrodes incorporated in BFC applications: It provides increased current densities and coulombic efficiencies for the conversion of substrates, resulting in synergistic effects and delivering enhanced electrochemical responses [136–138]. As shown for sensitive [NiFe] and [NiFeSe]-hydrogenases, additional protection is achieved by an enzymatic O$_2$-scavenging system immobilized on top of the active redox polymer/hydrogenase layer [139]. Thus, the unprecedented protection of enzymes from oxidative damage has enabled the implementation of high-performance bioelectrodes ready for application, for example, in energy conversion devices.

The very first feasibility of a high-performance bioanode that relied on an O$_2$-sensitive [FeFe]-hydrogenase as a biocatalyst was just recently demonstrated [36]. In fact, the combination of redox polymer (for O$_2$ protection) and gas-breathing electrodes (to enhance the H$_2$ diffusion kinetics) allowed the highly sensitive DdH enzyme to efficiently operate under 5% O$_2$, with a maximum power density of 5.4 mW cm^{-2} at 0.7 V when used as a dual-gas diffusion electrode (with a bilirubin oxidase at the cathode) under anode-limiting conditions. Figure 9A (right panel) shows a potential application of such a bioanode for the efficient generation of electricity from H$_2$ under air where the polymer protects HydA from damage and high potential inactivation. The generated electricity can subsequently be stored in a power capacitor. The reverse case of generating H_2 from electricity is likewise possible. Even without the usage of redox polymer this approach was successfully demonstrated for the [FeFe]-hydrogenase CpHydA1 from *Clostridium perfringens* in combination with a TiO_2 support. H_2 production rates were stable for several days with high Faradaic efficiency (~98%) [140]. However, exchanging the TiO_2 support for a suitable redox polymer may improve long-term stability, even in the presence of O_2 (Fig. 9A, left panel). With [FeFe]-hydrogenase CbA5H, the O_2 protection effect of the redox polymer would become dispensable due to the enzyme's inherent O_2 protection mechanism, highlighting the great application potential of this [FeFe] hydrogenase (Fig. 9A, middle panel). However, a reasonable approach using CbA5H as a H_2-producing power cell may still include redox polymer. The polymer acts as a supporting matrix for O_2 stability but also supports the electrochemical communication between enzyme and electrode material and enhances the enzyme coverage, thus increasing the total power output.

Another interesting strategy is to combine the advantage of integrating HydA into gas diffusion layers to overcome mass transport limitations of gaseous substrates while simultaneously protecting immobilized hydrogenases by a redox polymer matrix. This concept is shown for the bioanode in Fig. 9B and C. Note, however, that for these applications, CbA5H is most likely not able to provide sufficient electrons from H_2 oxidation due to its rapid inactivation at potentials above ~ −0.3 V (pH 7) (see Fig. 3A). In total, [FeFe]-hydrogenase-based BFCs have a great potential for future biotechnological applications due to a combination of enzyme re-design (see Section 3) and current advances in material design to overcome the inherent problems of O_2 damage and high potential inactivation.

The concept shown in Fig. 9C represents a new field of application for hydrogenase-based BFCs where at the cathode, a "natural" toolbox of enzyme cascades including a NAD(P)H-cofactor regenerating system [2, 141] could be fueled by hydrogenases via H_2 oxidation at the anode. To date, this process has only been tested with the less active class of [NiFe]-hydrogenases. Using [FeFe]-hydrogenases like CpI/DdH could further improve the efficiency of such systems. Moreover, the utilized electrode materials could be improved as the currently described H_2-driven enzyme cascades are mostly immobilized on carbon supports (e.g., MWCNT). While it provides them a suitable surface for immobilization [142, 143], it does not consider the individual needs for an enzyme to promote long-term stability and stereoselectivity, which is best achieved in more "native-like" environments. Due to its hydrogel character, solvated redox polymer may serve as a more suitable and biocompatible matrix for enzyme entrapment in this case, as it also includes the possibility to create a confined microenvironment of immobilized biocatalysts. Establishing semiartificial multi-enzymatic cascade reactions with the biocatalysts co-immobilized within a confined environment would allow to reconstitute the design of native metabolic pathways/systems [144–146]. Employing [FeFe]-hydrogenases for the synthesis of value-added chemicals and reaction intermediates represents a promising road for their effective usage in future applications [147, 148].

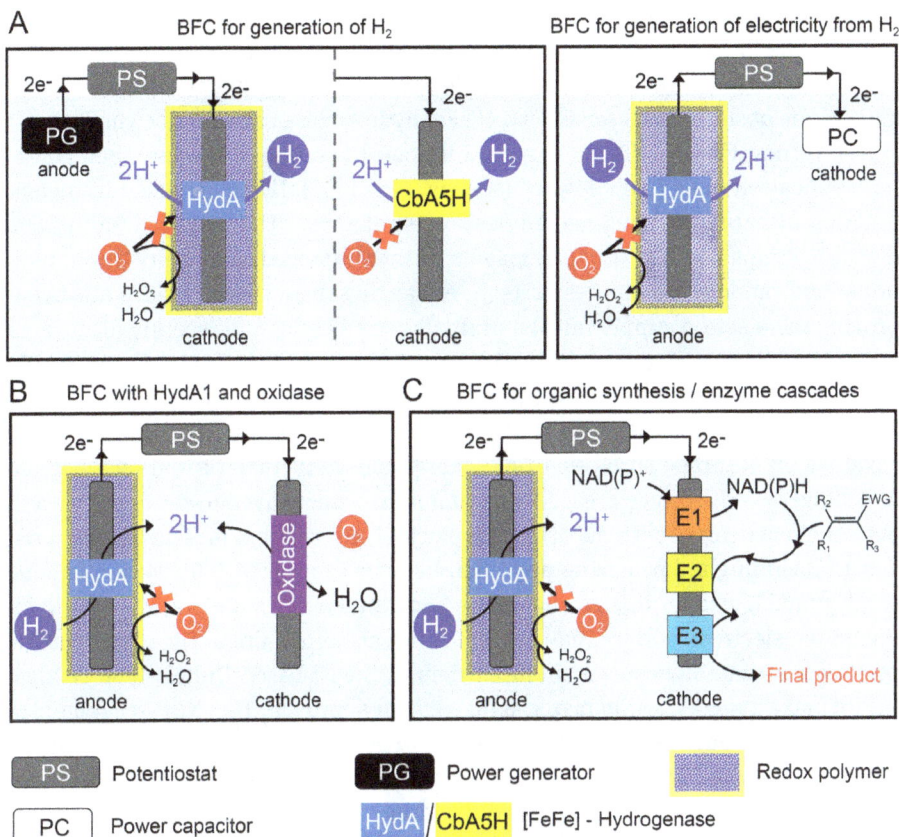

Fig. 9: Different application concepts for O$_2$-sensitive (HydA) and O$_2$-resistant (CbA5H) [FeFe]-hydrogenases in conjunction with a biofuel cell (BFC). In all cases except for CbA5H, a redox polymer (blue-white pattern highlighted in yellow) protects HydA from O$_2$ degradation. (A) Left panel shows a BFC using HydA/CbA5H at the cathode to form molecular hydrogen from protons and electrons, with the latter being supplied by a power generator. The reverse configuration (lower panel) can be used to split H$_2$ and inject the electrons into a power capacitor. (B) HydA at the anode provides electrons for an oxidase immobilized at the cathode in a membrane-free BFC. (C) The same anode as in (B) is used to supply electrons for the organic synthesis of fine chemicals via an enzyme cascade immobilized at the cathode.

4.4 Outlook and future perspectives

For more than 20 years, scientific efforts have contributed to gain invaluable in-sights into the molecular mechanisms underlying O$_2$-induced inhibition and/or deg-radation for both [NiFe] and [FeFe]-hydrogenases. Mechanisms involved have been elucidated by crystallographic, spectroscopic and kinetic methods. Overcoming the inherent drawbacks of O$_2$ sensitivity would pave the way for using hydrogenases in biotechnological and industrial applications. Due to their relative robustness toward O_2, the focus has been mainly on [NiFe]-hydrogenases as potential candidates for applications. However, due to considerable progress in understanding and removing the obstacle of O_2 sensitivity, [FeFe]-hydrogenases as supreme catalysts for $H_2/2H^+$ turnover [55, 56, 68, 99] have now become the center of interest. Their potential application in fuel cell devices (see Fig. 9) may finally be achieved by a combination of native and engineered "escape strategies." The most tolerant native [FeFe]-hydrogenase (e.g., CbA5H) may be further improved in stability through rational and random protein design (e.g., enhancing the molecular sieve effect/improving the electron supply rates) and finally be embedded in recently discovered nanomaterials (i.e., redox polymers) to maximize long-term stability against O_2 damage and high potential inactivation. This has already partly been realized but the enormous macromolecular size of these enzymes results in a relatively low coverage on electrode surfaces, even when porous "3D" materials are used to enhance the overall surface area. As this limits the achievable current densities and overall efficiencies of systems driven by biocatalysts, rational protein and material design including a downsizing approach have to be applied. Ideally, a synthetic catalyst much smaller in size and hence allowing for much greater catalyst loading on the electrode surface may overcome these shortcomings. The high catalytic performance and low overpotential barriers of the Ni-based DuBois-type catalyst would make a great starting material for such an approach. This $Ni(P_2N_2)_2$ complex can be further engineered by implementing peptide-based proton relay systems [149–152]. For the DuBois catalyst, a polymer-based electrode design with an inner electrocatalytic and an outer O_2 protection layer can prevent O_2-induced complex inactivation [153]. Still, a small peptide environment may be useful to implement design principles derived from biocatalyst research such as the O_2 cap mechanism of CbA5H. This may further push the limits of O_2 stability in design. Such semisynthetic hybrids may be the missing link in the development of inexpensive, noble-metal-free catalysts for efficient and long-term H_2 interconversion.

References

[1] Geider RJ, Delucia EH, Falkowski PG, Finzi AC, Grime JP, Grace J, et al. Primary productivity of planet earth: Biological determinants and physical constraints in terrestrial and aquatic habitats. Glob Chang Biol 2001, 7(8), 849–82.

[2] Morello G, Siritanaratkul B, Megarity CF, Armstrong FA. Efficient electrocatalytic CO2 fixation by nanoconfined enzymes via a C3-to-C4 reaction that is favored over H2 production. ACS Catal 2019, 9(12), 11255–62.

[3] Morello G, Megarity CF, Armstrong FA. The power of electrified nanoconfinement for energising, controlling and observing long enzyme cascades. Nat Commun 2021, 12(1), 340.

[4] Armstrong FA, Hirst J. Reversibility and efficiency in electrocatalytic energy conversion and lessons from enzymes. Proc Natl Acad Sci U S A 2011, 108(34), 14049.

[5]   Haumann M, Stripp ST. The molecular proceedings of biological hydrogen turnover. Accounts of Chemical Research. 2018.

[6]   Gallon JR. Reconciling the incompatible: N2 fixation And O2. New Phytol 1992, 122(4), 571–609.

[7]   Vincent KA, Parkin A, Lenz O, Albracht SP, Fontecilla-Camps JC, Cammack R, et al. Electrochemical definitions of O2 sensitivity and oxidative inactivation in hydrogenases. J Am Chem Soc 2005, 127(51), 18179–89.

[8]   Stripp ST, Goldet G, Brandmayr C, Sanganas O, Vincent KA, Haumann M, et al. How oxygen attacks [FeFe]-hydrogenases from photosynthetic organisms. Proc Natl Acad Sci U S A 2009, 106(41), 17331–6.

[9]   Can M, Armstrong FA, Ragsdale SW. Structure, Function, and. Mechanism of the nickel metalloenzymes, CO dehydrogenase, and acetyl-CoA synthase. Chem Rev 2014, 114(8). 4149–74.

[10]   Kroneck PM, Sosa Torres ME. Sustaining life on planet Earth: Metalloenzymes mastering dioxygen and other chewy gases. Met Ions Life Sci 2015, 15, vii–ix.

[11]   DiCosimo R, McAuliffe J, Poulose AJ, Bohlmann G. Industrial use of immobilized enzymes. Chem Soc Rev 2013, 42(15), 6437–74.

[12]   Itoh T, Hanefeld U. Enzyme catalysis in organic synthesis. Green Chem 2017, 19(2), 331–2.

[13]   Davis F, Higson SP. Biofuel cells – recent advances and applications. Biosens Bioelectron 2007, 22(7), 1224–35.

[14]   Guo KW. Chapter 15 – biofuel cells with enzymes as a catalyst. In: Srivastava N, Srivastava M, Mishra PK, Ramteke PW, Singh RL, eds. New and Future Developments in Microbial Biotechnology and Bioengineering. Elsevier, 2019, 261–82.

[15]   Bandodkar AJ, You J-M, Kim N-H, Gu Y, Kumar R, Mohan AMV, et al. Soft, stretchable, high power density electronic skin-based biofuel cells for scavenging energy from human sweat. Energy Environ Sci 2017, 10(7), 1581–9.

[16]   Barton SC, Gallaway J, Atanassov P. Enzymatic biofuel cells for implantable and microscale devices. Chem Rev 2004, 104(10), 4867–86.

[17]   Kwon CH, Ko Y, Shin D, Kwon M, Park J, Bae WK, et al. High-power hybrid biofuel cells using layer-by-layer assembled glucose oxidase-coated metallic cotton fibers. Nat Commun 2018, 9(1), 4479.

[18]   Xu L, Armstrong FA. Pushing the limits for enzyme-based membrane-less hydrogen fuel cells – achieving useful power and stability. RSC Adv 2015, 5(5), 3649–56.

[19]   Mazurenko I, Wang X, De Poulpiquet A, Lojou E. H2/O2 enzymatic fuel cells: From proof-of-concept to powerful devices. Sustainable Energy Fuels 2017, 1(7), 1475–501.

[20]   Lauterbach L, Lenz O, Vincent KA. H(2)-driven cofactor regeneration with NAD(P)(+)-reducing hydrogenases. FEBS J 2013, 280(13), 3058–68.

[21]   Al-Shameri A, Borlinghaus N, Weinmann L, Scheller PN, Nestl BM, Lauterbach L. Synthesis of N-heterocycles from diamines via H2-driven NADPH recycling in the presence of O2. Green Chem 2019, 21(6), 1396–400.

[22]   Volbeda A, Charon MH, Piras C, Hatchikian EC, Frey M, Fontecilla-Camps JC. Crystal structure of the nickel-iron hydrogenase from Desulfovibrio gigas. Nature 1995, 373(6515), 580–7.

[23]   Lubitz W, Ogata H, Rudiger O, Reijerse E. Hydrogenases. Chem Rev 2014, 114(8), 4081–148.

[24]   Pierik AJ, Roseboom W, Happe RP, Bagley KA, Albracht SP. Carbon monoxide and cyanide as intrinsic ligands to iron in the active site of [NiFe]-hydrogenases. NiFe(CN)2CO, Biology's way to activate H2. J Biol Chem 1999, 274(6), 3331–7.

[25]   Fontecilla-Camps JC, Amara P, Cavazza C, Nicolet Y, Volbeda A. Structure-function relationships of anaerobic gas-processing metalloenzymes. Nature 2009, 460(7257), 814–22.

[26]   Lamle SE, Albracht SP, Armstrong FA. Electrochemical potential-step investigations of the aerobic interconversions of [NiFe]-hydrogenase from *Allochromatium vinosum*: Insights into

the puzzling difference between unready and ready oxidized inactive states. J Am Chem Soc 2004, 126(45), 14899–909.

[27] Lamle SE, Albracht SP, Armstrong FA. The mechanism of activation of a [NiFe]-hydrogenase by electrons, hydrogen, and carbon monoxide. J Am Chem Soc 2005, 127(18), 6595–604.

[28] Ogata H, Hirota S, Nakahara A, Komori H, Shibata N, Kato T, et al. Activation process of [NiFe]-hydrogenase elucidated by high-resolution X-ray analyses: Conversion of the ready to the unready state. Structure 2005, 13(11), 1635–42.

[29] Volbeda A, Martin L, Cavazza C, Matho M, Faber BW, Roseboom W, et al. Structural differences between the ready and unready oxidized states of [NiFe]-hydrogenases. J Biol Inorg Chem 2005, 10(3), 239–49.

[30] Siegbahn PE, Tye JW, Hall MB. Computational studies of [NiFe] and [FeFe]-hydrogenases. Chem Rev 2007, 107(10), 4414–35.

[31] Ogata H, Kellers P, Lubitz W. The crystal structure of the [NiFe]-hydrogenase from the photosynthetic bacterium Allochromatium vinosum: Characterization of the oxidized enzyme (Ni-A state). J Mol Biol 2010, 402(2), 428–44.

[32] Abou Hamdan A, Burlat B, Gutierrez-Sanz O, Liebgott PP, Baffert C, De Lacey AL, et al. O2-independent formation of the inactive states of NiFe hydrogenase. Nat Chem Biol 2013, 9(1), 15–17.

[33] Fernandez VM, Rao KK, Fernandez MA, Cammack R. Activation and deactivation of the membrane-bound hydrogenase from Desulfovibrio desulfuricans, Norway strain. Biochimie 1986, 68(1), 43–8.

[34] Hexter SV, Grey F, Happe T, Climent V, Armstrong FA. Electrocatalytic mechanism of reversible hydrogen cycling by enzymes and distinctions between the major classes of hydrogenases. Proc Natl Acad Sci U S A 2012, 109(29), 11516–21.

[35] Ludwig M, Cracknell JA, Vincent KA, Armstrong FA, Lenz O. Oxygen-tolerant H2 oxidation by membrane-bound [NiFe]-hydrogenases of Ralstonia species. Coping with low level H2 in air. J Biol Chem 2009, 284(1), 465–77.

[36] Szczesny J, Birrell JA, Conzuelo F, Lubitz W, Ruff A, Schuhmann W. Redox-polymer-based high-current-density gas-diffusion H2-oxidation bioanode using [FeFe]-hydrogenase from Desulfovibrio desulfuricans in a membrane-free biofuel cell. Angewandte Chemie 2020, 59 (38), 16506–10.

[37] Hexter SV, Grey F, Happe T, Climent V, Armstrong FA. Electrocatalytic mechanism of reversible hydrogen cycling by enzymes and distinctions between the major classes of hydrogenases. Proc Natl Acad Sci U S A 2012, 109(29), 11516–21.

[38] Pandey K, Islam ST, Happe T, Armstrong FA. Frequency and potential dependence of reversible electrocatalytic hydrogen interconversion by [FeFe]-hydrogenases. Proc Natl Acad Sci U S A 2017, 114(15), 3843–8.

[39] Berggren G, Adamska A, Lambertz C, Simmons TR, Esselborn J, Atta M, et al. Biomimetic assembly and activation of [FeFe]-hydrogenases. Nature 2013, 499(7456), 66–9.

[40] Esselborn J, Lambertz C, Adamska-Venkates A, Simmons T, Berggren G, Noth J, et al. Spontaneous activation of [FeFe]-hydrogenases by an inorganic [2Fe] active site mimic. Nat Chem Biol 2013, 9(10), 607–9.

[41] Silakov A, Wenk B, Reijerse E, Lubitz W. (14)N HYSCORE investigation of the H-cluster of [FeFe]-hydrogenase: Evidence for a nitrogen in the dithiol bridge. Phys Chem Chem Phys 2009, 11(31), 6592–9.

[42] Lampret O, Adamska-Venkatesh A, Konegger H, Wittkamp F, Apfel UP, Reijerse EJ, et al. Interplay between CN(-) ligands and the secondary coordination sphere of the H-cluster in [FeFe]-hydrogenases. J Am Chem Soc 2017.

[43] Winkler M, Senger M, Duan JF, Esselborn J, Wittkamp F, Hofmann E, et al. Accumulating the hydride state in the catalytic cycle of [FeFe]-hydrogenases. Nat Commun 2017, 8.

[44] Duan J, Senger M, Esselborn J, Engelbrecht V, Wittkamp F, Apfel UP, et al. Crystallographic and spectroscopic assignment of the proton transfer pathway in [FeFe]-hydrogenases. Nat Commun 2018, 9(1), 4726.

[45] Morra S, Giraudo A, Di Nardo G, King PW, Gilardi G, Valetti F. Site saturation mutagenesis demonstrates a central role for cysteine 298 as proton donor to the catalytic site in CaHydA [FeFe]-hydrogenase. PloS One 2012, 7(10), e48400.

[46] Birrell JA, Pelmenschikov V, Mishra N, Wang H, Yoda Y, Tamasaku K, et al. Spectroscopic and computational evidence that [FeFe]-hydrogenases operate exclusively with CO-bridged intermediates. J Am Chem Soc 2020, 142(1), 222–32.

[47] Lorent C, Katz S, Duan J, Kulka CJ, Caserta G, Teutloff C, et al. Shedding light on proton and electron dynamics in [FeFe]-hydrogenases. J Am Chem Soc 2020, 142(12), 5493–7.

[48] Mebs S, Senger M, Duan J, Wittkamp F, Apfel UP, Happe T, et al. Bridging hydride at reduced H-cluster species in [FeFe]-hydrogenases revealed by infrared spectroscopy, isotope editing, and quantum chemistry. J Am Chem Soc 2017, 139(35), 12157–60.

[49] Wittkamp F, Senger M, Stripp ST, Apfel UP. [FeFe]-Hydrogenases: Recent developments and future perspectives. Chem Commun 2018, 54(47), 5934–42.

[50] Lampret O, Esselborn J, Haas R, Rutz A, Booth RL, Kertess L, et al. The final steps of [FeFe]-hydrogenase maturation. Proc Natl Acad Sci U S A 2019, 116(32), 15802–10.

[51] Silakov A, Kamp C, Reijerse E, Happe T, Lubitz W. Spectroelectrochemical characterization of the active site of the [FeFe]-hydrogenase HydA1 from Chlamydomonas reinhardtii. Biochemistry 2009, 48(33), 7780–6.

[52] Esselborn J, Muraki N, Klein K, Engelbrecht V, Metzler-Nolte N, Apfel UP, et al. A structural view of synthetic cofactor integration into [FeFe]-hydrogenases. Chem Sci 2016, 7(2), 959–68.

[53] Esselborn J, Kertess L, Apfel UP, Hofmann E, Happe T. Loss of specific active-site iron atoms in oxygen-exposed [FeFe]-hydrogenase determined by detailed X-ray structure analyses. J Am Chem Soc 2019, 141(44), 17721–8.

[54] Kubas A, Orain C, De Sancho D, Saujet L, Sensi M, Gauquelin C, et al. Mechanism of O2 diffusion and reduction in FeFe-hydrogenases. Nat Chem 2017, 9(1), 88–95.

[55] Morra S, Arizzi M, Valetti F, Gilardi G. Oxygen stability in the new [FeFe]-hydrogenase from Clostridium beijerinckii SM10 (CbA5H). Biochemistry 2016, 55(42), 5897–900.

[56] Winkler M, Duan J, Rutz A, Felbek C, Scholtysek L, Lampret O, et al. A safety cap protects hydrogenase from oxygen attack. Nat Commun 2021, 12(1), 756.

[57] Goldet G, Brandmayr C, Stripp ST, Happe T, Cavazza C, Fontecilla-Camps JC, et al. Electrochemical kinetic investigations of the reactions of [FeFe]-hydrogenases with carbon monoxide and oxygen: Comparing the importance of gas tunnels and active-site electronic/redox effects. J Am Chem Soc 2009, 131(41), 14979–89.

[58] Stripp ST, Goldet G, Brandmayr C, Sanganas O, Vincent KA, Haumann M, et al. How oxygen attacks [FeFe]-hydrogenases from photosynthetic organisms. Proc Natl Acad Sci U S A 2009, 106(41), 17331–6.

[59] Swanson KD, Ratzloff MW, Mulder DW, Artz JH, Ghose S, Hoffman A, et al. [FeFe]-hydrogenase oxygen inactivation is initiated at the H cluster 2Fe subcluster. J Am Chem Soc 2015, 137(5), 1809–16.

[60] Lautier T, Ezanno P, Baffert C, Fourmond V, Cournac L, Fontecilla-Camps JC, et al. The quest for a functional substrate access tunnel in FeFe hydrogenase. Faraday Discuss 2011, 148, 385–407. discussion 21-41.

[61] Baffert C, Demuez M, Cournac L, Burlat B, Guigliarelli B, Bertrand P, et al. Hydrogen-activating enzymes: Activity does not correlate with oxygen sensitivity. Angewandte Chemie 2008, 47(11), 2052–4.

[62] Lemon BJ, Peters JW. Binding of exogenously added carbon monoxide at the active site of the iron-only hydrogenase (CpI) from Clostridium pasteurianum. Biochemistry 1999, 38(40), 12969–73.

[63] Lambertz C, Leidel N, Havelius KG, Noth J, Chernev P, Winkler M, et al. O2 reactions at the six-iron active site (H-cluster) in [FeFe]-hydrogenase. J Biol Chem 2011, 286(47), 40614–23.

[64] Noth J, Kositzki R, Klein K, Winkler M, Haumann M, Happe T. Lyophilization protects [FeFe]-hydrogenases against O2-induced H-cluster degradation. Sci Rep 2015, 5, 13978.

[65] Adamska-Venkatesh A, Krawietz D, Siebel J, Weber K, Happe T, Reijerse E, et al. New redox states observed in [FeFe]-hydrogenases reveal redox coupling within the H-cluster. J Am Chem Soc 2014, 136(32), 11339–46.

[66] Foster CE, Kramer T, Wait AF, Parkin A, Jennings DP, Happe T, et al. Inhibition of [FeFe]-hydrogenases by formaldehyde and wider mechanistic implications for biohydrogen activation. J Am Chem Soc 134(17), 7553–7.

[67] Wait AF, Brandmayr C, Stripp ST, Cavazza C, Fontecilla-Camps JC, Happe T, et al. Formaldehyde--a rapid and reversible inhibitor of hydrogen production by [FeFe]-hydrogenases. J Am Chem Soc 133(5), 1282–5.

[68] Rodriguez-Macia P, Reijerse EJ, Van Gastel M, DeBeer S, Lubitz W, Rudiger O, et al. Sulfide protects [FeFe]-hydrogenases from O2. J Am Chem Soc 2018, 140(30), 9346–50.

[69] Rodriguez-Macia P, Galle LM, Bjornsson R, Lorent C, Zebger I, Yoda Y, et al. Caught in the Hinact: Crystal structure and spectroscopy reveal a sulfur bound to the active site of an O2-stable state of [FeFe]-hydrogenase. Angewandte Chemie 2020, 59(38), 16786–94.

[70] Orain C, Saujet L, Gauquelin C, Soucaille P, Meynial-Salles I, Baffert C, et al. Electrochemical measurements of the kinetics of inhibition of two FeFe-hydrogenases by O2 demonstrate that the reaction is partly reversible. J Am Chem Soc 2015, 137(39), 12580–7.

[71] Stiebritz MT, Reiher M. Theoretical study of dioxygen induced inhibition of [FeFe]-hydrogenase. Inorg Chem 2009, 48(15), 7127–40.

[72] Kubas A, De Sancho D, Best RB, Blumberger J. Aerobic damage to [FeFe]-hydrogenases: Activation barriers for the chemical attachment of O2. Angewandte Chemie 2014, 53(16), 4081–4.

[73] Bruska MK, Stiebritz MT, Reiher M. Binding of Reactive Oxygen Species at Fe-S Cubane Clusters. Chemistry 2015, 21(52), 19081–9.

[74] Mebs S, Kositzki R, Duan J, Senger M, Wittkamp F, Apfel UP, et al. Hydrogen and oxygen trapping at the H-cluster of [FeFe]hydrogenase revealed by site-selective spectroscopy and QM/MM calculations. Biochim Biophys Acta. 2017.

[75] Jang S, Imlay JA. Micromolar intracellular hydrogen peroxide disrupts metabolism by damaging iron-sulfur enzymes. J Biol Chem 2007, 282(2), 929–37.

[76] Pershad HR, Duff JLC, Heering HA, Duin EC, Albracht SPJ, Armstrong FA. Catalytic electron transport in Chromatium vinosum [NiFe]-hydrogenase: Application of voltammetry in detecting redox-active centers and establishing that hydrogen oxidation is very fast even at potentials close to the reversible H+/H2 Value. Biochemistry 1999, 38(28), 8992–9.

[77] Armstrong FA, Evans RM, Hexter SV, Murphy BJ, Roessler MM, Wulff P. Guiding principles of hydrogenase catalysis instigated and clarified by protein film electrochemistry. Acc Chem Res 2016, 49(5), 884–92.

[78] Vincent KA, Parkin A, Armstrong FA. Investigating and exploiting the electrocatalytic properties of hydrogenases. Chem Rev 2007, 107(10), 4366–413.

[79] Banks CE, Compton RG. New electrodes for old: From carbon nanotubes to edge plane pyrolytic graphite. Analyst 2006, 131(1), 15–21.

[80] Del Barrio M, Sensi M, Orain C, Baffert C, Dementin S, Fourmond V, et al. Electrochemical investigations of hydrogenases and other enzymes that produce and use solar fuels. Acc Chem Res 2018, 51(3), 769–77.

[81] Lampret O, Duan J, Hofmann E, Winkler M, Armstrong FA, Happe T. The roles of long-range proton-coupled electron transfer in the directionality and efficiency of [FeFe]-hydrogenases. Proc Natl Acad Sci U S A 2020, 117(34), 20520–9.

[82] Del Barrio M, Fourmond V. Redox (In)activations of Metalloenzymes: A Protein Film Voltammetry Approach. ChemElectroChem 2019, 6(19), 4949–62.

[83] Fourmond V, Baffert C, Sybirna K, Dementin S, Abou-Hamdan A, Meynial-Salles I, et al. The mechanism of inhibition by H2 of H2-evolution by hydrogenases. Chem Commun 2013, 49(61), 6840–2.

[84] Senger M, Mebs S, Duan J, Wittkamp F, Apfel UP, Heberle J, et al. Stepwise isotope editing of [FeFe]-hydrogenases exposes cofactor dynamics. Proc Natl Acad Sci U S A 2016, 113(30), 8454–9.

[85] Senger M, Mebs S, Duan J, Shulenina O, Laun K, Kertess L, et al. Protonation/reduction dynamics at the [4Fe-4S] cluster of the hydrogen-forming cofactor in [FeFe]-hydrogenases. Phys Chem Chem Phys 2018, 20(5), 3128–40.

[86] Roseboom W, De Lacey AL, Fernandez VM, Hatchikian EC, Albracht SP. The active site of the [FeFe]-hydrogenase from Desulfovibrio desulfuricans. II. Redox properties, light sensitivity and CO-ligand exchange as observed by infrared spectroscopy. J Biol Inorg Chem 2006, 11(1), 102–18.

[87] Albracht SP, Roseboom W, Hatchikian EC. The active site of the [FeFe]-hydrogenase from Desulfovibrio desulfuricans. I. Light sensitivity and magnetic hyperfine interactions as observed by electron paramagnetic resonance. J Biol Inorg Chem 2006, 11(1), 88–101.

[88] Akhtar MK, Jones PR. Deletion of iscR stimulates recombinant clostridial Fe-Fe hydrogenase activity and H2-accumulation in Escherichia coli BL21(DE3). Appl Microbiol Biotechnol 2008, 78(5), 853–62.

[89] Kuchenreuther JM, Grady-Smith CS, Bingham AS, George SJ, Cramer SP, Swartz JR. High-yield expression of heterologous [FeFe]-hydrogenases in Escherichia coli. PloS ne 2010, 5(11), e15491.

[90] Stapleton JA, Swartz JR. A cell-free microtiter plate screen for improved [FeFe]-hydrogenases. PloS One 2010, 5(5), e10554.

[91] Bingham AS, Smith PR, Swartz JR. Evolution of an [FeFe]-hydrogenase with decreased oxygen sensitivity. Int J Hydrogen Energy 2012, 37(3), 2965–76.

[92] Smith PR, Bingham AS, Swartz JR. Generation of hydrogen from NADPH using an [FeFe]-hydrogenase. Int J Hydrogen Energy 2012, 37(3), 2977–83.

[93] Vignais PM, Billoud B, Meyer J. Classification and phylogeny of hydrogenases. FEMS Microbiol Rev 2001, 25(4), 455–501.

[94] Vignais PM, Billoud B. Occurrence, classification, and biological function of hydrogenases: An overview. Chem Rev 2007, 107(10), 4206–72.

[95] Goldet G, Wait AF, Cracknell JA, Vincent KA, Ludwig M, Lenz O, et al. Hydrogen production under aerobic conditions by membrane-bound hydrogenases from Ralstonia species. J Am Chem Soc 2008, 130(33), 11106–13.

[96] Mohammadi M, Vashisth H. Pathways and thermodynamics of oxygen diffusion in [FeFe]-hydrogenase. J Phys Chem B 2017, 121(43), 10007–17.

[97] Koo J, Swartz JR. System analysis and improved [FeFe]-hydrogenase O2 tolerance suggest feasibility for photosynthetic H2 production. Metab Eng 2018, 49, 21–7.

[98] Koo J. Enhanced aerobic H2 production by engineering an [FeFe]-hydrogenase from Clostridium pasteurianum. Int J Hydrogen Energy 2020, 45(18), 10673–9.

[99] Corrigan PS, Tirsch JL, Silakov A. Investigation of the unusual ability of the [FeFe]-hydrogenase from Clostridium beijerinckii to access an O2-protected state. J Am Chem Soc 2020, 142(28), 12409–19.

[100] Kalms J, Schmidt A, Frielingsdorf S, Utesch T, Gotthard G, Von Stetten D, et al. Tracking the route of molecular oxygen in O2-tolerant membrane-bound [NiFe]-hydrogenase. Proc Natl Acad Sci U S A 2018, 115(10), E2229–E37.

[101] Fontecilla-Camps JC, Volbeda A, Cavazza C, Nicolet Y. Structure/function relationships of [NiFe]- and [FeFe]-hydrogenases. Chem Rev 2007, 107(10), 4273–303.

[102] Cohen J, Kim K, Posewitz M, Ghirardi ML, Schulten K, Seibert M, et al. Molecular dynamics and experimental investigation of H(2) and O(2) diffusion in [Fe]-hydrogenase. Biochem Soc Trans 2005, 33(Pt 1), 80–2.

[103] Cohen J, Kim K, King P, Seibert M, Schulten K. Finding gas diffusion pathways in proteins: Application to O2 and H2 transport in CpI [FeFe]-hydrogenase and the role of packing defects. Structure 2005, 13(9), 1321–9.

[104] Kalms J, Schmidt A, Frielingsdorf S, Van Der Linden P, Von Stetten D, Lenz O, et al. Krypton derivatization of an O2 -tolerant membrane-bound [NiFe]-hydrogenase reveals a hydrophobic tunnel network for gas transport. Angewandte Chemie 2016, 55(18), 5586–90.

[105] Liu Y, Mohammadi M, Vashisth H. Diffusion network of CO in FeFe-Hydrogenase. J Chem Phys 2018, 149(20), 204108.

[106] Liebgott PP, Leroux F, Burlat B, Dementin S, Baffert C, Lautier T, et al. Relating diffusion along the substrate tunnel and oxygen sensitivity in hydrogenase. Nat Chem Biol 2010, 6(1), 63–70.

[107] Leroux F, Dementin S, Burlat B, Cournac L, Volbeda A, Champ S, et al. Experimental approaches to kinetics of gas diffusion in hydrogenase. Proc Natl Acad Sci U S A 2008, 105 (32), 11188–93.

[108] Nadler W, Stein DL. Reaction–diffusion description of biological transport processes in general dimension. J Chem Phys 1996, 104(5), 1918–36.

[109] Sode O, Voth GA. Electron transfer activation of a second water channel for proton transport in [FeFe]-hydrogenase. J Chem Phys 2014, 141(22), 22D527.

[110] Fritsch J, Scheerer P, Frielingsdorf S, Kroschinsky S, Friedrich B, Lenz O, et al. The crystal structure of an oxygen-tolerant hydrogenase uncovers a novel iron-sulphur centre. Nature 2011, 479(7372), 249–52.

[111] Wulff P, Day C, Sargent F, Sargent F, Armstrong FA, Armstrong FA. How oxygen reacts with oxygen-tolerant respiratory [NiFe]-hydrogenases. (1091-6490 (Electronic)).

[112] Goris T, Wait AF, Saggu M, Fritsch J, Heidary N, Stein M, et al. A unique iron-sulfur cluster is crucial for oxygen tolerance of a [NiFe]-hydrogenase. Nat Chem Biol 2011, 7(5), 310–8.

[113] Pandelia ME, Fourmond V, Tron-Infossi P, Lojou E, Bertrand P, Leger C, et al. Membrane-bound hydrogenase I from the hyperthermophilic bacterium Aquifex aeolicus: Enzyme activation, redox intermediates and oxygen tolerance. J Am Chem Soc 2010, 132(20), 6991–7004.

[114] Saggu M, Teutloff C, Ludwig M, Brecht M, Pandelia ME, Lenz O, et al. Comparison of the membrane-bound [NiFe]-hydrogenases from R. eutropha H16 and D. vulgaris Miyazaki F in the oxidized ready state by pulsed EPR. Phys Chem Chem Phys 2010, 12(9), 2139–48.

[115] Volbeda A, Amara P, Darnault C, Mouesca JM, Parkin A, Roessler MM, et al. X-ray crystallographic and computational studies of the O2-tolerant [NiFe]-hydrogenase 1 from Escherichia coli. Proc Natl Acad Sci U S A 2012, 109(14), 5305–10.

[116] Mouesca JM, Amara P, Fontecilla-Camps JC. Electronic states of the O2-tolerant [NiFe]-hydrogenase proximal cluster. Proc Natl Acad Sci U S A 2013, 110(28), E2538.

[117] Fritsch J, Loscher S, Sanganas O, Siebert E, Zebger I, Stein M, et al. [NiFe] and [FeS] cofactors in the membrane-bound hydrogenase of Ralstonia eutropha investigated by X-ray absorption spectroscopy: Insights into O(2)-tolerant H(2) cleavage. Biochemistry 2011, 50(26), 5858–69.

[118] Lenz O, Ludwig M, Schubert T, Burstel I, Ganskow S, Goris T, et al. H2 conversion in the presence of O2 as performed by the membrane-bound [NiFe]-hydrogenase of Ralstonia eutropha. ChemPhysChem 2010, 11(6), 1107–19.

[119] Dementin S, Leroux F, Cournac L, De Lacey AL, Volbeda A, Leger C, et al. Introduction of methionines in the gas channel makes [NiFe]-hydrogenase aero-tolerant. J Am Chem Soc 2009, 131(29), 10156–64.

[120] Liebgott PP, De Lacey AL, Burlat B, Cournac L, Richaud P, Brugna M, et al. Original design of an oxygen-tolerant [NiFe]-hydrogenase: Major effect of a valine-to-cysteine mutation near the active site. J Am Chem Soc 2011, 133(4), 986–97.

[121] Del Barrio M, Guendon C, Kpebe A, Baffert C, Fourmond V, Brugna M, et al. Valine-to-cysteine mutation further increases the oxygen tolerance of Escherichia coli NiFe hydrogenase Hyd-1. ACS Catal 2019, 9(5), 4084–8.

[122] Zacarias S, Temporão A, Barrio M, Fourmond V, Léger C, Matias PM, et al. A hydrophilic channel is involved in oxidative inactivation of a [NiFeSe]-hydrogenase. ACS Catal 2019, 9(9), 8509–19.

[123] Hondal RJ, Ruggles EL. Differing views of the role of selenium in thioredoxin reductase. Amino Acids 2011, 41(1), 73–89.

[124] Evans RM, Brooke EJ, Wehlin SA, Nomerotskaia E, Sargent F, Carr SB, et al. Mechanism of hydrogen activation by [NiFe]-hydrogenases. Nat Chem Biol 2016, 12(1), 46–50.

[125] Rodríguez-Maciá P, Reijerse EJ, Van Gastel M, DeBeer S, Lubitz W, Rüdiger O, et al. Sulfide Protects [FeFe]-Hydrogenases From O2. J Am Chem Soc 2018, 140(30), 9346–50.

[126] Rodríguez-Maciá P, Galle LM, Bjornsson R, Lorent C, Zebger I, Yoda Y, et al. Caught in the hinact: Crystal structure and spectroscopy reveal a sulfur bound to the active site of an O2-stable State of [FeFe]-hydrogenase. Angew Chem Int Ed 2020, 59(38), 16786–94.

[127] Jones AK, Sillery E, Albracht SP, Armstrong FA. Direct comparison of the electrocatalytic oxidation of hydrogen by an enzyme and a platinum catalyst. Chem Commun 2002(8), 866–7.

[128] Goldet G, Brandmayr C, Stripp ST, Happe T, Cavazza C, Fontecilla-Camps JC, et al. Electrochemical kinetic investigations of the reactions of [FeFe]-hydrogenases with carbon monoxide and oxygen: Comparing the importance of gas tunnels and active-site electronic/redox effects. J Am Chem Soc 2009, 131(41), 14979–89.

[129] Krishnan S, Armstrong FA. Order-of-magnitude enhancement of an enzymatic hydrogen-air fuel cell based on pyrenyl carbon nanostructures. Chem Sci 2012, 3(4), 1015–23.

[130] Plumere N, Rudiger O, Oughli AA, Williams R, Vivekananthan J, Poller S, et al. A redox hydrogel protects hydrogenase from high-potential deactivation and oxygen damage. Nat Chem 2014, 6(9), 822–7.

[131] Oughli AA, Conzuelo F, Winkler M, Happe T, Lubitz W, Schuhmann W, et al. A redox hydrogel protects the O2 -sensitive [FeFe]-hydrogenase from Chlamydomonas reinhardtii from oxidative damage. Angewandte Chemie 2015, 54(42), 12329–33.

[132] Ruff A, Szczesny J, Zacarias S, Pereira IAC, Plumere N, Schuhmann W. Protection and Reactivation of the [NiFeSe]-hydrogenase from Desulfovibrio vulgaris Hildenborough under oxidative conditions. ACS Energy Lett 2017, 2(5), 964–8.

[133] Oughli AA, Velez M, Birrell JA, Schuhmann W, Lubitz W, Plumere N, et al. Viologen-modified electrodes for protection of hydrogenases from high potential inactivation while performing H2 oxidation at low overpotential. Dalton Trans 2018, 47(31), 10685–91.

[134] Ruff A, Szczesny J, Vega M, Zacarias S, Matias PM, Gounel S, et al. Redox-polymer-wired [NiFeSe]-hydrogenase variants with enhanced O2 stability for triple-protected high-current-density H2-oxidation bioanodes. ChemSusChem 2020, 13(14), 3627–35.

[135] Heller A. Electron-conducting redox hydrogels: Design, characteristics and synthesis. Curr Opin Chem Biol 2006, 10(6), 664–72.

[136] Zafar MN, Shao M, Ludwig R, Leech D, Schuhmann W, Gorton L. Improving the current density and the coulombic efficiency by a cascade reaction of glucose oxidizing enzymes. ECS Trans 2013, 53(2), 131–43.

[137] Shao M, Zafar MN, Sygmund C, Guschin DA, Ludwig R, Peterbauer CK, et al. Mutual enhancement of the current density and the coulombic efficiency for a bioanode by entrapping bi-enzymes with Os-complex modified electrodeposition paints. Biosens Bioelectron 2013, 40(1), 308–14.

[138] Hickey DP, Giroud F, Schmidtke DW, Glatzhofer DT, Minteer SD. Enzyme cascade for catalyzing sucrose oxidation in a biofuel cell. ACS Catal 2013, 3(12), 2729–37.

[139] Ruff A, Szczesny J, Markovic N, Conzuelo F, Zacarias S, Pereira IAC, et al. A fully protected hydrogenase/polymer-based bioanode for high-performance hydrogen/glucose biofuel cells. Nat Commun 2018, 9(1), 3675.

[140] Morra S, Valetti F, Sarasso V, Castrignano S, Sadeghi SJ, Gilardi G. Hydrogen production at high Faradaic efficiency by a bio-electrode based on TiO2 adsorption of a new [FeFe]-hydrogenase from Clostridium perfringens. Bioelectrochemistry 2015, 106((Pt B)), 258–62.

[141] Wan L, Heath RS, Siritanaratkul B, Megarity CF, Sills AJ, Thompson MP, et al. Enzyme-catalysed enantioselective oxidation of alcohols by air exploiting fast electrochemical nicotinamide cycling in electrode nanopores. Green Chem 2019, 21(18), 4958–63.

[142] Poznansky B, Thompson LA, Warren SA, Reeve HA, Vincent KA. Carbon as a simple support for redox biocatalysis in continuous flow. Org Process Res Dev 2020, 24(10), 2281–7.

[143] Thompson LA, Rowbotham JS, Reeve HA, Zor C, Grobert N, Vincent KA. Biocatalytic hydrogenations on carbon supports. Methods Enzymol 2020, 630, 303–25.

[144] Petroll K, Kopp D, Care A, Bergquist PL, Sunna A. Tools and strategies for constructing cell-free enzyme pathways. Biotechnol Adv 2019, 37(1), 91–108.

[145] Giannakopoulou A, Gkantzou E, Polydera A, Stamatis H. Multienzymatic nanoassemblies: Recent progress and applications. Trends Biotechnol 2020, 38(2), 202–16.

[146] Schoffelen S, Van Hest JC. Chemical approaches for the construction of multi-enzyme reaction systems. Curr Opin Struct Biol 2013, 23(4), 613–21.

[147] Kuchler A, Yoshimoto M, Luginbuhl S, Mavelli F, Walde P. Enzymatic reactions in confined environments. Nat Nanotechnol 2016, 11(5), 409–20.

[148] Huang G, Li F, Zhao X, Ma Y, Li Y, Lin M, et al. Functional and biomimetic materials for engineering of the three-dimensional cell microenvironment. Chem Rev 2017, 117(20), 12764–850.

[149] Helm ML, Stewart MP, Bullock RM, DuBois MR, DuBois DL. A synthetic nickel electrocatalyst with a turnover frequency above 100,000 s(-)(1) for H(2) production. Science 2011, 333(6044), 863–6.

[150] Dutta A, Ginovska B, Raugei S, Roberts JA, Shaw WJ. Optimizing conditions for utilization of an H2 oxidation catalyst with outer coordination sphere functionalities. Dalton Trans 2016, 45(24), 9786–93.

[151] Priyadarshani N, Dutta A, Ginovska B, Buchko GW, O'Hagan M, Raugei S, et al. Achieving reversible H2/H+ interconversion at room temperature with enzyme-inspired molecular complexes: A mechanistic study. ACS Catal 2016, 6(9), 6037–49.

[152] Boralugodage NP, Arachchige RJ, Dutta A, Buchko GW, Shaw WJ. Evaluating the role of acidic, basic, and polar amino acids and dipeptides on a molecular electrocatalyst for H2 oxidation. Catal Sci Technol 2017, 7(5), 1108–21.

[153] Oughli AA, Ruff A, Boralugodage NP, Rodríguez-Maciá P, Plumeré N, Lubitz W, et al. Dual properties of a hydrogen oxidation Ni-catalyst entrapped within a polymer promote self-defense against oxygen. Nat Commun 2018, 9(1), 864.

Tobias J. Erb, Jan Zarzycki, Marieke Scheffen

5 Synthetic enzymes and pathways for improved carbon capture and conversion

Abstract: Photosynthesis allows the sustainable capture and conversion of atmospheric CO_2 into multicarbon compounds, opening the way for a carbon-neutral, green economy. However, photosynthetic efficiency is currently limited by the inefficiencies of the CO_2-fixing enzymes and pathways that emerged during evolution, namely RubisCO and the Calvin–Benson–Bassham cycle. This chapter presents novel approaches in synthetic biology that aim at the design and realization of new-to-nature enzymes and synthetic pathways for the improved capture and conversion of CO_2. Future challenges and opportunities of these approaches will be critically discussed.

5.1 Goal

The capture and conversion of carbon dioxide (CO_2) is key to a sustainable, circular economy. Photosynthetic organisms allow the light-driven conversion of atmospheric CO_2 into multicarbon compounds. Being able to harness the synthetic capabilities of photosynthetic microorganisms would allow the direct production of value-added compounds from inorganic carbon, in contrast to current biotechnological processes that are mainly based on feeding photosynthetically fixed carbon (e.g., glucose) to heterotrophic microorganisms. However, the average light-to-carbon efficiency (i.e., the photosynthetic yield) in photosynthesis is low. For microalgae, annual yields of 3% have been reported in bioreactors, with short-term efficiencies of up to 5–7% [1, 2]. Albeit still substantially higher than the values reported for plants (typically 1%), the low light-to-carbon conversion ultimately limits volumetric productivity of microalgae. Important factors are the high energetic requirements and inefficiencies of the CO_2-fixing enzymes and pathways involved in photosynthesis.

The goal of synthetic biology is to create biological systems with improved or new-to-nature functions from defined parts, similar to the concept of engineering in physics. This synthetic constructive approach aims at expanding the biological solution space beyond what nature has invented during evolution, thus providing new approaches to overcome the natural limitations of living systems. In recent years, several approaches have focused on radically redesigning the metabolic enzymes and networks involved in carbon capture and conversion to increase photosynthetic yield. This chapter familiarizes the reader with the concept of synthetic metabolism, reports on the current state of the art and provides an outlook on future challenges and opportunities in creating designer enzymes and pathways for improved carbon capture and conversion.

https://doi.org/10.1515/9783110716979-005

5.2 Basic background

Photosynthesis can be divided into a light and a dark reaction. While the light reactions convert solar energy into chemical energy and redox equivalents (i.e., ATP and NADPH), the dark reactions use ATP and NADPH to capture and convert inorganic carbon (CO_2) into organic compounds, in particular sugar(phosphates). Naturally, cyanobacteria, algae and plants use the Calvin–Benson–Bassham cycle (CBB cycle) for CO_2 fixation for the dark reactions. This process relies on ribulose-1,5-bisphosphate carboxylase/oxygenase (RubisCO), the key enzyme in the CBB cycle, and the global carbon cycle in general, as the vast majority of carbon on the Earth is fixed through this carboxylase. Overall, it is estimated that about 0.7 Gt RubisCO are present in the biosphere, which makes RubisCO one of the most abundant enzymes on the Earth [3–5].

Photosynthetic efficiency directly depends on the capture and conversion of CO_2 and is dictated by two parameters: the energy required per CO_2 molecule captured and the maximum rate of conversion. In the CBB cycle, excluding photorespiration and carbon concentration mechanisms, nine ATP and six NADPH are required to convert three molecules of CO_2 into one molecule of glyceraldehyde-3-phosphate, which can be used in biosynthesis.

The rate of CO_2 fixation, which determines volumetric productivity, is set by the turnover frequency of the CBB cycle, and in particular RubisCO, which is one of the rate-limiting enzymes in the overall process. RubisCO is a relatively slow catalyst ($5–10$ s^{-1}) [6–8], which requires the allocation of a significant amount of protein resources to this enzyme to achieve sufficient CO_2 fixation rates in vivo [9]. In addition, RubisCO also shows an inherent side reaction with oxygen instead of CO_2, causing the phenomenon of photorespiration [10], which lowers carbon and energetic efficiency of carbon capture and conversion during photosynthesis [6, 11, 12] by more than 30%.

Based on the above considerations, the CBB cycle and RubisCO have been identified as prime targets for efforts of improving photosynthetic carbon capture and conversion. In the past, much effort has been focused on increasing photosynthetic productivity through overexpression of RubisCO or introduction of RubisCOs with improved kinetics [13, 14]. However, although these approaches have their own merits, they do not provide alternative solutions to address the actual problem and are thus restricted to improving an existing process. In other words, while classical metabolic engineering is able to optimize a given solution (i.e., finding a local maximum: improving flux through the CBB cycle), it is not necessarily able to identify the best solution (i.e., finding a global optimum: developing improved CO_2 fixation pathways).

With the advent of synthetic biology, it has become possible to design and realize new-to-nature enzymes and metabolic networks that nature has not explored during evolution [15, 16]. This has allowed the development of alternative, synthetic pathways for carbon fixation and photorespiration, which are predicted to increase energy and carbon efficiency of photosynthesis of up to 150% [6, 11, 12, 17], dramatically expanding the metabolic solution space of photosynthetic CO_2 fixation [15]. Although these pathways have been mainly demonstrated in vitro and only partially realized in vivo so far, the next decade will see the full implementation of the most promising designs in photosynthetic model organisms, to explore their actual potential for improving photosynthetic yield.

5.3 Methods involved

The design and realization of new-to-nature enzymes and metabolic pathways requires first a detailed understanding of the native metabolic and enzymatic repertoire of a given cell. Modern omics methods allow to assess the metabolome and proteome of cells under different conditions to provide a systems view on the endogenous genetic and metabolic network of cells. This forms the basis to understand, manipulate and extend the metabolism of cells by synthetic biological and metabolic engineering approaches.

Based on the ATP and redox cofactor (e.g., NAD(P)H, flavins, ferredoxins) requirements, the thermodynamic profile of metabolic pathways, such as the CBB cycle or photorespiration, are calculated, providing the basis for considerations to replace or introduce novel reactions or reaction sequences that are more efficient (i.e., consume less ATP or redox cofactors). Another important consideration is the question whether in these reaction segments carbon atoms are released or retained, which determines the carbon efficiency of these metabolic sequences (see photorespiration, below).

In a next step, alternative reaction sequences to naturally existing pathways (e.g., photorespiration) are developed using a process called metabolic retrosynthesis (Fig. 1) [6]. Provided with a metabolic starting point (e.g., a metabolite such as the photorespiratory metabolite 2-phosphoglycolate) and an endpoint (e.g., the CBB cycle intermediate 3-phosphoglycerate), the metabolic space is analyzed for different reactions. Like for organic synthesis, these retrosynthetic analysis can be performed manually [6] or with the support of computational algorithms [17–19]. Notably at that stage, these considerations are purely based on potential biochemical transformations, not actually known enzymes, to explore the full solution space from first principles and without any limitations a priori. Once different alternative reaction sequences are identified, their thermodynamic feasibility and carbon efficiencies are analyzed and compared to the natural sequence. Pathways with increased energetic and carbon efficiencies are further selected for realization with a special focus on the realization of short pathways with a strong overall thermodynamic profile.

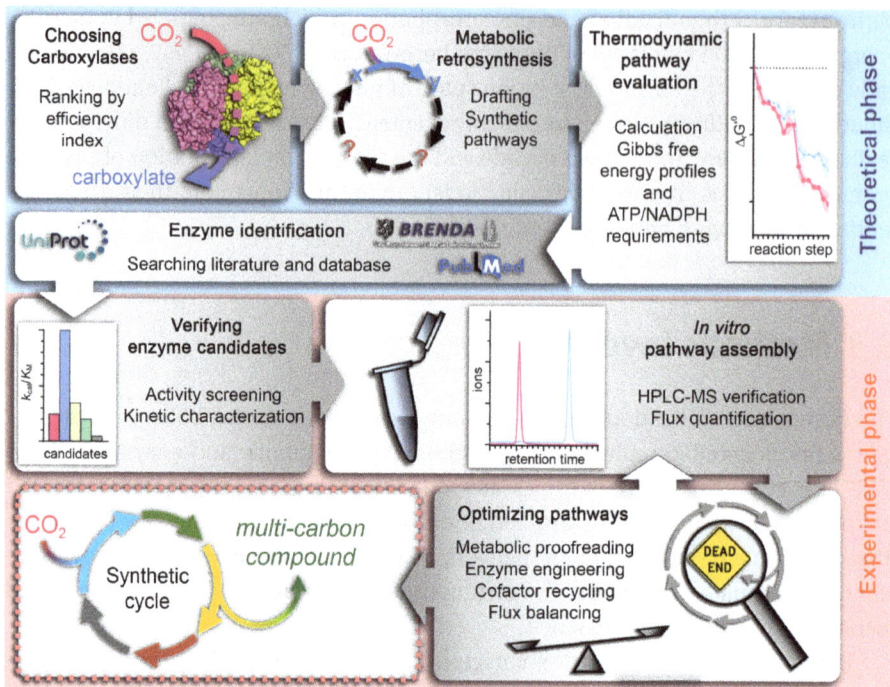

Fig. 1: Workflow scheme for the design and realization of new-to-nature pathways.

For the actual realization of a given pathway, enzymes that are able to catalyze each reaction in the designed network need to be identified. This is typically achieved through the screening of biological databases for suitable enzyme candidates and subsequent experimental validation. In case that a required reaction has not been described (yet), an enzyme catalyzing a similar reaction can be re-engineered for the new function. Based on a crystal structure or homology model of the enzyme, active site residues can be targeted to adopt the active site for new substrates [20] or reactions [21]. These rational efforts can be complemented by semirational or untargeted mutagenesis and high-throughput screens, both in vitro [12] and in vivo [22].

After all enzymes of a designed pathway have been identified, the reaction sequence is characterized, potential bottlenecks, side-reactions as well as dead-end products are identified and alleviated. Further pathway optimization may entail flux balancing between inputs and outputs including ATP and reducing equivalents to explore and improve the boundary conditions and robustness of the metabolic pathway, finally enabling its integration in vivo. This last part requires genetic tools for the expression of the corresponding genes in the photosynthetic target organism, which are reported in chapter 2 and 3 in more detail. Very often, complex pathways are transplanted stepwise into the target organism and native genes (e.g., genes involved in photorespiration) need to be knocked out to allow the implementation of the new reactions into the native metabolic network. Establishing selection schemes for the operation of the desired reaction sequence (e.g., replacing the native photorespiration by a synthetic photorespiration bypass) and applying selective pressure (e.g., lowering the CO_2 concentration or removing native carbon concentrating mechanisms (CCMs)) allows experimental evolution of the new-to-nature pathways to optimize their functionality in vivo.

5.4 State of the art

5.4.1 Natural and synthetic photorespiration pathways

Being central to the CBB cycle, the activity of RubisCO directly determines how much carbon can be captured and converted at a given time during photosynthesis. Average RubisCOs have turnover frequencies of $5-10$ s^{-1} in photosynthetic microorganisms and even less in plants. These catalytic rates are relatively low compared to those of enzymes in central carbon metabolism and are one to two orders of magnitude slower compared to other carboxylases. At the same time, RubisCO shows a side reaction with oxygen, which results in the production of 2-phosphoglycolate, a toxic metabolite that needs to be recycled in a process called photorespiration (Fig. 2) [10]. Overall, these unfavorable kinetic parameters and their direct consequences for photosynthetic yield have been a prime target for synthetic biological and protein engineering efforts.

Quite some effort has been put into discovering novel RubisCOs with improved kinetic parameters through (meta)genomic mining and/or system biochemical approaches [8, 23–25]. Indeed, several RubisCOs with increased activity were reported from these screens. Also, it has been shown that overexpressing RubisCO in cyanobacteria is able to improve photosynthetic yield and rate of growth [13, 14], which suggests that some of these newly isolated, faster RubisCOs could be used to achieve the same effects without allocating more protein resources to the carboxylation step. However, so far no RubisCO variants could be isolated that show both increased catalytic activity and decreased oxygenation side reaction. Similarly, efforts to engineer and/or experimentally evolve RubisCO [26–29] have not yielded a more active and – at the same time – more specific enzyme, supporting the notion that these two parameters are inversely coupled with each other and that photorespiration is an inevitable result of the enzyme's reaction mechanism [30]. Overall, this has shifted attention toward mitigating the effects of photorespiration to further improve photosynthetic yield.

In plants, photorespiration involves more than 12 steps, distributed over 3 organelles, chloroplast, peroxisome and mitochondria [Fig. 2]. In eukaryotic microalgae, the oxidation of glyoxylate, carried out by peroxisomal glycolate oxidase in

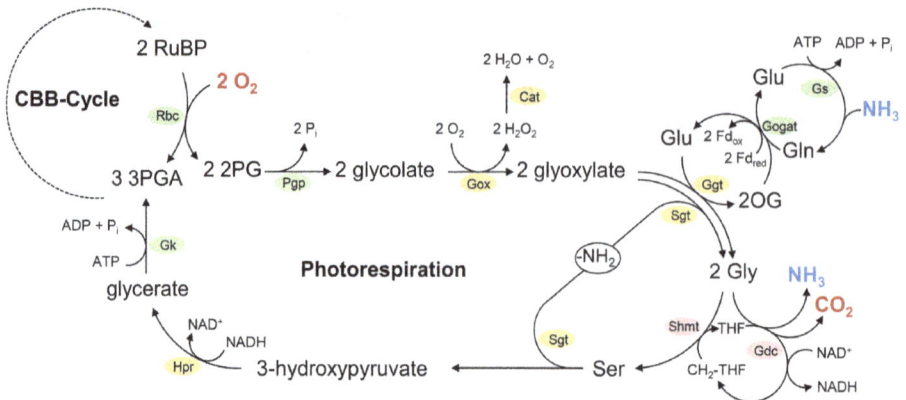

Fig. 2: Canonical photorespiration pathway as found in plants. Enzymes highlighted according to compartments in which they are located. Green, chloroplast; yellow, peroxisome; pink, mitochondria. Abbreviations: 2OG, 2-oxoglutarate; 2PG, 2-phosphoglycolate; 2PGA, 2-phosphoglycerate; 3PGA, 3-phosphoglycerate; Cat, catalase; CBB cycle, Calvin–Benson–Bassham cycle; Fd$_{ox}$, oxidized ferredoxin; Fd$_{red}$, reduced ferredoxin; Gdc, glycine decarboxylase; Ggt, glutamate/glyoxylate aminotransferase; Gk, glycerate kinase; Gln, glutamine; Gly, glycine; Gogat, glutamine/oxoglutarate aminotransferase; Gox, glycolate oxidase; Gs, glutamine synthetase; Hpr, hydroxypyruvate reductase; Pgp, phosphoglycolate phosphatase; Rbc, RubisCO; RuBP, ribulose-1,5-bisphosphate; Ser, serine; Sgt, serine/glyoxylate aminotransferase; Shmt, serine hydroxymethyl transferase; THF, tetrahydrofolate.

plants, is redirected to the mitochondria, where glycolate dehydrogenase is catalyzing the reaction [31]. In addition, there is evidence for extensive metabolic feedback between the CBB cycle and photorespiration, within and among the different involved organelles, controlling the flow of carbon [32]. In cyanobacteria, photorespiration is not subdivided into different compartments and is achieved by three different pathways that operate in parallel, as shown in *Synechocystis* [33, 34]. However, in all cases, previously fixed CO_2 is re-released and additional energy is consumed during photorespiration, which both lowers photosynthetic yield by up to 30% in different photosynthetic organisms.

To improve the carboxylation efficiency of RubisCO, photosynthetic microorganisms have developed CCMs, which increase the intracellular concentration of inorganic carbon, thus repressing the oxygenation reaction of RubisCO [35, 36]. In cyanobacteria, this includes the active transport of inorganic carbon to the cytoplasm through a sodium gradient, ATP- and redox-driven processes, resulting in up to 20–40 mM intracellular inorganic carbon. Cyanobacteria also produce carboxysomes, which are bacterial microcompartments that are composed of a polyhedral shell encapsulating and concentrating RubisCO and carbonic anhydrase inside. Notably, the shell is selectively permeable. Specific pores in the shell have been shown to transport bicarbonate but exclude oxygen [37], thus creating optimal working conditions for RubisCO (i.e., high CO_2 and low O_2 concentrations) inside of the microcompartment. In green algae, such as *Chlamydomonas*, RubisCO also becomes spatially concentrated in the so-called pyrenoid at low CO_2 concentrations. However, in this case, RubisCO is not packed into a protein shell but rather forms liquid-like condensates that undergo liquid–liquid phase separation [38]. This process is driven by complex coacervation, likely through electrostatic interactions of the negatively charged *Chlamydomonas* RubisCO with a positively charged linker protein called EPCY1 (essential pyrenoid component 1) [39]. Notably, the cyanobacterial enzyme shows also a large negative surface area, suggesting that in the future functional pyrenoids could be realized in cyanobacteria replacing the need for carboxysomes.

5.4.1.1 Synthetic photorespiration bypasses based on natural pathway modules

Although CCMs are quite efficient in mitigating photorespiration, they use a lot of cellular resources and still do not solve the actual problem of the wasteful photorespiration process. This has inspired the development of photorespiration pathway bypasses of increased carbon and energy efficiency. The first generation of photorespiration bypasses was based on naturally existing glycolate or glyoxylate converting pathways, respectively (Tab. 1). One prominent example is the transplantation of the (cyano)bacterial glycerate pathway that converts two molecules of glyoxylate into the C_3 compound glycerate. Compared to canonical photorespiration, the glycerate pathway requires only four enzymes and less energy (Fig. 3A). In tobacco, the glycerate pathway showed up to 13% increase in biomass, demonstrating that providing alternative photorespiration pathways is indeed able to improve photosynthetic yield [40]. Very recently, the beta-hydroxyaspartate cycle for the conversion of two molecules of glycolate into oxaloacetate was described in proteobacteria [41]. This pathway also requires only four enzymes [Fig. 3B] and has been transplanted into the peroxisomes of a photorespiratory mutant *Arabidopsis*, where it was able to restore wild-type-like growth [42]. Having demonstrated the function of these photorespiratory bypasses in plants opens up the use of these pathways also in photosynthetic microalgae in the future to increase photosynthetic yield. Another project recently aimed at installing the 3-hydroxypropionate pathway as photorespiratory bypass in *Synechococcus elongatus* PCC 7942 [Fig. 3C]. This pathway would notably not release CO_2 during glycolate recycling but allow the additional capture of CO_2, thus representing a "carbon-positive" pathway that would not only be more energy but also carbon efficient. While all the required heterologous enzymes of the 3-hydroxypropionate photorespiratory bypass were successfully expressed together in vivo, a major bottleneck was identified in the endogenous acetyl-CoA carboxylase [43]. Its activity was below the required level to sustain the cycle and will require additional overexpression or experimental evolution of the strain, before the pathway becomes fully operational.

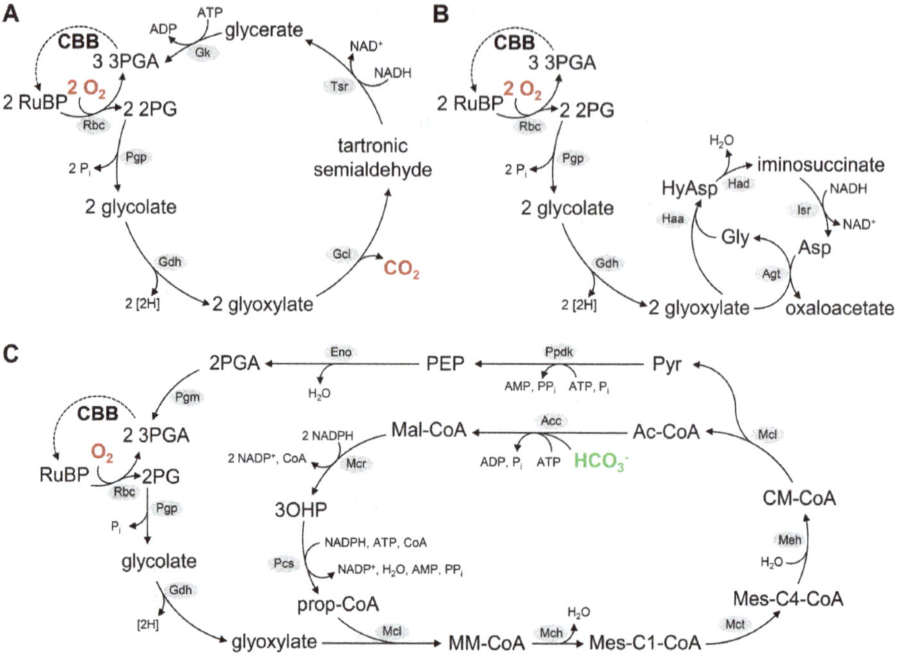

Fig. 3: Photorespiratory bypasses based on natural metabolic pathways. (A) Glycerate pathway as found in cyanobacteria [32, 33]. (B) Beta-hydroxyaspartate cycle as photorespiratory bypass [40, 41]. (C) Half-cycle of the 3-hydroxypropionate bicycle [42]. Abbreviations: 2PG, 2-phosphoglycolate; 2PGA, 2-phosphoglycerate; 3OHP, 3-hydroxypropionate; 3PGA, 3-phosphoglycerate; Acc, acetyl-CoA carboxylase; Ac-CoA, acetyl-CoA; Agt, aspartate/glyoxylate aminotransferase; Asp, aspartate; CBB, Calvin–Benson–Bassham cycle; CM-CoA, citramalyl-CoA; Eno, enolase; Gcl, glyoxylate carboligase; Gdh, glycolate dehydrogenase; Gk, glycerate kinase; Gly, glycine; Haa, β-hydroxyaspartate aldolase; Had, β-hydroxyaspartate dehydratase; HyAsp, β-hydroxyaspartate; Isr, iminosuccinate reductase; Mal-CoA, malonyl-CoA; Mch, mesaconyl-C1-CoA hydratase; Mcl, malyl-CoA/β-methylmalyl-CoA/citramalyl-CoA lyase; Mcr, malonyl-CoA reductase; Mct, mesaconyl-CoA transferase; Meh, mesaconyl-C4-CoA hydratase; Mes-C1-CoA, mesaconyl-C1-CoA; Mes-C4-CoA, mesaconyl-C4-CoA; MM-CoA, β-methylmalyl-CoA; Pcs, propionyl-CoA synthase; PEP, phosphoenolpyruvate; Pgm, phosphoglycerate mutase; Pgp, phosphoglycolate phosphatase; Ppdk, pyruvate phosphate dikinase; prop-CoA, propionyl-CoA; Pyr, pyruvate; Rbc, RubisCO; RuBP, ribulose-1,5-bisphosphate; Tsr, tartronic semialdehyde reductase.

5.4.1.2 New-to-nature photorespiration bypasses

While above photorespiration bypasses rely on naturally existing pathway sequences, the natural solution space of photorespiration was recently enlarged by systematic designs of potential pathways starting from 2-phosphogylcolate and considering specific biochemical reactions, such as C–C bond formation, activation of carboxyl groups into phosphoanhydrides and CoA esters, as well as their subsequent reduction [11]. One important boundary condition was to avoid the release of CO_2 to create "carbon-neutral" or even "carbon-positive" photorespiration bypasses. The design phase was followed by a thermodynamic feasibility analysis to discard potential pathways that were thermodynamically limited. Several potential carbon-neutral designs were identified that were all based on the reduction of glycolate into glycolaldehyde and the subsequent assimilation of glycolaldehyde through condensation reactions. The analysis also identified a carbon-positive pathway, the so-called tartronyl-CoA pathway that is based on the carboxylation of glycolyl-CoA into tartronyl-CoA [11, 12].

Tab. 1: Photorespiratory bypasses.

Pathway	Classification	Characteristics/implementation
Glycerate bypass	Natural	CO_2 release implemented into *Arabidopsis thaliana* [44], *Camelina sativa* [45], *Nicotiana tabacum* [40]
Beta-hydroxyaspartate pathway	Natural	CO_2 neutral implemented into *Arabidopsis thaliana* [42]
3-Hydroxypropionate bypass	Natural	CO_2 fixing implemented into *Synechococcus elongatus* [43]
Arabinose-5-phosphate shunt	New-to-nature	CO_2 neutral demonstrated in vitro [11]
Tartronyl-CoA pathway	New-to-nature	CO_2 fixing demonstrated in vitro [12]

Although the analysis identified several promising photorespiration pathways, a challenge has been the fact that all these pathways contain metabolites and reactions that are not known to take part in natural metabolism, which required the engineering of new-to-nature enzymes to realize these pathways. In a proof of principle, the arabinose-5-phosphate pathway (Fig. 4A), a carbon neutral pathway, was demonstrated, for which a glycolyl-CoA synthetase and a glycolyl-CoA reductases were engineered [11]. The tartronyl-CoA pathway (Fig. 4B) likewise required engineering of a glycolyl-CoA synthetase but notably also the development of a new-to-nature CO_2-fixing enzyme, glycolyl-CoA carboxylase. This was achieved by combining rational design, high-throughput microfluidics and microplate screens to create an engineered enzyme that matches the properties of natural carboxylases [12].

The arabinose-5-phosphate and the tartronyl-CoA pathway were both reconstituted in vitro and could be successfully coupled to RubisCO demonstrating that both new-to-nature reaction sequences are indeed compatible with the native carbon capture metabolism of photosynthesis. Theoretical calculations show that carbon-neutral and carbon-positive pathways should be able to increase photosynthetic yield by up to 60% [11, 12], requiring much less energy than natural photorespiration (Fig. 5).

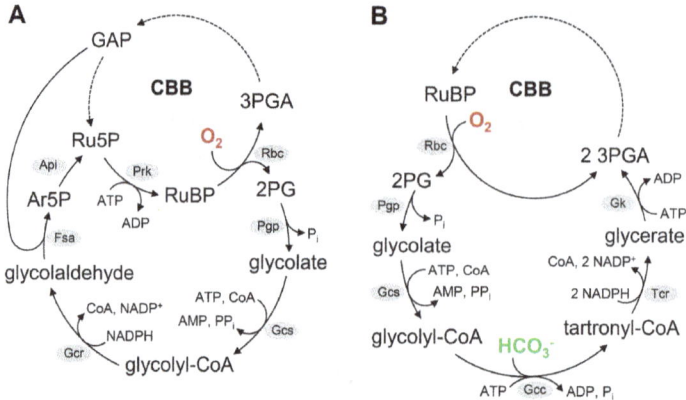

Fig. 4: Artificial new-to-nature photorespiratory bypasses. (A) Arabinose-5-phosphate pathway. (B) Tartronyl-CoA pathway. Abbreviations: 2PG, 2-phosphoglycolate; 3PGA, 3-phosphoglycerate; Api, arabinose-5-phosphate isomerase; Ar5P, arabinose-5-phosphate; CBB, Calvin–Benson–Bassham cycle; Fsa, fructose-6-phosphate aldolase; GAP, glyceraldehyde-3-phosphate; Gcc, glycolyl-CoA carboxylase; Gcr, glycolyl-CoA reductase; Gcs, glycolyl-CoA synthetase; Gk, glycerate kinase; Pgp, phosphoglycolate phosphatase; Prk, phosphoribulokinase; Rbc, RubisCO; Ru5P, ribulose-5-phosphate; RuBP, ribulose-1,5-bisphosphate; Tcr, tartronyl-CoA reductase.

Current efforts are underway to transplant these routes into the chloroplasts of plants and different photosynthetic microorganisms to demonstrate their feasibility in vivo and allow their experimental evolution to further integrate and optimize function of these synthetic photorespiration pathways.

5.4.2 New-to-nature CO_2 fixation pathways

Although the CBB cycle has evolved over millions of years it is notably not the only and, by far, not the most efficient CO_2-fixing pathway in nature. Over the last three decades, six additional CO_2 fixation pathways – and several variants thereof – were discovered in different autotrophic microorganisms (i.e., microorganisms that can grow with CO_2 as a sole carbon source) [46–50]. These pathways all differ with respect to their energetic requirements, the efficiencies of their carboxylases (i.e., CO_2-fixing enzymes), their operating conditions (e.g., anaerobic and aerobic) and their host organism [51–53], which directly influences the efficiency of light-to-carbon conversion.

As an example, while the CBB cycle consumes seven ATP per CO_2 converted into pyruvate (excluding photorespiration, see below), the reverse citric acid cycle, for instance, requires only two ATP per CO_2 fixed into pyruvate, which notably allows organisms using this pathway to capture more CO_2 at the same energy input. Thus, if introduced successfully into microalgae, the reverse citric acid cycle would theoretically increase photosynthetic yield by several fold. A challenge, however, is

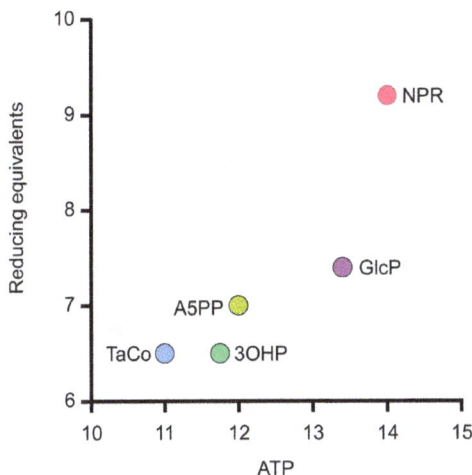

Fig. 5: Energetic requirements comparison between natural and synthetic photorespiration (bypass) pathways. Presented values derived from a flux-balance analysis according to Scheffen et al. [12] for net fixation of three CO_2 and formation of one molecule of 3-phosphoglycerate via CBB cycle in combination with respective photorespiratory pathway. Beta-hydroxyaspartate cycle omitted here as it does not directly arrive at 3-phosphoglycerate but at oxaloacetate. NPR, natural photorespiration; GlcP, glycerate pathway; A5PP, arabinose 5-phosphate pathway; 3OHP, 3-hydroxypropionate bypass; TaCo, tartronyl-CoA pathway.

the oxygen sensitivity of the key enzymes of the reverse citric acid cycle, α-ketoglutarate:ferredoxin and pyruvate:ferredoxin oxidoreductase, respectively, which limit the realization of this pathway in oxygenic photosynthetic organisms. Another practical problem is the limited capability of reprogramming the native metabolic and genetic networks of cells, which has prevented the simple transplantation of naturally existing CO_2 fixation pathways between microorganisms, although some progress has been made in expanding the genetic toolbox for microalgae, recently [54–56].

However, as mentioned before, a big drawback is that such "classical" metabolic engineering efforts are restricted to the handful of naturally existing enzymes and pathways that emerged during evolution. In contrast, synthetic biology opens new ways to improve biological CO_2 fixation through the design and realization of new-to-nature enzymes and pathways to explore novel, more efficient CO_2-fixing routes that nature has not explored yet [16].

In pioneering studies, the space of naturally existing enzymes was screened for new combinations to create alternative CO_2-fixing pathways that are RubisCO independent. Several solutions were identified, among them the malonyl-CoA-oxaloacetate-glyoxylate pathway, the reductive glycine cleavage (rGly) pathway and the crotonyl-CoA/ethylmalonyl-CoA/hydroxybutyryl-CoA (CETCH) cycle that are shorter and require less energy than the CBB cycle [6, 17, 57].

The rGly pathway [Fig. 6A] is a linear pathway that builds on the reversal of the glycine cleavage system that (usually) operates in glycine catabolism. Interestingly, while the rGly has been initially proposed as a synthetic pathway, it has in the meanwhile been described as a naturally existing CO_2 fixation pathway [58, 59]. In the last years, several successful attempts to transplant the core reaction sequence into several heterotrophic model organisms have been reported, including yeast and *Escherichia coli* [60, 61]. This paves the way for the future implementation of the rGly into microalgae and cyanobacteria. One challenge for the realization is the requirement of the rGly pathway for high CO_2 concentrations, which are necessary to drive the reaction of the enzyme complex in the direction of carbon fixation and not the preferred direction (i.e., CO_2 release). Sufficient supply of CO_2 to the culture and/or further adoption of the glycine cleavage system toward lower CO_2 concentrations might be required for a successful realization of the rGly in photosynthetic microorganisms.

The CETCH cycle, on the other hand (Fig. 6B), is designed around enoyl-CoA carboxylases/reductases (ECRs), a recently discovered new class of CO_2-fixing enzymes that belong to the most efficient carboxylases described to date [62, 63]. Notably, ECRs are up to 20-fold faster than an average RubisCO; and unlike RubisCOs, ECRs are able to selectively interact with CO_2 and do not show a side reaction with O_2, which together allows to operate CO_2 fixation at higher rate and efficiency. The CETCH cycle was drafted by metabolic retrosynthesis and established in several rounds in vitro by enzyme engineering and metabolic proofreading [6]. The complete cycle involves 12 core and 5 helper enzymes originating from 9 different organisms of all 3 domains of life, including 3 re-engineered enzymes.

To demonstrate the general feasibility of the CETCH cycle beyond an in vitro application, first attempts have been made to test the synthetic pathway in a more realistic in vivo scenario. In a recent study, a "synthetic chloroplast" was created by combining the CETCH cycle with photosynthetic membranes of spinach (and *Chlamydomonas*) in microfluidic droplets to understand whether the synthetic CO_2 fixation cycle can be coupled to the natural photochemical machinery of the cell [64]. After several optimizations and exchange of some enzyme components, the thylakoid extracts were indeed able to power the CETCH cycle in the microfluidic droplets, demonstrating that the pathway itself is in principle compatible with the native light-harvesting machinery of the cell. Current efforts focus on establishing the CETCH cycle in different hetero- and autotrophic model systems. The conversion of glyoxylate produced by the CETCH cycle into biomass can be achieved by using existing pathways, such as the glycerate pathway, the beta-hydroxyaspartate cycle [41] or eventually by coupling to the tartronyl-CoA pathway, as already demonstrated in vitro [12].

A

CO$_2$

Cyt$_{red}$
Fco
Cyt$_{ox}$

formate

ATP, THF
Ftl
ADP, P$_i$

10-HCO-THF

H$_2$O
Fch

5,10-CH=THF

NADH
MtdA
NAD$^+$

5,10-CH$_2$-THF

NH$_3$ — CO$_2$

NADH
Gcv
NAD$^+$ — THF

Gly

Shmt

Ser

NH$_3$
Sda
Pyr

B

CO$_2$

crotonyl-CoA
Ccr
NADPH   NADP$^+$
EM-CoA

H$_2$O
Hbd

Epi
Ecm

4-OHB-CoA

methylsuccinyl-CoA

ADP, P$_i$
Hbs
ATP, CoA

O$_2$
Mco
H$_2$O$_2$

4-OHB

mesaconyl-CoA

NADP$^+$
Ssr
NADPH

H$_2$O
Mch

SSA

methylmalyl-CoA

NADP$^+$
Scr
NADPH

Mcl

succinyl-CoA

glyoxylate

Mcm

propionyl-CoA

Epi
NADP$^+$
MM-CoA
Ccr
NADPH

Pco
O$_2$

acrylyl-CoA
H$_2$O$_2$

CO$_2$

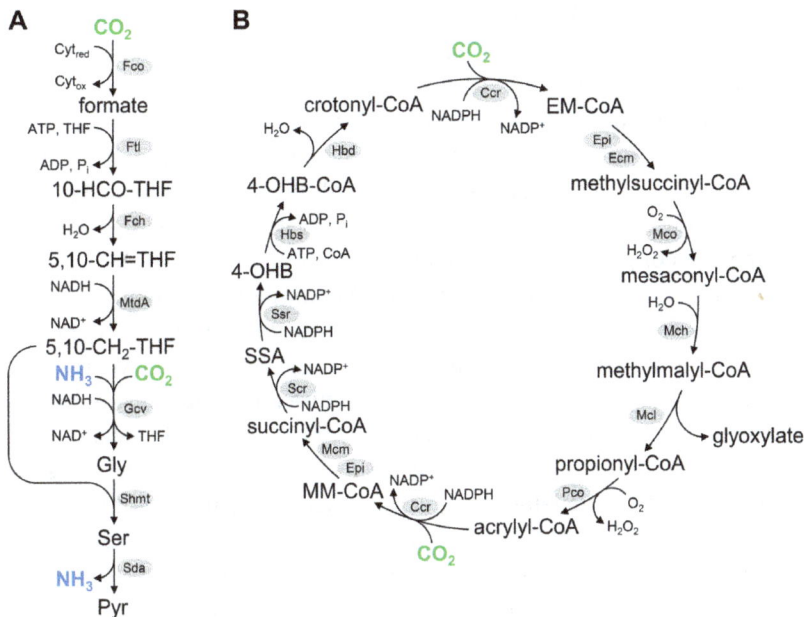

Fig. 6: New-to-nature CO$_2$ fixation pathways. (A) Reductive glycine cleavage (rGly) pathway and
(B) crotonyl-CoA/ethylmalonyl-CoA/hydroxybutyryl-CoA (CETCH) cycle. Abbreviations: 4-OHB,
4-hydroxybutyrate; 4-OHB-CoA, 4-hydroxybutyryl-CoA; Ccr, crotonyl-CoA carboxylase/reductase;
Cyt$_{ox}$, oxidized cytochrome; Cyt$_{red}$, reduced cytochrome; Ecm, ethylmalonyl-CoA mutase; EM-CoA,
ethylmalonyl-CoA; Epi, epimerase; Fch, methenyl-THF cyclohydrolase; Fco, formate dehydrogenase;
Ftl, formate-THF ligase; Gcv, glycine cleavage system; Gly, glycine; Hbd, 4-hydroxybutyryl-CoA
dehydratase; Hbs, 4-hydroxybutyryl-CoA synthetase; Mch, mesaconyl-CoA hydratase; Mcl, beta-
methylmalyl-CoA lyase; Mcm, methylmalonyl-CoA mutase; Mco, methylsuccinyl-CoA oxidase;
MM-CoA, methylmalonyl-CoA; MtdA, methenyl-THF dehydrogenase; Pco, propionyl-CoA oxidase;
Pyr, pyruvate; Scr, succinyl-CoA reductase; Sda, serine deaminase; Ser, serine; Shmt, serine
hydroxymethyl transferase; SSA, succinic semialdehyde; ssr, succinic semialdehyde reductase;
THF, tetrahydrofolate.

5.5 Outlook: future perspective and economic feasibility

A limiting factor in photosynthesis is the dark reaction, that is, the capture and conversion of CO$_2$ through the CBB cycle and in particular its key enzyme, RubisCO,
which causes the wasteful process of photorespiration. It has been shown that photosynthetic efficiency can be increased by modifying natural metabolism through
the integration of alternative enzymes or reaction sequences into the native metabolic
network of photosynthetic organisms. Beyond first approaches that were based on integrating naturally existing reaction sequences (such as the glycerate pathway or the

beta-hydroxyaspartate cycle), the advent of synthetic biology has allowed the design of new-to-nature enzymes and metabolic pathways that carry the potential to increase photosynthetic efficiency significantly.

While some of the promising new-to-nature designs, such as the arabinose-5-phosphate pathway, the tartronyl-CoA pathway or the CETCH cycle have been demonstrated in vitro, their in vivo implementation is still pending. The next step will require the successful integration of these designs into the endogenous metabolic and genetic networks of photosynthetic organisms. This is not only technically demanding (e.g., the coordinated expression of up to 14 enzymes at the required level) but also a biological challenge (e.g., unpredictable interactions of the new-to-nature metabolites with several thousand different proteins in the cell and vice versa). Several rounds and different model systems will be required to establish suitable selection systems to bring these pathways finally (in)to life. Inevitably, some of the designs will not work in some of the host organisms because some of the rules for the successful realization of new-to-nature pathways in cells still remain to be discovered. However, the increasing use of computational methods and artificial intelligence in biology, as well as new methods to assemble complex DNA pieces up to completely synthetic chromosomes will help to decipher these "biosynthetic" rules and guide experimental efforts to success.

Notably, these efforts in prokaryotes and eukaryotic microalgae are not only useful to increase efficiency in photosynthetic microorganisms but might also pave the way for the successful transplantation of these new-to-nature pathways into higher plants in the future. To allow the integration of these new-to-nature pathways into the native metabolism of photosynthetic organisms, experimental evolution systems will be required. The short generation times, chemostat cultivation, as well as the ease to assert selection pressure in microalgae or cyanobacteria make the experimental evolution in these "chloroplast-like" model system feasible. Thus, cyanobacteria and microalgae will have a great future and be an important building block in creating a photosynthesis 2.0.

References

[1] Blankenship RE, Tiede DM, Barber J, Brudvig GW, Fleming G, Ghirardi M, Gunner MR, Junge W, Kramer DM, Melis A, Moore TA, Moser CC, Nocera DG, Nozik AJ, Ort DR, Parson WW, Prince RC, Sayre RT. Comparing photosynthetic and photovoltaic efficiencies and recognizing the potential for improvement. Science 2011, 332, 805–9.

[2] Wijffels RH, Barbosa MJ. An outlook on microalgal biofuels. Science 2010, 329, 796–9.

[3] Bar-On YM, Milo R. The global mass and average rate of rubisco. Proc Natl Acad Sci USA 2019, 116, 4738–43.

[4] Phillips R, Milo R. A feeling for the numbers in biology. Proc Natl Acad Sci USA 2009, 106, 21465–71.

[5] Ellis RJ. Most Abundant Protein in the World. Trends Biochem Sci 1979, 4, 241–4.

[6] Schwander T, Schada Von Borzyskowski L, Burgener S, Cortina NS, Erb TJ. A synthetic pathway for the fixation of carbon dioxide *in vitro*. Science 2016, 354, 900–4.

[7] Trudeau DL, Edlich-Muth C, Zarzycki J, Scheffen M, Goldsmith M, Khersonsky O, Avizemer Z, Fleishman SJ, Cotton CAR, Erb TJ, Tawfik DS, Bar-Even A. Design and in vitro realization of carbon-conserving photorespiration. Proc Natl Acad Sci USA 2018, 115, E11455–E64.

[8] Scheffen M, Marchal DG, Beneyton T, Schuller SK, Klose M, Diehl C, Lehmann J, Pfister P, Carrillo M, He H, Aslan S, Cortina NS, Claus P, Bollschweiler D, Baret JC, Schuller JM, Zarzycki J, Bar-Even A, Erb TJ. A new-to-nature carboxylation module to improve natural and synthetic CO_2 fixation. Nat Catal 2021, 105–15.

[9] Bar-Even A, Noor E, Savir Y, Liebermeister W, Davidi D, Tawfik DS, Milo R. The moderately efficient enzyme: Evolutionary and physicochemical trends shaping enzyme parameters. Biochemistry 2011, 50, 4402–10.

[10] Davidi D, Shamshoum M, Guo Z, Bar-On YM, Prywes N, Oz A, Jablonska J, Flamholz A, Wernick DG, Antonovsky N, De Pins B, Shachar L, Hochhauser D, Peleg Y, Albeck S, Sharon I, Mueller-Cajar O, Milo R. Highly active rubiscos discovered by systematic interrogation of natural sequence diversity. EMBO J 2020, 39, e104081.

[11] Jahn M, Vialas V, Karlsen J, Maddalo G, Edfors F, Forsström B, Uhlén M, Käll L, Hudson EP. Growth of cyanobacteria is constrained by the abundance of light and carbon assimilation proteins. Cell Rep 2018, 25, 478–86.e8.

[12] Bowes G, Ogren WL, Hageman RH. Phosphoglycolate production catalyzed by ribulose diphosphate carboxylase. Biochem Biophys Res Commun 1971, 45, 716–22.

[13] Liang F, Englund E, Lindberg P, Lindblad P. Engineered cyanobacteria with enhanced growth show increased ethanol production and higher biofuel to biomass ratio. Metab Eng 2018, 46, 51–9.

[14] Liang F, Lindblad P. *Synechocystis* PCC 6803 overexpressing RuBisCO grow faster with increased photosynthesis. Metab Eng Commun 2017, 4, 29–36.

[15] Wurtzel ET, Vickers CE, Hanson AD, Millar AH, Cooper M, Voss-Fels KP, Nikel PI, Erb TJ. Revolutionizing agriculture with synthetic biology. Nat Plants 2019, 5, 1207–10.

[16] Erb TJ, Jones PR, Bar-Even A. Synthetic metabolism: Metabolic engineering meets enzyme design. Curr Opin Chem Biol 2017, 37, 56–62.

[17] Bar-Even A, Noor E, Lewis NE, Milo R. Design and analysis of synthetic carbon fixation pathways. Proc Natl Acad Sci USA 2010, 107, 8889–94.

[18] Carbonell P, Parutto P, Baudier C, Junot C, Faulon JL. Retropath: Automated pipeline for embedded metabolic circuits. ACS Synth Biol 2014, 3, 565–77.

[19] Koch M, Duigou T, Faulon JL. Reinforcement Learning for Bioretrosynthesis. ACS Synth Biol 2020, 9, 157–68.

[20] Peter DM, Von Borzyskowski LS, Kiefer P, Christen P, Vorholt JA, Erb TJ. Screening and engineering the synthetic potential of carboxylating reductases from central metabolism and polyketide biosynthesis. Angew Chem Int Ed 2015, 54, 13457–61.

[21] Schwander T, McLean R, Zarzycki J, Erb TJ. Structural basis for substrate specificity of methylsuccinyl-CoA dehydrogenase, an unusual member of the acyl-CoA dehydrogenase family. J Biol Chem 2018, 293, 1702–12.

[22] Calzadiaz-Ramirez L, Calvo-Tusell C, Stoffel GMM, Lindner SN, Osuna S, Erb TJ, Garcia-Borras M, Bar-Even A, Acevedo-Rocha CG. *In vivo* selection for formate dehydrogenases with high efficiency and specificity toward NADP. ACS Catal 2020, 10, 7512–25.

[23] Bohnke S, Perner M. Seeking active RubisCOs from the currently uncultured microbial majority colonizing deep-sea hydrothermal vent environments. ISME J 2019, 13, 2475–88.

[24] Bohnke S, Perner M. A function-based screen for seeking RubisCO active clones from metagenomes: Novel enzymes influencing RubisCO activity. ISME J 2015, 9, 735–45.

[25] Varaljay VA, Satagopan S, North JA, Witte B, Dourado MN, Anantharaman K, Arbing MA, Hoeft Mccann S, Oremland RS, Banfield JF, Wrighton KC, Tabita FR. Functional metagenomic selection of ribulose 1, 5-bisphosphate carboxylase/oxygenase from uncultivated bacteria. Environ Microbiol 2016, 18, 1187–99.

[26] Greene DN, Whitney SM, Matsumura I. Artificially evolved *Synechococcus* PCC6301 Rubisco variants exhibit improvements in folding and catalytic efficiency. Biochem J 2007, 404, 517–24.

[27] Wilson RH, Martin-Avila E, Conlan C, Whitney SM. An improved *Escherichia coli* screen for Rubisco identifies a protein-protein interface that can enhance CO_2-fixation kinetics. J Biol Chem 2018, 293, 18–27.

[28] Satagopan S, Tabita FR. RubisCO selection using the vigorously aerobic and metabolically versatile bacterium *Ralstonia eutropha*. FEBS J 2016, 283, 2869–80.

[29] Mueller-Cajar O, Morell M, Whitney SM. Directed evolution of rubisco in *Escherichia coli* reveals a specificity-determining hydrogen bond in the form II enzyme. Biochemistry 2007, 46, 14067–74.

[30] Flamholz AI, Prywes N, Moran U, Davidi D, Bar-On YM, Oltrogge LM, Alves R, Savage D, Milo R. Revisiting trade-offs between Rubisco kinetic parameters. Biochemistry 2019, 58, 3365–76.

[31] Timm S, Florian A, Fernie AR, Bauwe H. The regulatory interplay between photorespiration and photosynthesis. J Exp Bot 2016, 67, 2923–9.

[32] Nakamura Y, Kanakagiri S, Van K, He W, Spalding MH. Disruption of the glycolate dehydrogenase gene in the high-CO_2-requiring mutant HCR89 of *Chlamydomonas reinhardtii*. Can J Bot 2005, 83, 820–33.

[33] Eisenhut M, Kahlon S, Hasse D, Ewald R, Lieman-Hurwitz J, Ogawa T, Ruth W, Bauwe H, Kaplan A, Hagemann M. The plant-like C2 glycolate cycle and the bacterial-like glycerate pathway cooperate in phosphoglycolate metabolism in cyanobacteria. Plant Physiol 2006, 142, 333–42.

[34] Eisenhut M, Ruth W, Haimovich M, Bauwe H, Kaplan A, Hagemann M. The photorespiratory glycolate metabolism is essential for cyanobacteria and might have been conveyed endosymbiontically to plants. Proc Natl Acad Sci USA 2008, 105, 17199–204.

[35] Badger MR, Price GD. CO_2 concentrating mechanisms in cyanobacteria: Molecular components, their diversity and evolution. J Exp Bot 2003, 54, 609–22.

[36] Badger MR, Price GD. The CO_2 concentrating mechanism in cyanobactiria and microalgae. Physiol Plant 1992, 84, 606–15.

[37] Mahinthichaichan P, Morris DM, Wang Y, Jensen GJ, Tajkhorshid E. Selective permeability of carboxysome shell pores to anionic molecules. J Phys Chem B 2018, 122, 9110–8.

[38] Freeman Rosenzweig ES, Xu B, Kuhn Cuellar L, Martinez-Sanchez A, Schaffer M, Strauss M, Cartwright HN, Ronceray P, Plitzko JM, Förster F, Wingreen NS, Engel BD, Mackinder LCM, Jonikas MC. The eukaryotic CO_2-concentrating organelle is liquid-like and exhibits dynamic reorganization. Cell 2017, 171, 148–62 e19.

[39] Wunder T, Cheng SLH, Lai SK, Li HY, Mueller-Cajar O. The phase separation underlying the pyrenoid-based microalgal Rubisco supercharger. Nat Commun 2018, 9, 5076.

[40] South PF, Cavanagh AP, Liu HW, Ort DR. Synthetic glycolate metabolism pathways stimulate crop growth and productivity in the field. Science 2019, 363.

[41] Schada Von Borzyskowski L, Severi F, Krüger K, Hermann L, Gilardet A, Sippel F, Pommerenke B, Claus P, Cortina NS, Glatter T, Zauner S, Zarzycki J, Fuchs BM, Bremer E, Maier UG, Amann RI, Erb TJ. Marine Proteobacteria metabolize glycolate via the beta-hydroxyaspartate cycle. Nature 2019, 575, 500–4.

[42] Roell MS, SvB L, Westhoff P, Paczia N, Claus P, Erb TJ, Weber APM. A synthetic C4 shuttle via the β-hydroxyaspartate cycle in C3 plants. Under review.

[43] Shih PM, Zarzycki J, Niyogi KK, Kerfeld CA. Introduction of a Synthetic CO_2 fixing Photorespiratory Bypass into a Cyanobacterium. J Biol Chem 2014, 289, 9493–500.

[44] Kebeish R, Niessen M, Thiruveedhi K, Bari R, Hirsch HJ, Rosenkranz R, Stabler N, Schonfeld B, Kreuzaler F, Peterhansel C. Chloroplastic photorespiratory bypass increases photosynthesis and biomass production in *Arabidopsis thaliana*. Nat Biotechnol 2007, 25, 593–9.

[45] Dalal J, Lopez H, Vasani NB, Hu Z, Swift JE, Yalamanchili R, Dvora M, Lin X, Xie D, Qu R, Sederoff HW. A photorespiratory bypass increases plant growth and seed yield in biofuel crop *Camelina sativa*. Biotechnol Biofuels 2015, 8, 175.

[46] Berg IA, Kockelkorn D, Buckel W, Fuchs G. A 3-hydroxypropionate/4-hydroxybutyrate autotrophic carbon dioxide assimilation pathway in Archaea. Science 2007, 318, 1782–6.

[47] Zarzycki J, Brecht V, Müller M, Fuchs G. Identifying the missing steps of the autotrophic 3-hydroxypropionate CO_2 fixation cycle in *Chloroflexus aurantiacus*. Proc Natl Acad Sci USA 2009, 106, 21317–22.

[48] Huber H, Gallenberger M, Jahn U, Eylert E, Berg IA, Kockelkorn D, Eisenreich W, Fuchs G. A dicarboxylate/4-hydroxybutyrate autotrophic carbon assimilation cycle in the hyperthermophilic Archaeum *Ignicoccus hospitalis*. Proc Natl Acad Sci USA 2008, 105, 7851–6.

[49] Evans MCW, Buchanan BB, Arnon DI. A new ferredoxin-dependent carbon reduction cycle in a photosynthetic bacterium. Proc Natl Acad Sci USA 1966, 55, 928-&.

[50] Ragsdale SW. Enzymology of the Wood-Ljungdahl pathway of acetogenesis. Ann N Y Acad Sci 2008, 1125, 129–36.

[51] Berg IA, Ramos-Vera WH, Petri A, Huber H, Fuchs G. Study of the distribution of autotrophic CO_2 fixation cycles in Crenarchaeota. Microbiology 2010, 156, 256–69.

[52] Berg IA. Ecological aspects of the distribution of different autotrophic CO_2 fixation pathways. Appl Environ Microbiol 2011, 77, 1925–36.

[53] Berg IA, Kockelkorn D, Ramos-Vera WH, Say RF, Zarzycki J, Hügler M, Alber BE, Fuchs G. Autotrophic carbon fixation in archaea. Nat Rev Microbiol 2010, 8, 447–60.

[54] Crozet P, Navarro FJ, Willmund F, Mehrshahi P, Bakowski K, Lauersen KJ, Perez-Perez ME, Auroy P, Gorchs Rovira A, Sauret-Gueto S, Niemeyer J, Spaniol B, Theis J, Trosch R, Westrich LD, Vavitsas K, Baier T, Hubner W, De Carpentier F, Cassarini M, Danon A, Henri J, Marchand CH, De Mia M, Sarkissian K, Baulcombe DC, Peltier G, Crespo JL, Kruse O, Jensen PE, Schroda M, Smith AG, Lemaire SD. Birth of a photosynthetic chassis: A MoClo Toolkit enabling synthetic biology in the microalga *Chlamydomonas reinhardtii*. ACS Synth Biol 2018, 7, 2074–86.

[55] Baier T, Jacobebbinghaus N, Einhaus A, Lauersen KJ, Kruse O. Introns mediate post-transcriptional enhancement of nuclear gene expression in the green microalga *Chlamydomonas reinhardtii*. PLoS Genet 2020, 16, e1008944.

[56] Baier T, Wichmann J, Kruse O, Lauersen KJ. Intron-containing algal transgenes mediate efficient recombinant gene expression in the green microalga *Chlamydomonas reinhardtii*. Nucleic Acids Res 2018, 46, 6909–19.

[57] Cotton CA, Edlich-Muth C, Bar-Even A. Reinforcing carbon fixation: CO_2 reduction replacing and supporting carboxylation. Curr Opin Biotechnol 2018, 49, 49–56.

[58] Sanchez-Andrea I, Guedes IA, Hornung B, Boeren S, Lawson CE, Sousa DZ, Bar-Even A, Claassens NJ, Stams AJM. The reductive glycine pathway allows autotrophic growth of *Desulfovibrio desulfuricans*. Nat Commun 2020, 11, 5090.

[59] Figueroa IA, Barnum TP, Somasekhar PY, Carlstrom CI, Engelbrektson AL, Coates JD. Metagenomics-guided analysis of microbial chemolithoautotrophic phosphite oxidation yields evidence of a seventh natural CO_2 fixation pathway. Proc Natl Acad Sci USA 2018, 115, E92–E101.

[60] Yishai O, Bouzon M, Döring V, Bar-Even A. *In Vivo* assimilation of one-carbon via a synthetic reductive glycine pathway in *Escherichia coli*. ACS Synth Biol 2018, 7, 2023–8.

[61] Gonzalez De La Cruz J, Machens F, Messerschmidt K, Bar-Even A. Core catalysis of the reductive glycine pathway demonstrated in yeast. ACS Synth Biol 2019, 8, 911–7.

[62] Erb TJ, Berg IA, Brecht V, Müller M, Fuchs G, Alber BE. Synthesis Of C5-dicarboxylic acids from C2-units involving crotonyl-CoA carboxylase/reductase: The ethylmalonyl-CoA pathway. Proc Natl Acad Sci USA 2007, 104, 10631–6.

[63] Erb TJ, Brecht V, Fuchs G, Müller M, Alber BE. Carboxylation mechanism and stereochemistry of crotonyl-CoA carboxylase/reductase, a carboxylating enoyl-thioester reductase. Proc Natl Acad Sci USA 2009, 106, 8871–6.

[64] Miller TE, Beneyton T, Schwander T, Diehl C, Girault M, McLean R, Chotel T, Claus P, Cortina NS, Baret JC, Erb TJ. Light-powered CO_2 fixation in a chloroplast mimic with natural and synthetic parts. Science 2020, 368, 649–54.

Part 2: **Environment and photobioreactor design**

Christian Wilhelm, Heiko Wagner

6 Rate-limiting steps in algal energy conversion from sunlight to products – the role of photosynthesis

Abstract: Photosynthesis is the driving force in the conversion of airborne CO_2 to biomass. Unlike higher plants, in microalgae cultures all cells perform photosynthesis and are photosynthetically active during their whole life cycle. Moreover, algae can be cultivated under very controlled conditions to achieve optimal growth rates, resulting in much higher productivities and much better control under varying environmental conditions in comparison with higher plants. However, in real cells, even under optimal conditions, the efficiency of biomass production is far lower than the efficiency of photosynthetic reactions. Here we present the *state of the art* on the physiological processes that are responsible for the downregulation of photosynthesis and how they can be triggered by environmental and/or genetic control.

6.1 Goal

The maximum amount of algal biomass that can be produced at a given place is highly controversial in the literature [1–3]. However, running photobioreactor units produce much less than the theoretical amount [4–6]. Intensive work is focused on closing this gap between theory and practice and many proposals are made to overcome these limitations, especially with modifications in the photosynthetic apparatus [7, 8]. The aim of this chapter is to introduce the physiology of the algal cells and to evaluate their intrinsic physiological features under the conditions of an industrial bioreactor. It is shown how energy balances from photon to biomass can identify bottlenecks under reactor conditions and the most promising targets for optimizing algal product formation. Optimized product synthesis can be achieved either by increasing the content of the metabolite of interest by means of synthetic biology or by new pathways of excretion which will allow product formation without continuous biomass formation. Such a new concept is presented on how to overcome the intrinsic restrictions of biomass production by microalgae.

Acknowledgment: The financial support by the Sächsische Aufbaubank (SAB) grant number 100330646 for Dr. H. Wagner is greatly acknowledged.

https://doi.org/10.1515/9783110716979-006

6.2 Current methods for microalgae photosynthesis optimization

The assessment of potential improvements in the photosynthetic apparatus requires methods that specifically allow to evaluate the performance of partial reactions in the biochemical production machinery. At best, this can be done by energy balances from photon to biomass or related products, which requires a complex set of methods and a well-defined experimental design. In order to understand the physiology of growing cells, experiments have to be designed for a constant growth rate and constant environmental conditions during growth. Otherwise, the physiological methods are only a snapshot that cannot be extrapolated to long-term production conditions. This is optimally realized in a chemostat culture with low biomass concentration to maintain constant light and nutrient capture. Such an analytical system should include the following elements: (1) quantification of the amount of photons per cell and time to quantify the energy uptake per cell as a starting point for the energy balance, referred to as Q_{Phar}. (2) Pulse-amplitude-modulated fluorescence analysis during growth to measure the operative quantum yield of photosystem 2 (PS2) and the electron transport rate (ETR) per absorbed photon. This also provides data for the so-called non-photochemical quenching (NPQ). (3) Quantification of the oxygen production/consumption per cell or chlorophyll and time. Comparison of absolute ETR and oxygen production then provides the rates of alternative electron transport pathways. This fraction of electrons pumped by PS2 does not contribute to the formation of NADPH because the electrons are consumed by other acceptors or processes. (4) The CO_2 assimilation rate per cell/chlorophyll which in combination with the oxygen rate yields the photosynthetic quotient (PQ). The O_2/CO_2 ratio (PQ) always exceeds 1 in growing cells, depending on the required electrons for the reduction/assimilation of nitrogen and the biosynthesis of cellular compounds that are more reduced than carbohydrates (e.g., lipids). (5) The macromolecular composition of the newly synthesized biomass on the level of quantitative amounts of lipids, proteins and carbohydrates which sum up to more than 90% of the organic cellular biomass. (6) The daily growth rate including the light and dark phases. Details of these methods are described elsewhere [9–11]. The complete data set gained by all these methods allows a quantitative estimate of photon losses for CO_2 fixed in the biomass. In addition, it also quantitatively controls the validity of these methods, as the sum of the quantitative estimates of the electrons/carbon molecules should yield the biomass which can be harvested and weighed. This type of "modeling" differs from modeling, which is done on flux estimates in the major pathway which cannot be validated by independent experimental results [12].

Due to the high data density, the described set of methods can ultimately be used to optimize the quantum efficiency of biomass production. In addition to optimizing the individual physiological processes, for example, by changing the cultivation conditions, there are some promising genetic engineering approaches to improve the quantum-use efficiency of photosynthesis. Thus, in the last decade, progress has been made in manipulating the genomes of phototrophic organisms (cyanobacteria and eukaryotic algae), enabling the production of new strains specifically modified in the key elements in photosynthesis. As the molecular tools required for such a strain optimization are not topic of this chapter, please refer to [13–15]. Genome editing in diatoms is reviewed in [16] and functional genomics for *Chlamydomonas* in [17].

6.3 Basic background

6.3.1 The relationship between light and photosynthesis is the key for algal biomass formation

Figure 1 shows a typical light saturation curve of photosynthesis based on oxygen (lower curve) and fluorescence measurements (upper curve). While the first x-axis shows the intensity of the incident light (PAR), the second x-axis refers to the amount of absorbed quanta (Q_{Phar}). This experiment was performed in a diluted suspension (about 2 mg chlorophyll a L^{-1}), which only absorbed about 5% of the incident light. This indicates a only weakly pronounced light gradient in the suspension, suggesting an exposition of all cells to more or less identical light conditions. As the light climate (dynamics in intensity and wavelength, day length) strongly influences the physiological state of the cells, this is an appropriate condition for physiological measurements.

In a photobioreactor, the biomass loading must be much higher. Depending on the thickness of the vessel or flat panel, respectively, the biomass loading is between 2 and 20 g L^{-1} [3], corresponding to about 100–1000 mg chlorophyll a L^{-1} assuming that about 1–5% of the biomass is bound to chlorophylls [9–11]. This high biomass loading is necessary to increase the light absorption. Ideally, light absorption is about 90% of the incident light to capture as much energy as possible. These differences in biomass loading between physiological and biotechnical applications have drastic consequences for the photosynthetic performance of the cells. First, cells in a bioreactor are no longer exposed to continuous light, but to flickering light with a frequency below 1 s, depending on the aeration of the reactor. Flickering light can enhance the photosynthetic performance [19]. However, under high biomass loads, the aeration also strongly influences the CO_2 availability for the cells and the oxygen concentration in the culture medium. High O_2/CO_2 ratios inhibit the biomass production by the mechanism of photorespiration [20]. Based on these differences between high and low biomass-loaded suspensions, physiological performance data cannot easily be extrapolated to bioreactor conditions. In most cases, independent photobioreactor experiments must be performed to obtain reliable production data.

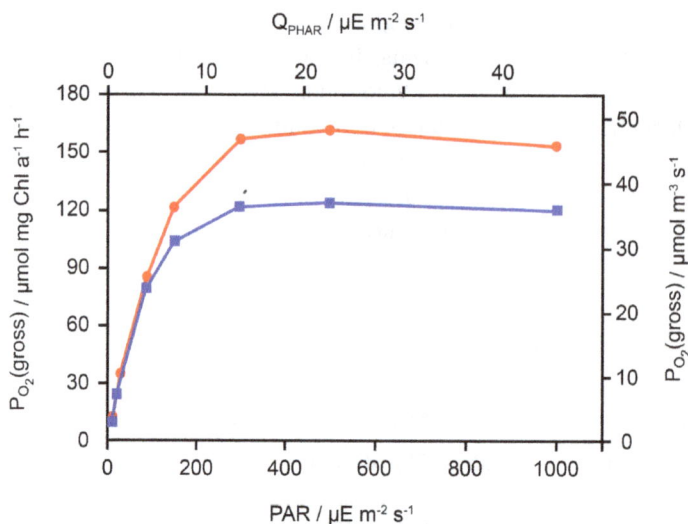

Fig. 1: P–I curve from the green alga *Chlorella*. Oxygen evolution rates (P_{O2}) as a function of photosynthetic active radiation (PAR) and photosynthetically effective absorbed radiation (Q_{Phar}). Comparison of oxygen evolution rates directly measured (blue) and calculated from fluorescence data (PAM, red). Absolute values for fluorescence-based electron transport based on multiplication of efficient PS2 quantum yields and absorbed quanta per time (Q_{Phar}). The gross rates include the losses by respiration. Figure taken from [18].

In Fig. 1, a second *y*-axis is labeled, indicating the oxygen production per volume. In biotechnology the volumetric activity is the most used reference parameter, however, in photoautotrophic systems it is very difficult to calculate the physiological activity of the cells from volumetric data. This is because volumetric activity depends on biomass concentration. However, biomass concentration strongly alters the availability of substrates such as light (Q_{Phar}), CO_2, O_2 and can change the pH value. In Fig. 1, the upper curve is based on fluorescence data, whereas the lower one is based on direct oxygen measurements. The oxygen production rate can be calculated from fluorescence data as follows: The fluorescence data indicate the apparent quantum yield of PS2. If this quantum yield is recorded after a dark phase, it corresponds to the optimum activity of PS2 and reflects the physiological intactness of the photochemistry. The "operative" quantum yield is measured during illumination, which "closes" a defined fraction of PS2 reaction centers (RC) and lowers the yield. Under saturating light conditions, all RCs are closed, which minimizes the variable fluorescence. The quantum yield is then determined as the ratio of absorbed photons (Q_{Phar}) to photochemically used photons in PS2. If this ratio is multiplied with the number of absorbed photons (which is equivalent to Q_{Phar}), the number of used photons can be converted into photosynthetic electrons [3, 21]. The whole electron transport from water to NADPH (carrying two electrons more than NADP$^+$) requires two light reactions, which means that the number of photons per electron must be divided by 2. Since the emission of one oxygen molecule requires two water molecules, eight photosynthetic electrons are statistically equivalent to one molecule NADPH formed. Figure 1 shows that both curves match perfectly at the linear part of the P–I curve. However, at saturating light intensities, fluorescence-based oxygen rates exceed the rate of oxygen evolution. This is due to the fact that electrons can take alternative pathways returning to oxygen and producing water again [22, 23]. This reaction is considered as a protective mechanism against photoinhibition.

A typical P–I curve shows a linear part at low light intensities, indicating that the process of photosynthesis is light limited. Under these conditions, the light reactions (absorption, photochemistry and electron transport) operate with high efficiency (photon per molecule oxygen or electron, respectively) and the slope of the curve provides the operative quantum yield. At optimal growth conditions, the slope is in the range of about 10 absorbed photons per molecular oxygen, which in a first approximation is equivalent to 1 carbon molecule. In contrast, at light saturation, the dark reactions are limiting and the photosynthetic capacity cannot be further increased by components of the light reactions. Under full sunlight, the growth rate of an algal cell is therefore primarily determined by the cellular capacity of the dark reactions. There is consensus that the most important bottleneck in dark reactions is the amount and activity of Rubisco, that is, the CO_2-assimilating enzyme [24, 25]. However, the electron sink capacity is not only determined by the CO_2-assimilating activity but also by other electron-consuming reactions such as the Mehler reaction, nitrate/nitrite reduction, sulfate reduction and electron consumption by reductive reactions in the chloroplast like fatty acid biosynthesis. The turnover times of the light reactions under full sunlight are about 100 times faster than the enzymatic dark reactions. Photosynthetic electron transport thus exceeds the capacity of the electron consumption resulting in a reduction of the components in the thylakoid membrane and a blocking of the electron transport [26]. The reduced acceptors in PS2 and PS1 can generate toxic chlorophyll excited states (Chl^3) or toxic oxygen species like singlet oxygen or superoxide anions, which can oxidize sensitive proteins (e.g., the D1 subunit of PS2) or lipid molecules by forming lipid radicals [27, 28]. These destructive reactions must be prevented by the so-called photosynthetic control [30].

6.3.2 Natural light is not optimal for autotrophic growth

Natural light is characterized by dynamic changes in light intensity and light quality (the spectrum). Here we focus on the changes in light intensity because this is the major factor for biomass production. Light intensity changes during the day either sinusoidally when the sky is clear, or the intensity drops down for unpredictable time intervals when it is cloudy. During evolution, different mechanisms have been developed to adapt the cells to these conditions. Three light conditions are distinguished in a P–I curve: (1) light limitation below the light saturation parameter (E_k), (2) full sunlight (above E_k) and (3) darkness. The annual "light usage dilemma" reveals the problem: In Tab. 1, the light–dark distribution in the natural light climate is compared between two regions at different latitudes. The light usage dilemma shows that on the one side most of the time light is limiting or absent, whereas on the other side the light intensity is often too high. Calculating the photon distribution based on these data reveals that 82% of the incident photosynthetically active photons reach the surface with a light intensity above E_k; however, 80% of the total time the cells are exposed to limiting light or darkness. Therefore, under natural light, the cells cannot operate under optimal physiological conditions. If the physiological growth conditions are optimized for full sunlight, for example, by decreasing their absorption capacity [29], the cells will suffer from energy limitation in the morning and afternoon. Since most of the light energy can be harvested under high light intensities, the culture must be optimally managed for this condition. For instance, it is essential to know the regulatory processes called "photosynthetic control" [30] quantitatively under full sunlight. Also, conditions have to be established to minimize photon losses as efficiently as possible.

Tab. 1: Annual light distribution in two different European locations (data from the local data base of the Institute of Geography, University of Leipzig and Almeria, Spain).

	Leipzig 50 (LAT)	Almeria 36.8 (LAT)
Hours per year with incident light above E_k	1594	3000
Hours per year with incident light below E_k	2542	1136
Hours of darkness per year	4224	4136

6.3.3 Photosynthetic control and metabolic regulation in response to light

One of the most important control mechanisms of the photosynthetic electron transport chain can be visualized by enhanced quenching of chlorophyll fluorescence, representing a regulated increase of heat dissipation: it is called NPQ [31]. In eukaryotic algae, this process is regulated by an extremely well-controlled mechanism [32]. Under high light intensities, the proton gradient across the thylakoid membrane is high, causing the antenna complexes to aggregate. This induces a drop in luminal pH and activates the enzyme violaxanthin de-epoxidase, thus, converting violaxanthin into zeaxanthin. Zeaxanthin binds again to a specific light-harvesting protein and induces a conformational change. The resulting complex can absorb excess excitons from chlorophyll a and convert their energy into heat. This regulatory mechanism is known to be present in green algae but also in heterokontophytes – in the latter case using diadinoxanthin and diatoxanthin as functional light protection pigments [33, 34]. This regulatory process lowers the quantum yield and converts the photosynthetic apparatus from a photochemical to a heat-dissipating complex. The upper part of Fig. 2 shows the buildup of NPQ as a function of incident light in two different diatom species, indicating that the degree of photosynthetic control differs between species. Exposing the cells to an oscillating light regime (simulating a cloudy day or mixing in a slow running river) reveals that the xanthophyll cycle operates in response to the intensity of the incident light (Fig. 2B and C). It follows that cells exposed to high light intensity have a lower photosynthetic efficiency than cells in flickering light or permanent low light. Comparing Figs. 1 and 2 reveals that NPQ, as a mechanism of the photosynthetic control, is more pronounced above the transition between the linear increase of the $P–I$ curve and the phase of light saturation. This transition is defined as the light saturation parameter E_k defining the incident light intensity at which photosynthesis starts to saturate. It is evident that below E_k, the ETR of photosynthesis is below the potential biological capacity of the cells but above E_k the higher the light, the more photosynthetic efficiency is impaired. In this case, the quantum-use efficiency of biomass production expressed as oxygen evolution of PS2 (Φ_{Onet}) is reduced by the influence of NPQ under high light conditions. Therefore, the cells are in a conflicting situation: When high light delivers sufficient energy for high growth rates, the quantum yield is low. The optimal state of light conversion into biomass is given when single cells absorb the amount of energy that corresponds to the E_k. However, both flat panel and tubular bioreactors designed for optimal sunlight utilization and with a generally high biomass loading will not maintain an optimal physiological state. This means that if the conditions resulting in a physiological state near the E_K are to be transferred to a bioreactor with a high biomass load, a physiological state near the E_K can only be achieved by "light dilution," that is the incident photons are distributed over a larger area. This can be accomplished by vertical orientation of the reactor by an optimal management of the cell concentration and an optimized mixing regime [3]. As a result, the number of photons per cell and time is then close to the E_k value under non-self-shading conditions.

Quantum-use efficiency of biomass production can also be impaired by a variety of defined physiological processes that lead to losses of absorbed quanta. These losses can be divided into constitutive losses (entropy of physical and chemical reactions, referred to as ($\Phi_{f,D}$)), regulated heat emission (xanthophyll cycle-dependent NPQ, Φ_{NPQ}), alternative ETRs (mainly the Mehler reaction, which does not deliver reductants but produces ATP, denoted as Φ_{alt}), photorespiration and dark respiration (Φ_{resp}). The biological variability of these processes was studied in two different diatoms which possess different NPQ characteristics. It was hypothesized that *Cyclotella meneghiniana* would benefit from its large and flexible NPQ potential under fluctuating light conditions when photoprotection is vital, whereas the small NPQ capacity in *Skeletonema costatum* (Fig. 2A) should have a negative impact on growth. However, resulting growth rates that are higher in *S. costatum* clearly contradict this expectation (Tab. 2).

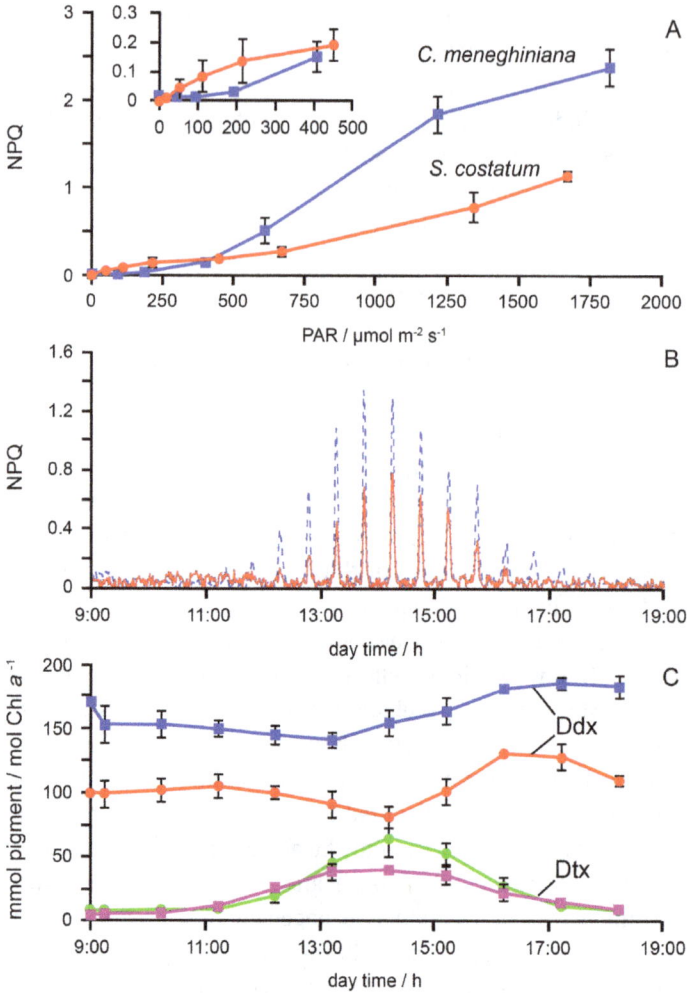

Fig. 2: Non-photochemical quenching (NPQ) potential as a function of light intensity in *Cyclotella meneghiniana* and *Skeletonema costatum* grown under fluctuating light conditions mimicking cloudy days or mixed reactor conditions. (A) Samples taken at 09:00 before the start of the light period. NPQ values derived from photosynthetic irradiance curves via fluorescence measurements (inset shows the initial slope). (B) NPQ values in *C. meneghiniana* (blue line) and *S. costatum* (red line) over the course of the light period under fluctuating light conditions. NPQ peaks refer to peaks in the light intensity during the day. (C) Changes in the sum of the amounts of the xanthophyll cycle pigments diadinoxanthin (Ddx; blue and red symbols) and diatoxanthin (Dtx; green and magenta) in *C. meneghiniana* (blue and magenta) and *S. costatum* (red and green circles) over the course of the light period under fluctuating light conditions. Data and figure taken from Su et al. [9].

Tab. 2: Physiological parameters of *Cyclotella meneghiniana* and *Skeletonema costatum* observed under exponentially fluctuating light conditions according to Su et al. [9].

Parameter	*C. meneghiniana*	*S. costatum*
Q_{Phar} (mmol mg^{-1} Chl a × day)	34.3 (0.85)	35.9 (0.71)
O_2 production (mmol O_2 mg^{-1} Chl a × day)	1.49 (0.1)	1.67 (0.1)
ETR_{alt} (mmol e$^-$ mg^{-1} Chl a × day)	1.26 (0.26)	2.34 (0.33)
Growth rate (day^{-1})	0.33 (0.04)	0.44 (0.01)
Respiration (mmol O_2 mg^{-1} Chl a × day)	0.63 (0.02)	0.97 (0.18)
Lipid:protein ratio (mg mg^{-1})	1:2.79	1:5.57
C:N ratio (mol mol^{-1})	7.13 (0.13)	6.5 (0.28)
PQ (mol O_2 mol^{-1} CO_2)	2.26 (0.14)	1.43 (0.27)
Φ_C (mol C mol^{-1} quanta)	0.011	0.014

Q_{Phar}, photosynthetically absorbed radiation; Chl a, cellular content of chlorophyll a; ETR_{alt}, amount of alternative electrons; growth rate, growth rate based on Chl a content; C:N, ratio of carbon: nitrogen; PQ, photosynthetic quotient; Φ_C, carbon-related biomass production.

The results show that the main loss of quanta is due to unregulated heat emission ($\Phi_{f,D}$) as physically inevitable metabolic cost (Fig. 3). Other loss components show unexpected variability and extent. The higher NPQ in *C. meneghiniana* is compensated by a lower activity of alternative ETRs (Φ_{alt}) and its finally higher quantum yield for oxygen production (Φ_{Onet}). From the better photosynthetic performance in *C. meneghiniana*, one would expect better growth. Although both species absorb the same amount of quanta per time and chlorophyll, *S. costatum* shows lower Φ_{Onet} and higher quantum losses in total (Σ of $\Phi_{f,D}$, Φ_{NPQ}, Φ_{alt} and Φ_{resp}; Fig. 3). However, *S. costatum* has a significantly higher growth rate than *C. meneghiniana* and a clearly better quantum-use efficiency of carbon assimilation (Tab. 2). This in turn leads to a clearly different biomass composition. *S. costatum* contains much more protein in relation to lipid (Tab. 2). Since protein synthesis requires more ATP than lipid synthesis, the higher respiration rate and the alternative electron cycling contribute to generating energy for anabolic processes. However, the biomass of *C. meneghiniana* is more reduced and therefore more electrons from water splitting are needed to synthesize cellular carbon backbones. From these data, it is clear that the efficiency of biomass formation and thus biotechnological product yields are strongly regulated in the metabolic network and that photosynthetic control adapts the photosynthetic apparatus to the metabolic needs [9].

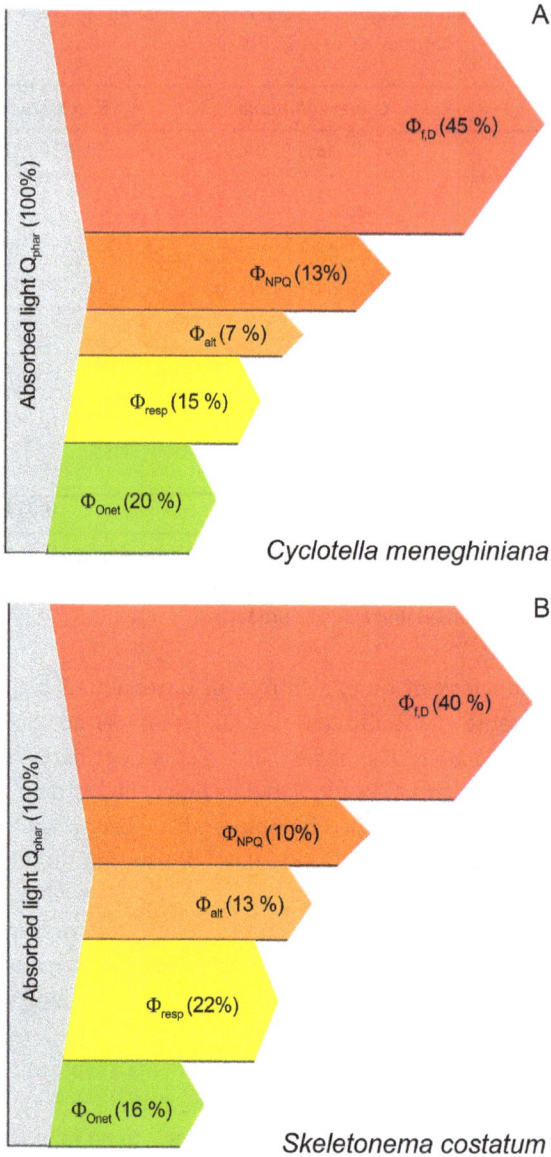

Fig. 3: Distribution of absorbed light energy in *Cyclotella meneghiniana* and *Skeletonema costatum* under exponentially fluctuating light conditions. Q_{Phar}, total photosynthetically absorbed radiation; $\Phi_{f,D}$, sum of fluorescence dissipation and constitutive unregulated heat emission; Φ_{alt}, energy for alternative electron transport; Φ_{resp}, energy loss via respiration; Φ_{Onet}, fraction of absorbed light used for net oxygen evolution. Figure modified after Su et al. [9].

6.3.4 Carbon sequestration under optimal and non-optimal nutrient conditions

The variation in energy demand for carbon assimilation into harvestable biomass is of great importance in the context of biofuel production based on the formation of lipids. From the data presented above, there is a trade-off situation for optimum growth and high lipid yield. This matches perfectly with the observation that fast-growing cells contain only little amounts of neutral lipids but contain abundant proteins to catalyze the biochemical reactions needed for growth [35]. Therefore, recent approaches proposed a two-step cultivation with an initial replete growth phase followed by a second lipid accumulation phase [36, 37]. In the latter phase, the contribution of photosynthesis to carbon acquisition is only of minor importance. Since the lipid productivity is the product of growth rate and lipid content, it is not clear if the optimum lipid production rate is primarily controlled by the growth rate or by the lipid content [38]. In the literature, quantitative studies that have followed the fate of the electrons in lipid-accumulating cells are scarce. However, an extended energy balance study by Jakob et al. [39] compared the energetic losses of N-replete and N-limited cultures which still perform growth. The results in Fig. 4 show that in N-replete growing cells the losses by NPQ are much lower (9% vs. 31%) and the relative amounts of photosynthetic electrons are higher (15% vs. 7% in N-limited cells). Despite the lower losses in NPQ and the higher photochemical efficiency Φ_{Onet}, the quantum-use efficiency of carbon fixation (Φ_C) is lower in fast-growing cells compared to N-depleted ones (Fig. 4). The reason becomes obvious by analyzing the macromolecular composition of the cells: fast-growing cells contain very high levels of proteins, whereas the storage compounds lipids and carbohydrates are low. In contrast, N-limited cells contain only low amounts of proteins and much more carbohydrates. Since carbohydrates are less reduced compounds, the electron and energy demand for one carbon molecule embedded in the biomass is much lower and less photons are needed to fix a CO_2 into an organic C-molecule as part of the biomass. As protein biosynthesis always has a very high ATP demand which cannot be covered by photophosphorylation alone, mitochondrial respiration is needed for cellular protein biosynthesis. Therefore, respiration losses are higher in protein-rich cells, supporting the linear relationship in phytoplankton between mitochondrial respiratory activity and growth [40]. As a result, the inclusion of biomass reduction efficiency, general quantum losses and respiration rates in a general quantum-use balance as a function of nutrient conditions is required and very helpful for the optimization of cultivation conditions in bioreactors.

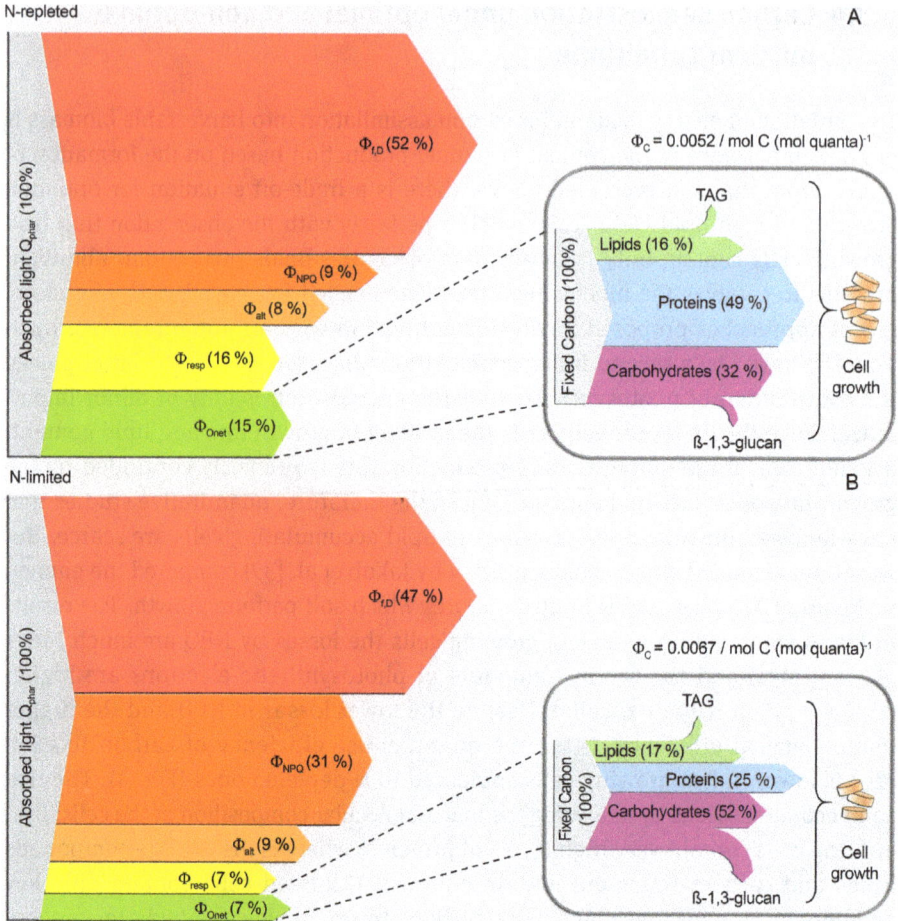

Fig. 4: Schematic representation of energy and carbon allocation in model diatoms based on nutrient conditions: (A) optimal growth conditions or (B) nitrogen limitation. Proportions of the different pathways of photosynthetic energy calculated based on the amount of absorbed radiation Q_{Phar}. Distribution of fixed carbon among the individual macromolecular pools including lipid storage (TAG, triacylglycerol) and polysaccharide storage (b-1,3-glucan, chrysolaminarin) shown on the right side. $\Phi_{f,D}$, sum of fluorescence dissipation and constitutive unregulated energy dissipation; Φ_{alt}, energy for alternative electron transport; Φ_{resp}, energy loss via respiration; Φ_{Onet}, fraction of absorbed light used for net oxygen evolution; Φ_C, quantum-use efficiency of carbon fixation. Figure modified according to [40], original data in [39].

6.3.5 Changing carbon sequestration patterns by temperature

Each algal species has a specific temperature optimum for growth. Figure 5 shows the relationship between growth and temperature in different taxa of freshwater algae. While it is obvious that most species isolated from waters of temperate re-

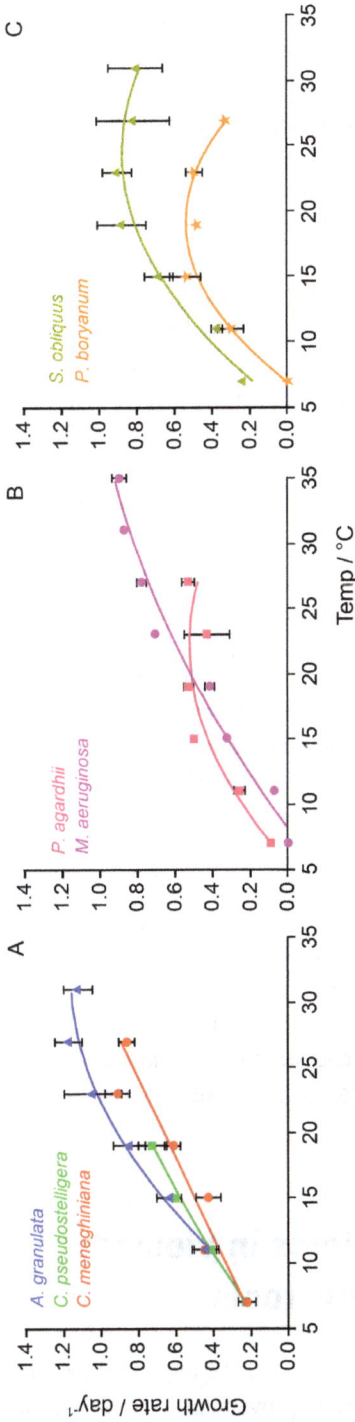

Fig. 5: Growth rate response to different growth temperatures of (a) *Cyclotella meneghiniana*, *Cyclotella pseudostelligera* and *Aulacoseira granulata*; (b) *Microcystis aeruginosa* and *Planktothrix agardhii*; (c) *Pediastrum boryanum* and *Scenedesmus obliquus*. The solid lines represent a second-order polynomial function fitted to the measured data. Figure taken from [41].

gions perform well under 20 °C, growth reduction beyond the optimal growth temperature is clearly different between the species.

At higher temperatures, the growth stops completely in some cases after a critical threshold. The reasons for this behavior are complex and not completely understood. In theory, the photosynthetic light reactions (light absorption, photochemistry and photosynthetic electron transport) are very insensitive to temperature, whereas the C-assimilation and the metabolic reactions for biomass formation follow the van't Hoff law with a few exceptions. This results in a duplication of the reaction kinetics of biochemical reactions upon increasing the temperature by 10 K and leads to an overexcited state of the cells when they are exposed to lower temperatures under constant light. An open question is the fate of excessive energy that cannot be used for growth due to temperature limitations. Using again the energy-balancing approach, the results clearly show a diversity among different species (Fig. 6). As expected, in most species the amount of the absorbed energy invested into growth (Φ_{Onet}) increases with temperature (with the exception of *C. pseudostelligera*; Fig. 6).

However, the energy partitioning shows that the quantum yield of the unregulated heat emission ($\Phi_{f,D}$) is more or less constant at different temperatures, whereas the differences between species are high (in the range of 20% up to 55%). The fraction of NPQ (Φ_{NPQ}) again has a different extent among the species but in general decreases with higher temperature. A prominent factor of regulation is the alternative electron transport (Φ_{alt}). This example clearly shows again that the link between photosynthesis and growth is rather weak under dynamic environmental conditions.

The conclusion from the study [41] matches perfectly with previously findings that the growth rate of algae is not only necessarily dependent on the species but rather on temperature [42], light [43] and nutrient [44] conditions that always correlate with the amount of storage compounds. This seems to be a general law for algal carbon sequestration strategies. Carbon obtained from photosynthetic CO_2 assimilation can only be directed to growth, if the higher growth rate is co-regulated with the synthesis of membrane lipids and proteins for cell structure formation. If these processes are limited due to nutrient limitation or temperature, the carbon will preferably be directed to storage products.

6.4 State of the art to improve yields in biomass or products via photosynthetic tools

Recent progress in the generation of transgenic algal lines with improved features for biotechnological applications has stimulated the improvement of photosynthesis by addressing targets identified as bottlenecks in the conversion of light to bio-

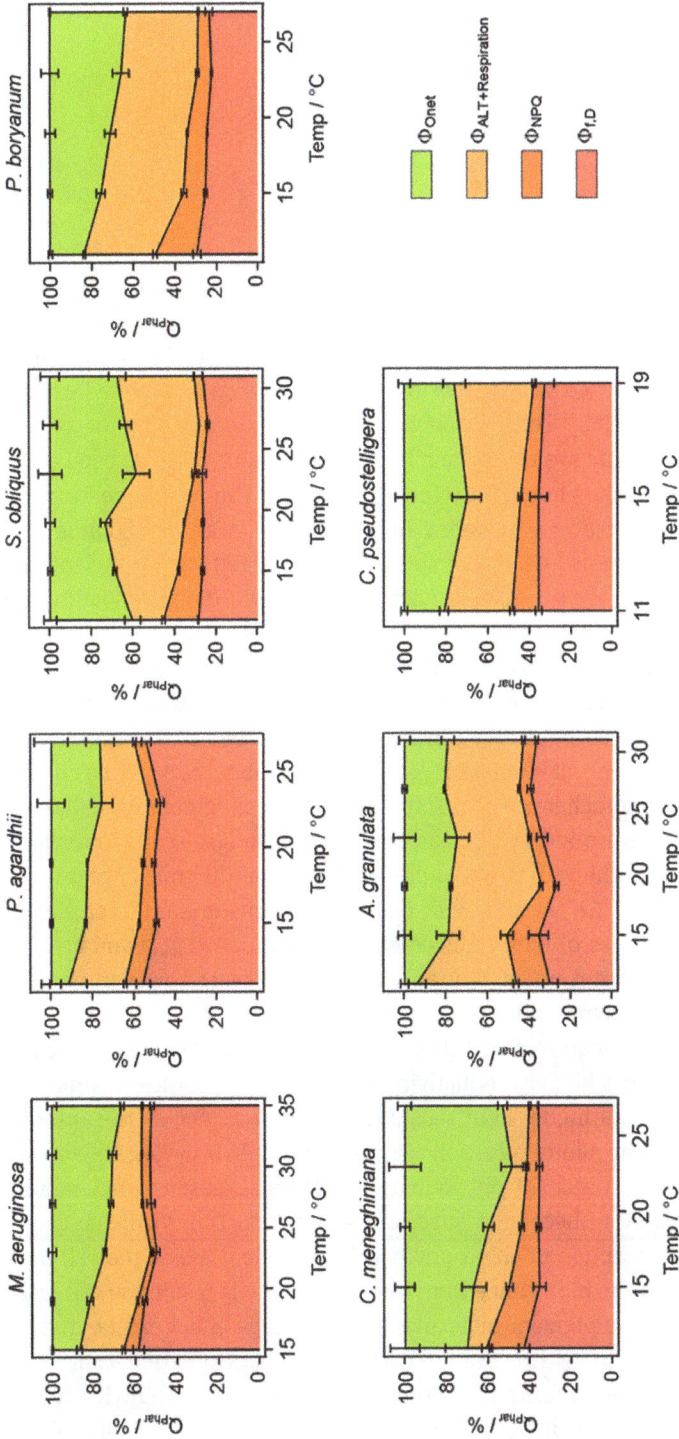

Fig. 6: Absorbed photosynthetic energy (Q_{Phar}) partitioning in seven phytoplankton species acclimated to temperature. Data as percentage of Q_{Phar}; $\Phi_{f,D}$, sum of fluorescence dissipation and constitutive unregulated energy dissipation; Φ_{alt}, energy for alternative electron transport and respiration; Φ_{Onet}, fraction of absorbed light used for net oxygen evolution. Figure modified according to [41].

mass. Here we focus on (1) truncated light-harvesting antennae, (2) engineered Rubisco, (3) learning from highly productive strains and (4) using natural metabolic pathways to produce carbon-based compounds. For further targets and more details, see recent reviews [46–48].

6.4.1 Antenna size reduction

In the last decade, intensive work has been done with *Chlamydomonas* [45], *Chlorella* [46] and *Nannochloropsis* [47] strains to optimize the light to biomass conversion efficiency by reducing the light-harvesting antenna system. Minimization of the phycobiliprotein-composed light-harvesting system has also been tested experimentally in cyanobacteria [48]. This approach is based on the following theoretical consideration: As microalgae are very strong light absorbers, cells shade each other at higher cell densities in the photobioreactor. This shading can only be reduced by intensive mixing, which creates a flickering light regime. However, mixing is also one of the major running costs [49], which should be minimized for an efficient biotechnological application. Even under intensive mixing, the biomass loading by self-shading limits the volumetric activity, which is a critical parameter in bioreactor technology. For comparison, the volumetric activity of algal cultures is about 20–100 times lower than that of heterotrophic bacteria or yeast cultures. This disadvantage of algal cultures can be reduced by an increase of the volumetric activity, that is, if the absorption per unit biomass is reduced. Since about 50% of the chlorophyll is bound to the photochemically inactive light-harvesting complexes (LHC), reduction of the light absorption by decreasing the antenna size seems promising [46, 50, 51]. This can be achieved, for example, by inhibiting the transcription process of LHC-related genes [52]. Results with *Chlorella vulgaris* mutant lines clearly show higher maximum rates of light-induced oxygen evolution (P_{max}) and also higher biomass production rates under saturating light conditions; however, production is clearly below wild type under low light conditions [53]. Also, the achieved improvements are rather small (around 30%) and under simulated environmental light climate conditions the higher productivity disappears more or less completely [45]. This could be due to a higher light sensitivity of the mutants, as the antenna system is also involved in photoprotection via the xanthophyll cycle (see above). Based on data, Schramm et al. [54] contributed physiological reasons why antenna size reduction is unlikely to become a successful approach: Due to the so-called package effect, light absorption of chlorophylls is less efficient inside the cell than in solution [29, 55], with the high concentration of chlorophylls inside the cell leading to a self-shading of the pigment molecules. This package effect can be quantified by the so-called Chl *a* in vivo absorption coefficient a^*_{phy} expressed as $m^2 \, mg^{-1}$ Chl *a*, which is the scatter-corrected absorption divided by the chlorophyll concentration, describing the area available for the population of all chlorophyll molecules. Consequently, a higher packing results in a lower light absorption efficiency while a chlorophyll reduction increases the efficiency of light absorption by each chlorophyll molecule. Blache et al. [29] showed this relationship between a^*_{phy} and the cellular chlorophyll content (Fig. 7).

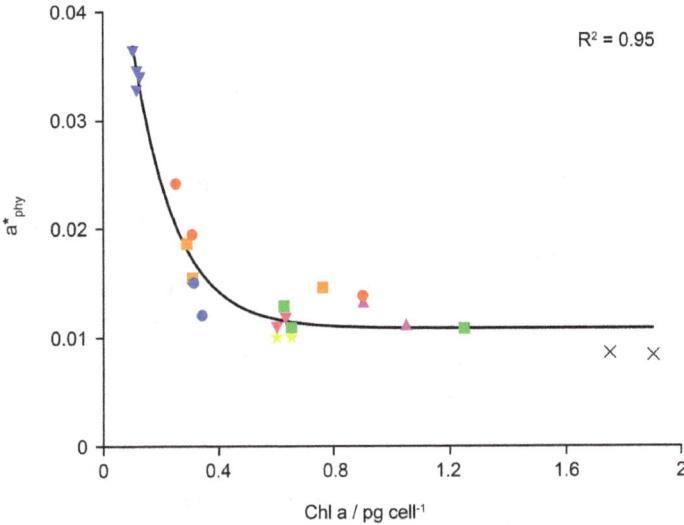

Fig. 7: Correlation between chlorophyll-specific absorption coefficient a^*_{phy} [m^2 (mg Chl a)$^{-1}$] and cellular chlorophyll content in *Chlorella vulgaris* under nutrient-replete conditions (■),1/5 (●) and 1/10 (■) nitrogen-limited conditions, in *Nannochloropsis salina* (▽), *Skeletonema costatum* (△) and *C. meneghiniana* under nutrient-replete conditions (X), in *C. vulgaris* (★) and *Phaeodactylum tricornutum* under fluctuating light (▼) conditions in combination with nitrogen limitation (●). Figure modified according to [29].

It should be emphasized that the strong change in light absorption efficiency by a factor of about 3 occurs exactly in that range where the truncated antennae strategy is operating, that is, the absorption efficiency of the remaining chlorophyll is strongly enhanced in the cells with reduced antenna. Therefore, a clear analysis of the quantitative effects of chlorophyll-reduced cells can only be done under conditions of equal energy uptake per volume. Such an analysis was carried out by Schramm et al. [54], showing that the biomass-specific volumetric respiration is also increased by antenna-truncated cells. Here, two processes counteract the positive effect of the reduced antenna size under natural light conditions: Increased respiration and enhanced chlorophyll absorptivity. Overall, this led to the conclusion that pale cells perform better than wild-type cells only in a small window of conditions: Long periods of oversaturating light combined with low temperature during short nights. Under real conditions, however, the light is only consistently high under warm night conditions. In addition, the study clearly shows that the best wild-type strain has a better performance than the mutants, because the wild type adapts itself to high light conditions. Adaptation to high light conditions, however, is not only accompanied by antenna size reduction but also by improved photoprotection and an enhanced capacity of dark reactions. Therefore, successful attempts have also been made to isolate genetic mutants with enhanced resistance to reactive oxygen species [52, 56]. However, the applicability of these strains in an industrial process has not yet been proven.

6.4.2 Rubisco engineering

Rubisco is considered as the rate-limiting step in the energy conversion from light to sugar in plants [57] and microalgae [26]. The reason for the low carboxylation capacity of the cells is the kinetic feature of this enzyme. With an in vivo turnover rate of about $0.3 \ s^{-1}$, Rubisco is about 50–60 times slower than the turnover rate of the rate-limiting step in photosynthetic electron transport [26]. As a consequence of this kinetic imbalance, cells accumulate high amounts of Rubisco protein inside the chloroplast stroma, reaching concentrations of up to $240 \ mg \ mL^{-1}$ [58]. Manipulations of Rubisco, therefore, aim to increase its concentration, its specificity, its kinetics (k_{cat}) and its activation status at low CO_2 concentrations [59]. As Rubisco enzymes in algae and plants are the result of a long phylogenetic optimization process, it is hard to improve its catalytic activity and specificity. In addition, increased amounts of this protein failed to increase photosynthesis in *Chlamydomonas* [60]. By contrast, gene overexpression of Rubisco activase, which catalyzes the activation of the enzyme in response to changed CO_2 concentrations, resulted in higher photosynthetic activity, growth and biomass accumulation in *Nannochloropsis* cells at air-level CO_2 [61]. One of the major constraints of Rubisco in algal cultures is its sensitivity to oxygen, especially at elevated temperatures due to the different solubility of CO_2 and O_2. Under high O_2/CO_2 ratios, the enzyme starts the process of photorespiration, which leads to a significant loss of CO_2 or organic carbon through the excretion of glycolate [62]. Besides Rubisco and activase, other Calvin cycle enzymes and enzymes in CO_2 transport or pyrenoid organization have also been used as targets for improving the CO_2 assimilation process. Pyrenoid engineering, for example, has been proposed as a new mechanism to enhance crop yields in the future [63]. However, none of these improvements could so far be transferred to the next level of technical readiness through system biology and the state of the art remains at the laboratory scale.

6.4.3 The case study of *Chlorella ohadii*

The isolation of the green alga *Chlorella ohadii* from a desert crust has opened new perspectives for algal biotechnology. This alga shows the highest division rates of an autotrophic eukaryotic cell ever reported of up to 4 h per cell cycle. However, the fastest growth rates were not accompanied by highest photosynthetic rates [64]. The reasons for this extremely high growth rate are not completely clear. Firstly, the cells show extreme resistance to photoinhibition when exposed to ultra-high light intensities. The authors propose a specific electron recombination pathway in PS2 for resistance to photo damage that effectively prevents the formation of harmful singlet oxygen [65]. A second difference to other algae is the observation of multi-phasic growth kinetics, with phases of high growth rates exhibiting an unusual metabolic status. In this context, high concentrations of polyamines are suggested as candidate signals for metabolic shifts and culture synchronization [64]. Furthermore, recent data indicate that under excessive reduction pressure, *C. ohadii* possesses an unusual redox-triggered potential to accumulate storage products that can later be transformed into biomass. However, the underlying mechanism is not yet clear [66].

6.4.4 Using metabolites of photosynthetic pathways as products

The excretion of metabolites directly from algal biosynthetic pathways can be used biotechnologically and is referred to as milking [67]. For instance, in the green alga *Botryococcus braunii*, hydrocarbons are directly excreted into the medium and can be easily recovered as a final product for gasoline production [68]. There are, however, also less noticed metabolic intermediates being excreted by the cell that may be used for potential biotechnological applications. For instance, under certain circumstances, glycolate is released into the medium as a product of photorespiration and may serve as organic carbon source in chemical applications [69]. The biosynthetic origin of this glycolate production is the double function of Rubisco as carboxylase or oxygenase [70], with the balance of both reactions depending on temperature and the CO_2/O_2 ratio in the environment. High temperatures and low CO_2/O_2 ratios favor the oxygenation reaction, leading to the production of one molecule of glycolate and glycerate each. Glycerate is reduced in the Calvin cycle and can be used either for the regeneration of Ribulose-1,5-bisphosphate or for sugar biosynthesis. Since glycolate is toxic, the cells can only survive if they metabolize glycolate in the so-called C2 cycle, or excrete the compound into the medium [69, 71, 72]. For this reason, a high yield of excreted glycolate is achieved by inactivation of the C2 cycle, for instance, by blocking the first glycolate-metabolizing enzyme, glycolate dehydrogenase [73]. Under these conditions, a continuous production of glycolate can be achieved with a 2:1 ratio of carboxylation/oxygenation. In this case, two fixed carbon atoms are excreted as glycolate with a zero net carbon gain for the cell, that is, the photosynthetically fixed organic carbon accumulates in the medium without any cell growth. Such carbon loss can be reduced if the so-called carbon concentrating mechanisms (CCM) are activated [62]. CCMs increase the availability of CO_2 at the Rubisco, resulting in an effective suppression of photorespiration. Recently, a stable production of glycolate without growth and nutrient consumption at an appropriate ratio of CO_2/O_2 in the air stream, combined with inhibition of the C2 cycle and the CCM mechanism by the addition of a carboanhydrase inhibitor could be shown [74].

Glycolate concentrations of approx. 3 g L^{-1} as achieved in the algae medium are sufficient for direct use in anaerobic fermentation to convert glycolate into methane [75]. The mixture of methane and CO_2 resulting from the fermentation of glycolate can then be used directly for further processing into clean methane, which is applicable for C1 chemistry without extensive purification. Harvesting and refinement are easy to manage with this method and have only a minor impact on the energy balance of the overall process. Surprisingly, these glycolate excreting cells showed high photosynthetic activity without significant growth for weeks. Fourier transform infrared spectroscopy analysis revealed that the macromolecular composition is stable during glycolate production and shows no accumulation of storage products during the light phase, as observed under limiting conditions (see above). Furthermore, about 80% of the fixed carbon can finally be found in the glycolate accumulated in the medium [74]. The physiological stability of the cells under glycolate excreting conditions suggests that the application of the process can be stabilized for an entire growing season. Also, the lacking cell growth opens up the perspective for an alternate bioreactor concept with the cells as living catalysts – possibly engineered for optimized photosynthetic performance. Advantages of such a concept include reduced nutrient consumption, high CO_2 use efficiency, as well as the low effort required for cell harvesting and processing. The development of those new types of photobioreactors is now under work.

6.4.5 Conclusions

Although photosynthesis is the energetic driver of growth, it has become evident over the last 10 years that it is not directly linked to growth. Energy conversion from light to biomass or products is rather uncoupled from the activity of the photosynthetic apparatus by many metabolic control steps that have not yet been elucidated at the mechanistic level. Critical regulatory elements for the efficiency of biomass formation per absorbed photon are photon losses through heat dissipation, NPQ, alternative electron transport pathways and respiration, as well as the pattern for carbon allocation to macromolecules. This is stoichiometrically evident because the reduction degree of biomass and the ATP consumption of the different biosynthetic pathways strongly influence the energy balance from photon to product. Under outdoor reactor conditions, cells are exposed to dynamic and unpredictable changes in light, temperature and nutrient conditions (mainly CO_2). Therefore, the real situation of the cells in an outdoor reactor is far from permanently optimal homoeostatic conditions. This induces systemic acclimation costs, which are an inevitable load on the photon to biomass conversion efficiency.

6.5 Outlook: future perspective and economic feasibility

It is now accepted that algal biomass production through sophisticated photobio-reactor management has a great potential for providing high-value products which cannot otherwise be produced in a climate-friendly manner by synthetic chemistry. The large variety of potential products includes pigments, polyunsaturated fatty acids, sterols, phenolic compounds and peptides with a healthcare application. The unsolved challenge is to scale up to production areas of millions of km^2 in order to make a significant contribution to the global demand for organic carbon by CO_2 se-questration. This goal can only be achieved if the efficiency of photon to biomass conversion can be improved by at least an order of magnitude. Although the biolo-gical potential for algal biomass production has not yet been completely exploited, current research does open up only limited perspectives for a real breakthrough. Considerable progress can only be achieved if the following bottleneck reactions can be overcome: (1) reduction of energy for mixing and harvesting, (2) efficient re-cycling of nutrients (mainly, N, P, Mg, K), (3) a real CO_2 sequestration balance, (4) a complete conversion of algal biomass into products of interest and (5) the reduction of investment costs for large-scale photobioreactor plants. At best, these restrictions can be overcome by preventing biomass accumulation and using the living cells as photocatalysts in cell factories. This can be achieved, for example, by a "milking" approach, for example, using glycolate as an excretion product. Such a carbon-based compound can then be used either in conventional technical biotechnology or in chemical conversion facilities to produce organic carbon for the chemical in-dustry in realistic amounts to replace crude oil.

Abbreviations

a^*_{phy}	Chl a in vivo absorption coefficient (m^2 mg^{-1} Chl a)
CCM	Carbon concentrating mechanisms
E_K	Inclination point in the light saturation curve (μmol quanta m^{-2} s^{-1})
ETR	Electron transport rate (mmol e^- mg^{-1} Chl a h)
Φ_{alt}	Quantum yield of alternative ETR
Φ_{NPQ}	Quantum yield of the xanthophyll cycle-dependent NPQ
Φ_{Onet}	Quantum yield of the photosynthetic oxygen production
Φ_C	Quantum yield of the photosynthetic carbon assimilation
$\Phi_{f,D}$	Quantum yield of the fluorescence-based heat emission
Φ_{resp}	Quantum yield of the e^- loss by mitochondrial respiration
NPQ	Non-photochemical quenching (dimensionless)
P–I curve	Photosynthetic irradiance curve (incident light)
RC	Photosynthetic reaction center
PS2	Photosystem 2

Q_{phar} Amount of absorbed PAR (μmol quanta mg^{-1} Chl a h)
PAR   Photosynthetic active radiation (μmol quanta m^{-2} s^{-1})
PQ   Photosynthetic quotient (mmol O$_2$ released per mmol CO$_2$ assimilated)

References

[1]   Chisti Y. Biodiesel from microalgae. Biotechnol Adv 2007, 25, 294–306.
[2]   Brennan L, Owende P. Biofuels from microalgae – A review of technologies for production, processing, and extractions of biofuels and co-products. Renew Sustain Energy Rev 2010, 14, 557–77.
[3]   Posten C. Design principles of photo-bioreactors for cultivation of microalgae. Eng Life Sci 2009, 9, 165–77.
[4]   Fabris M, Abbriano RM, Pernice M, Sutherland DL, Commault AS, Hall CC, et al. Emerging technologies in algal biotechnology: Toward the establishment of a sustainable, algae-based bioeconomy. Front Plant Sci 2020, 11.
[5]   Jorquera O, Kiperstok A, Sales EA, Embiruçu M, Ghirardi ML. Comparative energy life-cycle analyses of microalgal biomass production in open ponds and photobioreactors. Biores Techn 2010, 101, 1406–13.
[6]   Tredici MR, Bassi N, Prussi M, Biondi N, Rodolfi L, Chini Zittelli G, et al. Energy balance of algal biomass production in a 1-ha "Green Wall Panel" plant: How to produce algal biomass in a closed reactor achieving a high Net Energy Ratio. Appl Energy 2015, 154, 1103–11.
[7]   Work VH, D'Adamo S, Radakovits R, Jinkerson RE, Posewitz MC. Improving photosynthesis and metabolic networks for the competitive production of phototroph-derived biofuels. Cur Opin Biotech 2012, 23, 290–7.
[8]   Vecchi V, Barera S, Bassi R, Dall'Osto L. Potential and challenges of improving photosynthesis in algae. Plants 2020, 9, 67.
[9]   Su W, Jakob T, Wilhelm C. The impact of nonphotochemical quenching of fluorescence on the photon balance in diatoms under dynamic light conditions. J Phycol 2012, 48, 336–46.
[10] Kunath C, Jakob T, Wilhelm C. Different phycobilin antenna organisations affect the balance between light use and growth rate in the cyanobacterium Microcystis aeruginosa and in the cryptophyte Cryptomonas ovata. Photosynth Res 2012, 111, 173–83.
[11] Wilhelm C, Jakob T. From photons to biomass and biofuels: Evaluation of different strategies for the improvement of algal biotechnology based on comparative energy balances. Appl Microbiol Biotechnol 2011, 92, 909–19.
[12] Del Rio-Chanona EA, Liu J, Wagner JL, Zhang D, Meng Y, Xue S, et al. Dynamic modeling of green algae cultivation in a photobioreactor for sustainable biodiesel production. Biotechnol Bioeng 2018, 115, 359–70.
[13] Stephenson PG, Moore CM, Terry MJ, Zubkov MV, Bibby TS. Improving photosynthesis for algal biofuels: Toward a green revolution. Trends Biotechnol 2011, 29, 615–23.
[14] Benedetti M, Vecchi V, Barera S, Dall'Osto L. Biomass from microalgae: The potential of domestication towards sustainable biofactories. Microbial Cell Factories 2018, 17, 173.
[15] Georgianna DR, Mayfield SP. Exploiting diversity and synthetic biology for the production of algal biofuels. Nature 2012, 488, 329–35.
[16] Kroth PG, Bones AM, Daboussi F, Ferrante MI, Jaubert M, Kolot M, et al. Genome editing in diatoms: Achievements and goals. Plant Cell Rep 2018, 37, 1401–8.
[17] Dent RM, Han M, Niyogi KK. Functional genomics of plant photosynthesis in the fast lane using Chlamydomonas reinhardtii. Trends Plant Sci 2001, 6, 364–71.

[18] Gilbert M, Wilhelm C, Richter M. Bio-optical modelling of oxygen evolution using in vivo fluorescence: Comparison of measured and calculated photosynthesis/irradiance (P-I) curves in four representative phytoplankton species. J Plant Physiol 2000, 57, 307–14.

[19] Sato T, Yamada D, Hirabayashi S. Development of virtual photobioreactor for microalgae culture considering turbulent flow and flashing light effect. Energy Convers Manag 2010, 51, 1196–201.

[20] Hagemann M, Fernie AR, Espie GS, Kern R, Eisenhut M, Reumann S, et al. Evolution of the biochemistry of the photorespiratory C2 cycle. Plant Biol 2013, 15, 639–47.

[21] Gilbert M, Domin A, Becker A, Wilhelm C. Estimation of primary productivity by chlorophyll a in vivo fluorescence in freshwater phytoplankton. Photosynthetica 2000, 38, 111–26.

[22] Asada K. The water-water cycle as alternative photon and electron sinks. Philos Trans R Soc Lond B Biol Sci 2000, 355, 1419–31.

[23] Miyake C. Alternative electron flows (water-water cycle and cyclic electron flow around PSI) in photosynthesis: Molecular mechanisms and physiological functions. Plant Cell Physiol 2010, 51, 1951–63.

[24] Whitney SM, Houtz RL, Alonso H. Advancing our understanding and capacity to engineer nature's CO2-Sequestering Enzyme, Rubisco. Plant Physiol 2011, 155, 27–35.

[25] Raines CA. Increasing photosynthetic carbon assimilation in C3 plants to improve crop yield: Current and future strategies. Plant Physiol 2011, 155, 36–42.

[26] Wilhelm C, Selmar D. Energy dissipation is an essential mechanism to sustain the viability of plants: The physiological limits of improved photosynthesis. J Plant Physiol 2011, 168, 79–87.

[27] Foyer CH, Shigeoka S. Understanding oxidative stress and antioxidant functions to enhance photosynthesis. Plant Physiol 2011, 155, 93–100.

[28] Pospíšil P. Production of reactive oxygen species by photosystem ii as a response to light and temperature stress. Front Plant Sci 2016, 7.

[29] Blache U, Jakob T, Su W, Wilhelm C. The impact of cell-specific absorption properties on the correlation of electron transport rates measured by chlorophyll fluorescence and photosynthetic oxygen production in planktonic algae. Plant Physiol Biochem 2011, 49, 801–8.

[30] Foyer CH, Neukermans J, Queval G, Noctor G, Harbinson J. Photosynthetic control of electron transport and the regulation of gene expression. J Exp Bot 2012, 63, 1637–61.

[31] Goss R, Ann Pinto E, Wilhelm C, Richter M. The importance of a highly active and delta pH-regulated diatoxanthin epoxidase for the regulation of the PS II antenna function in diadinoxanthin cycle containing algae. J Plant Physiol 2006, 63, 1008–21.

[32] Goss R, Lepetit B. Biodiversity of NPQ. J Plant Physiol 2015, 172, 13–32.

[33] Schaller S, Latowski D, Jemioła-Rzemińska M, Dawood A, Wilhelm C, Strzałka K, et al. Regulation of LHCII aggregation by different thylakoid membrane lipids. Biochim Biophys Acta (Bioenergetics) 2011, 1807, 326–35.

[34] Jahns P, Latowski D, Strzalka K. Mechanism and regulation of the violaxanthin cycle: The role of antenna proteins and membrane lipids. Biochim Biophys Acta (Bioenergetics) 2009, 1787, 3–14.

[35] Sharma J, Kumar SS, Bishnoi NR, Pugazhendhi A. Enhancement of lipid production from algal biomass through various growth parameters. J Mol Liqu 2018, 269, 712–20.

[36] Lv J-M, Cheng L-H, Xu X-H, Zhang L, Chen H-L. Enhanced lipid production of Chlorella vulgaris by adjustment of cultivation conditions. Biores Technol 2010, 101, 6797–804.

[37] Aziz MMA, Kassim KA, Shokravi Z, Jakarni FM, Liu HY, Zaini N, et al. Two-stage cultivation strategy for simultaneous increases in growth rate and lipid content of microalgae: A review. Renew Sustain Energy Rev 2020, 119, 109621.

[38] Ryu KH, Kim B, Lee JH. A model-based optimization of microalgal cultivation strategies for lipid production under photoautotrophic condition. Comput Chem Eng 2019, 121, 57–66.

[39] Jakob T, Wagner H, Stehfest K, Wilhelm C. A complete energy balance from photons to new biomass reveals a light- and nutrient-dependent variability in the metabolic costs of carbon assimilation. J Exp Bot 2007, 58, 2101–12.

[40] Wagner H, Jakob T, Fanesi A, Wilhelm C. Towards an understanding of the molecular regulation of carbon allocation in diatoms: The interaction of energy and carbon allocation. Phil Trans R Soc B 2017, 372, 20160410.

[41] Fanesi A, Wagner H, Becker A, Wilhelm C. Temperature affects the partitioning of absorbed light energy in freshwater phytoplankton. Freshw Biol 2016, 61, 1365–78.

[42] Fanesi A, Wagner H, Wilhelm C. Phytoplankton growth rate modelling: Can spectroscopic cell chemotyping be superior to physiological predictors? Proc R Soc B 2017, 284, 20161956.

[43] Jebsen C, Norici A, Wagner H, Palmucci M, Giordano M, Wilhelm C. FTIR spectra of algal species can be used as physiological fingerprints to assess their actual growth potential. Physiol Plant 2012, 146, 427–38.

[44] Wagner H, Jebsen C, Wilhelm C. Monitoring cellular C:N ratio in phytoplankton by means of FTIR-spectroscopy. J Phycol 2019, 55, 543–51.

[45] De Mooij T, Janssen M, Cerezo-Chinarro O, Mussgnug JH, Kruse O, Ballottari M, et al. Antenna size reduction as a strategy to increase biomass productivity: A great potential not yet realized. J Appl Phycol 2015, 27, 1063–77.

[46] Cazzaniga S, Dall'Osto L, Szaub J, Scibilia L, Ballottari M, Purton S, et al. Domestication of the green alga Chlorella sorokiniana: Reduction of antenna size improves light-use efficiency in a photobioreactor. Biotechnol Biofuels 2014, 7, 157.

[47] Perin G, Bellan A, Segalla A, Meneghesso A, Alboresi A, Morosinotto T. Generation of random mutants to improve light-use efficiency of Nannochloropsis gaditana cultures for biofuel production. Biotechnol Biofuels 2015, 8, 161.

[48] Kwon J-H, Bernát G, Wagner H, Rögner M, Rexroth S. Reduced light-harvesting antenna: Consequences on cyanobacterial metabolism and photosynthetic productivity. Algal Res 2013, 2, 188–95.

[49] Norsker N-H, Barbosa MJ, Vermuë MH, Wijffels RH. Microalgal production – A close look at the economics. Biotechnol Adv 2011, 29, 24–7.

[50] Ort DR, Zhu X, Melis A. Optimizing antenna size to maximize photosynthetic efficiency. Plant Physiol 2011, 155, 79–85.

[51] Melis A. Solar energy conversion efficiencies in photosynthesis: Minimizing the chlorophyll antennae to maximize efficiency. Plant Sci 2009, 177, 272–80.

[52] Polle JEW, Kanakagiri S-D, Melis A. tla1, a DNA insertional transformant of the green alga Chlamydomonas reinhardtii with a truncated light-harvesting chlorophyll antenna size. Planta 2003, 217, 49–59.

[53] Dall'Osto L, Cazzaniga S, Guardini Z, Barera S, Benedetti M, Mannino G, et al. Combined resistance to oxidative stress and reduced antenna size enhance light-to-biomass conversion efficiency in Chlorella vulgaris cultures. Biotechnol Biofuels 2019, 12, 221.

[54] Schramm A, Jakob T, Wilhelm C. The impact of the optical properties and respiration of algal cells with truncated antennae on biomass production under simulated outdoor conditions. Cur Biotechnol 2016, 5, 142–53.

[55] Baird ME, Timko P, Wu L. The effect of packaging of chlorophyll within phytoplankton and light scattering in a coupled physical–biological ocean model. Mar Freshw Res 2007, 58, 966–81.

[56] Schierenbeck L, Ries D, Rogge K, Grewe S, Weisshaar B, Kruse O. Fast forward genetics to identify mutations causing a high light tolerant phenotype in Chlamydomonas reinhardtii by whole-genome-sequencing. BMC Genomics 2015, 16, 57.

[57] Farquhar GD, Von Caemmerer S, Berry JA. A biochemical model of photosynthetic CO_2 assimilation in leaves of C3 species. Planta 1980, 149, 78–90.

[58] Wildner GF. Ribulose-1,5-bisphosphate carboxylase-oxygenase: Aspects and prospects. Physiol Plant 1981, 52, 385–9.

[59] Parry MAJ, Andralojc PJ, Mitchell RAC, Madgwick PJ, Keys AJ. Manipulation of Rubisco: The amount, activity, function and regulation. J Exp Bot 2003, 54, 1321–33.

[60] Li C, Salvucci ME, Portis AR. Two residues of rubisco activase involved in recognition of the Rubisco substrate. J Biol Chem 2005, 280, 24864–9.

[61] Wei L, Wang Q, Xin Y, Lu Y, Xu J. Enhancing photosynthetic biomass productivity of industrial oleaginous microalgae by overexpression of RuBisCO activase. Algal Res 2017, 27, 366–75.

[62] Hagemann M, Kern R, Maurino VG, Hanson DT, Weber APM, Sage RF, et al. Evolution of photorespiration from cyanobacteria to land plants, considering protein phylogenies and acquisition of carbon concentrating mechanisms. J Exp Bot 2016, 67, 2963–76.

[63] Freeman Rosenzweig ES, Xu B, Kuhn Cuellar L, Martinez-Sanchez A, Schaffer M, Strauss M, et al. The Eukaryotic CO2-Concentrating Organelle Is Liquid-like and Exhibits Dynamic Reorganization. Cell 2017, 171, 148–162.e19.

[64] Treves H, Murik O, Kedem I, Eisenstadt D, Meir S, Rogachev I, et al. Metabolic flexibility underpins growth capabilities of the fastest growing alga. Cur Biol 2017, 27, 2559–2567.e3.

[65] Treves H, Raanan H, Kedem I, Murik O, Keren N, Zer H, et al. The mechanisms whereby the green alga Chlorella ohadii, isolated from desert soil crust, exhibits unparalleled photodamage resistance. New Phytol 2016, 210, 1229–43.

[66] Treves H, Siemiatkowska B, Luzarowska U, Murik O, Fernandez-Pozo N, Moraes TA, et al. Multi-omics reveals mechanisms of total resistance to extreme illumination of a desert alga. Nature Plants 2020, 6, 1031–43.

[67] Hejazi MA, Wijffels RH. Milking of microalgae. Trends Biotechnol 2004, 22, 189–94.

[68] Jackson BA, Bahri PA, Moheimani NR. Repetitive non-destructive milking of hydrocarbons from Botryococcus braunii. Renew Sustain Energy Rev 2017, 79, 1229–40.

[69] Vílchez C, Galván F, Vega JM. Glycolate photoproduction by free and alginate-entrapped cells of Chlamydomonas reinhardtii. Appl Microbiol Biotechnol 1991, 35, 716–9.

[70] Portis AR, Parry MAJ. Discoveries in Rubisco (Ribulose 1,5-bisphosphate carboxylase/oxygenase): A historical perspective. Photosynth Res 2007, 94, 121–43.

[71] Dellero Y, Jossier M, Schmitz J, Maurino VG, Hodges M. Photorespiratory glycolate–glyoxylate metabolism. J Exp Bot 2016, 67, 3041–52.

[72] Rademacher N, Kern R, Fujiwara T, Mettler-Altmann T, Miyagishima S, Hagemann M, et al. Photorespiratory glycolate oxidase is essential for the survival of the red alga Cyanidioschyzon merolae under ambient CO_2 conditions. J Exp Bot 2016, 67, 3165–75.

[73] Günther A, Jakob T, Goss R, Koenig S, Spindler D, Raebiger N, et al. Methane production from glycolate excreting algae as a new concept in the production of biofuels. Bioresour Technol 2012, 121, 454–7.

[74] Taubert A, Jakob T, Wilhelm C. Glycolate from microalgae: An efficient carbon source for biotechnological applications. Plant Biotechnol J 2019, 17, 1538–46.

[75] Günther S, Becker D, Hübschmann T, Reinert S, Kleinsteuber S, Müller S, et al. Long-term biogas production from glycolate by diverse and highly dynamic communities. Microorganisms 2018, 6, 103.

Rosa Rosello Sastre, Clemens Posten

7 Optimization of photosynthesis by reactor design

Abstract: This chapter describes the process of designing photobioreactors (PBR). The most important thing is the mindset that a PBR transforms the given environmental conditions into favorable ones for the microalgae inside the reactor with respect to light, dissolved gases, nutrients and temperature. The tools for calculating this transformation are briefly outlined. The conditions for the microalgae are described in terms of kinetics in the first part of the chapter. Kinetics considers the relation between concentrations and turnover rates, giving the total amounts of the different energy and mass fluxes to supply the cells. In the central part of this treatise, a range of current solutions for PBR design are described and discussed. This includes numbers for regularly achieved productivities but also various pros and cons. Actually, there are still a lot of disadvantages that need to be stated. Innovative approaches to cope with these challenges are outlined. This supports the expectation that sustainable and economically viable microalgae production using closed PBRs will be possible in the near future.

7.1 Goal

Photobioreactors (PBRs) are the production engine of microalgal biomass. The first modern scientific investigations with respect to engineering go back to [1, 2]. Many different designs have been reported and patented; nevertheless, there is no concordance on the most efficient type. Flat plate and tubular reactors are widely used and are the workhorses of production. The first design criteria were elaborated based on engineering considerations [3] and later were increasingly based on model approaches for energy and material balances [4, 5]. Reliable data based on outdoor pilot-scale reactors came from [6, 7]. As the environment changes from location to location and cost issues change from product to product, the optimum reactor design also has to change, following product- and site-specific issues. Despite the employment of various computer simulations, there are no clear statements on design rules for PBRs. The stage is still open for more or less well-elaborated new ideas.

This chapter will provide a basic construction framework to design and assess PBRs. The starting point in every case is microalgal physiology. An architect designs a house based on the needs and wishes of a family and the conditions of the environment. In analogy, the engineer needs to understand the physiology of microalgal cells and has to design the reactor as a compromise between these physiological facts, the possibilities of the environment and the needs of the market. The reactor

https://doi.org/10.1515/9783110716979-007

transforms environmental variables such as light intensities and temperature into feasible physiological conditions (Fig. 1).

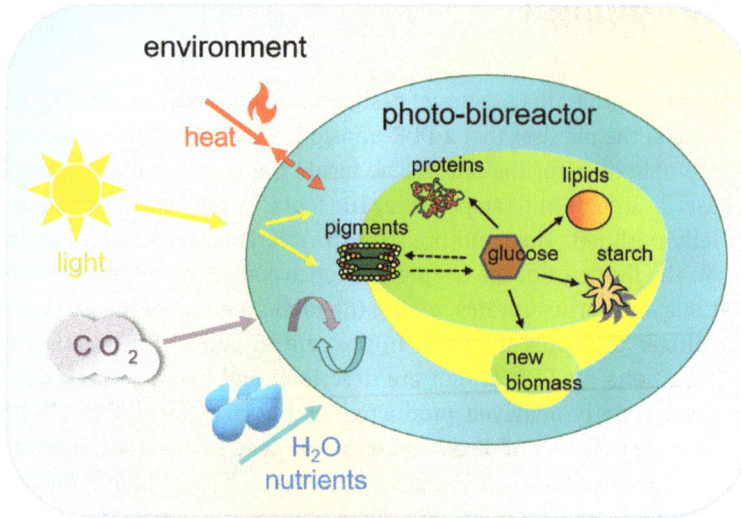

Fig. 1: The inner kernel of reactor design is microalgal physiology. The most important items are kinetic constants providing information about the concentration and fluxes of light, CO_2 and other energy or material items. The reactor itself is a device to convert environmental conditions to optimum physiological conditions.

Reactor design means understanding cell needs and reactor transformations. The art of microalgal production lies in a reasonably simplified description of the growth processes. This can include light and uptake kinetics, growth curves or time-varying intracellular composition. These items need to be coupled with the transport equations of the reactor concerning light and gas transfer. The goal of this chapter is to outline this approach for the most important environmental variables, microalgal kinetics and common reactor types. This allows us to finally assess and optimize photosynthesis with respect to energy efficiency, areal and volumetric productivity.

7.2 Basic background

Every attempt to design PBRs has to start from an understanding of the reactor and the cell as two systems on two scales. The PBR itself has the tasks of transporting energy and materials from the environment to the cell suspension and the cells themselves. The PBR, therefore, needs to be understood as a transport system of matter and energy. In terms of classic reactor systems, it is a "four-phase system," namely biomass (solid), medium including dissolved nutrients (fluid), CO_2, O_2 and N_2 as bubbles (gas), and last but not least photons (light). This transport has to take place in sufficient quantities, but also by keeping a sufficient concentration in the medium to allow for optimum uptake and reaction rates by the cells. The individual cells can be understood as bioreaction systems. The interface between transport and reaction is called kinetics. Understanding the cell's needs on a quantitative basis means basically to know the respective kinetics and their stoichiometric relations. The PBR has to be designed around the cells, as an architect does by first finding out the needs of the family who will live in the apartment. Only then can they assign areas to different purposes and play with light or the shortest connections between rooms. In the following paragraph, this approach is briefly outlined for the most important environmental variables.

7.2.1 Light transfer and the consequences for reactor geometry

The most unique component of PBRs is, of course, the light transfer into the reactor and the absorption by the biomass. The energy of the photons is then converted into biomass with a specific efficiency. This relation can be used to calculate the light demand for achieving a given biomass growth. While in photosynthesis, approximately ten mols of photons are used to fix one mole of carbon, some energy is lost during light respiration, but mainly dark respiration. This delivers an ATP for anabolic synthesis from respiration on starch produced in photosynthesis. After all, the real measurements are between 20 and 30 mole photons per mole biomass (calculated as C1) or, vice versa; 1–2 g biomass needs 1 mol of photons.

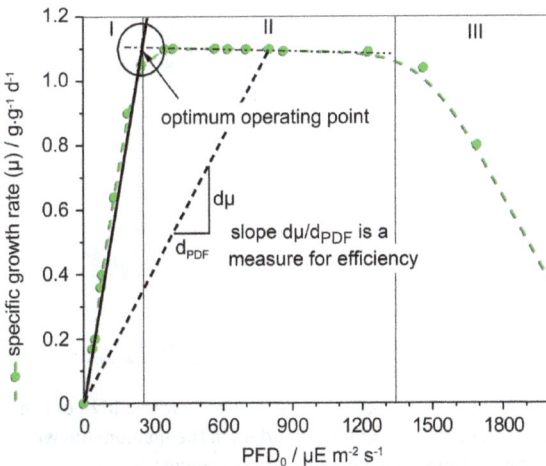

Fig. 2: The growth kinetics of a typical microalga (*Chlorella* sp.) as a function of the specific growth rate µ from light intensity after [8]. The biomass concentration was low allowing for an equal PFD everywhere in the reactor. Measurements were taken after one day of acclimation time.

To get an indication of the light intensity, at which the light should be supplied to the microalgal cells, growth kinetics can be measured. A typical example under ideal conditions is shown in Fig. 2. The specific growth rate $\mu/g \cdot g^{-1} \cdot h^{-1}$ is determined against incident light intensity measured as photon flux density $PFD_0/\mu E \cdot m^{-2} \cdot s^{-1}$, which is understood in the context of this chapter as being in the photosynthetically active range (PAR). During the measurement, it was ensured through low biomass concentrations that this light intensity is the same everywhere in the reactor. In production reactors, light intensity changes inside the reactor mainly along the light path, but also during the daily cycle. Therefore, the evaluation of this figure has to be done in the ideal case as timely and spatial discretization.

Light is scattered in the medium and absorbed by the microalgal cells. This leads, for example, in flat geometries, to an exponential decrease of light intensity along the light path. This means in practice that the cells experience high light conditions close to the illuminated side and lower light conditions on the averted side. As long as the biomass concentration is low, the gradient is low and there are nearly optimum growth conditions everywhere in the reactor. The cells can grow with the maximum specific growth rate depending on incident light intensity. With increasing biomass concentration but constant light-supply, the number of photons per cell becomes smaller leading to a smaller specific growth rate. In sum, the exponential growth curve transforms into a linear growth curve as depicted in Fig. 3. The slope of the biomass curve results next to the light intensity from light conversion efficiency.

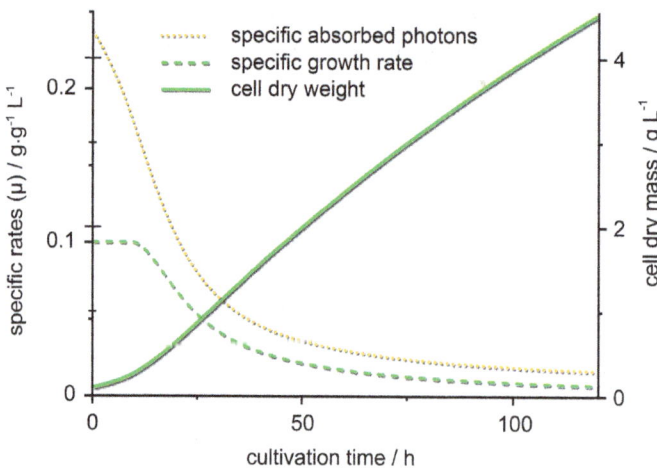

Fig. 3: Typical growth curve of microalgae: At low concentrations, cells can grow with a high or even the maximum growth rate; at high concentrations, all the light is absorbed in the medium but with a reduced biomass-specific photon availability. The specific growth rate consequently goes down. Assuming constant yields (new biomass per photon), the result is not an exponential but a nearly linear growth curve.

This linear growth has several consequences for an optimal working point. At low biomass concentrations, the growth rate is high, but photons are lost in the saturation range or may leave the reactor on the averted side. But higher biomass concentrations lead to increasing gradients. The light path length becomes shorter. The dark part will also increase with the effect of unproductive volume, which nevertheless requires mixing energy. In addition, a low light intensity means low or even negative growth rates due to respiration for maintenance. This has to be avoided by choosing thinner reactors to obtain higher biomass concentrations. As a consequence, the first design rule can be formulated:

Rule 1: A photobioreactor must have at least one geometric dimension that is no longer than the effective light path. Practical dimensions are between 2 and 20 mm.

A second aspect comes up while operating PBRs in sunlight. This is about 10 times more intensive than the saturation point during the day, meaning that many photons are absorbed but not used by the algae in case of direct insolation. This is the case, for example, in open ponds. A PBR should consequently be constructed and oriented so that the incident light intensity does not exceed the saturation point, see Fig. 4. This ability of the reactor is called "light dilution," which is realized by "dilution" of the light falling through a higher transparent area of the PBR than the footprint area. This practical implementation is concluded in a second design rule:

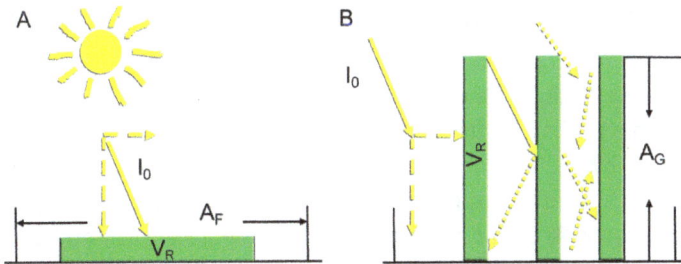

Fig. 4: The effect of light dilution: (A) Light from the Sun impacting in time-varying angles and intensities on the surface (yellow arrows, length corresponding to intensity). Only spatial components (dashed arrows) with a perpendicular (normal) impact on the reactor surface penetrating into the medium which is decisive for light saturation, here shown for an open pond. (B) Vertical reactor reducing the angle, thereby leading to lower light intensity in the first layer of the suspension and helping to reduce light saturation or inhibition (left). In a first approximation, this "dilution" factor is the ratio of the transparent reactor surface A_R (both sides) and the footprint area A_F. A secondary effect on light dilution is mutual shading and surface reflection (middle). Furthermore, diffuse light (dotted) is distributed over the reactor surface by surface and ground reflection. I_0, light intensity/$\mu E \cdot m^{-2} \cdot s^{-1}$; V_R, reactor volume/m^3; A_G, transparent area/m^2; A_F footprint area/m^2.

Rule 2: The active transparent surface area of a PBR should be several times greater than the footprint area. Practical values of this surface/footprint ratio (SFR = A_G/A_F) are in the range of 2–5 for flat plate reactors or more than 10 for tubular reactors.

Trees have the same problem. Here, the total area of the leaves is about 200 times greater than the covered ground area. Nevertheless, the surface area has to be large enough for a given incident light intensity to supply the desired biomass concentration with enough photons to reach a reasonably high specific growth rate according to the kinetics. This is specified in the next rule:

Rule 3: The ratio between the active transparent surface area and the reactor volume (SVR) should be high. Typical values range from 20 to 200 m^{-1}.

Usually, all the transparent surfaces of a PBR – front and back, light and dark – are counted to avoid discussions about considering, for example, the changing positions of the Sun. This rule 3 can to some extent be understood as a combination of rules 1 and 2. All three rules consider the actual biomass concentration in the reactor. Design and operation can therefore only be optimum for a pre-determined biomass concentration. Deviations lead to more or less significant losses in volumetric productivity and energetic efficiency. As this cannot be guaranteed for all strains or environmental conditions, reactor design has to live with compromises.

7.2.2 Gas transfer and its consequences for the energy balance

Next to light, the supply of CO_2 and the removal of O_2 are important transport tasks of PBRs. This happens in a similar way to heterotrophic cultures through air supply via bubbling, in phototrophic cultures optionally enriched with carbon dioxide. The next transport steps are the diffusion of the gases through the gas–liquid interface followed by convective transport supported by mixing. Then both oxygen and carbon dioxide can leave or enter the cells by diffusion.

The need for carbon dioxide can be estimated by the carbon content of the biomass. This can vary between 45% and 50% depending on the macromolecular composition of the cells. Cells rich in carbohydrates also have lower values while the values for lipid-rich cells can exceed the given range. The necessary net transport of CO_2 into the reactor is therefore dependent on the growth rate. It has to be mentioned that the turnover rate in the cells is nearly twice as high, as about half of the fixed carbon is lost by respiration. In phototrophic cultures, the biomass concentration is only about 10% of the value in heterotrophic cultivation. The same holds for growth rates. That means that the necessary gas transfer rates are typically two orders of magnitude lower. Additional mixing, for example by a stirrer, is almost always dispensable. Gassing rates can go down to 0.05 vvm (volume gas per volume liquid and minute), without a serious limitation in carbon supply or mixing efficiency. With respect to energy efficiency, bubbling is an expensive way of gas transfer and mixing. The energy needed for gas compression can easily reach the amount of energy that is fixed in the biomass. So, energy for aeration is one sticking point in achieving algae production for staple food or still less for energy in an economically sensible way.

The next item to adjust gassing is the value of the necessary gas concentration in the medium. To allow for the maximum carbon uptake by the cells, the partial pressure for dissolved CO_2 (pCO_2) is in the range of 0.5–1% saturation. As practical processes are usually light-limited, a value of $pCO_2 = 0.2\%$ is also reported to be sufficient. Actually, the role of pH, the uptake of carbonate or (energy-dependent) intracellular transport mechanisms are reported in references as not comprehensive enough to allow for setting up quantitative kinetics. It has to be said that the value of 0.2% is still five times higher than the current air concentration. Even strong shaking in flasks or heavy aeration cannot support high specific growth rates, but may be sufficient for lab cultures. To reduce aeration rates and make use of feasible and sustainable CO_2 sources, the molar fraction x_{CO2} in the feeding gas has to be increased. Available CO_2 sources are, for example, biogas plants, where methane has either been separated or burned in a thermal power station. Gas from combustion plants usually has a CO_2 fraction of $x_{CO2} = 10\%$ in the gas phase. A difference in the gas phase and the desired partial pressure in the liquid phase is needed for fast diffusion through the gas–liquid interface. The p_{CO2} in the medium results from the balance between CO_2 uptake and CO_2 diffusion. As both parameters may change with changing environmental conditions, for example light available for growth, p_{CO2} has to be controlled. This is usually done by measuring and controlling the pH value via changing the aeration rate or the molar fraction of CO_2. Additional care needs to be taken as other compounds in the medium such as ammonia also have an influence on the pH-value.

Rule 4: For a given gas composition, a PBR should be low to reduce hydraulic pressure at the bottom.

This will reduce the necessary compression of gas and saves thick plastic walls. Rapid drifting through high-pressure gradients may also not be good for the algal cells. This concept has the drawback of complicating gas transfer, as it reduces the duration of stay of the bubbles. More on this point is given in the outlook paragraph.

7.3 Methods involved

Like all chemical and bioreactors, simulation tools are the most important methods during the design process. Commonly they include computational fluid dynamics (CFD) to simulate and optimize gas transfer and fluid movement, as well as the necessary energy input. Gas transfer was studied at the beginning of microalgal photo-bioengineering by [9]. Light transfer simulation is a special case for PBR to observe the light distribution on the surface and inside the medium in dependency on the solar trajectory and design parameters. Beside the gauges of the single reactors, their orientation with respect to the cardinal directions and latitude can also be decided. Results have firstly been published by [10]. Of course, many parameters play a role and these have to be experimentally tested. An exemplary study compares, for example, productivity for reactors with the same surface/footprint area but different heights of the reactor modules [11].

7.3.1 Computational fluid dynamics

Unlike the usual stirred tank reactor, three different axes are to be considered in PBRs. These are the axis with the main gas transport (height of the plate reactor), the scaling axis (width of the plate reactor) and the axis of the light path (thickness). Driven by turbulence between the bright and the dark parts of the reactor, microalgal cells experience a rapid change of light intensity. Their reaction is described as the "intermittent" or "flashing" light effect [12]. The higher the fluctuation frequencies are (e.g., >5 Hz), the better the cells can compensate for the dark parts of a cycle using intracellular stored metabolites (Fig. 5).

Particle tracking options in CFD programs can produce typical results of light/dark cycles, for example Fig. 5. Note that in principle, each cell has its own light "history." Slow light/dark cycles have to be considered as well. These are enforced, for example, by gas exchange vessels without illumination. Unfortunately, they lead to an over-proportional loss in specific growth rate [14]. To support the flashing light effect using vortices with defined frequencies (Dean, Taylor vortices) along the light axis is an option [15].

7.3.2 Light field simulation

The most specific method in the design of PBRs is the simulation of the light field, see Fig. 6. This is not possible with the usual CFD programs, as the light field is more complicated than a vector field. While in fluid dynamics, the fluid velocity at each point in space can be described by a vector, in the case of a light field, photons can come independently from different directions at the same time. The basic idea of light simulation is to calculate single photons as particles according to the optical laws of absorption, scattering and reflection. Like CFD, the geometry has to be represented in the computer including the optical parameters. The evaluation of billions of calculated photons will then deliver a good picture of the light field. Some programs generate these virtual photons using stochastic assumptions according to

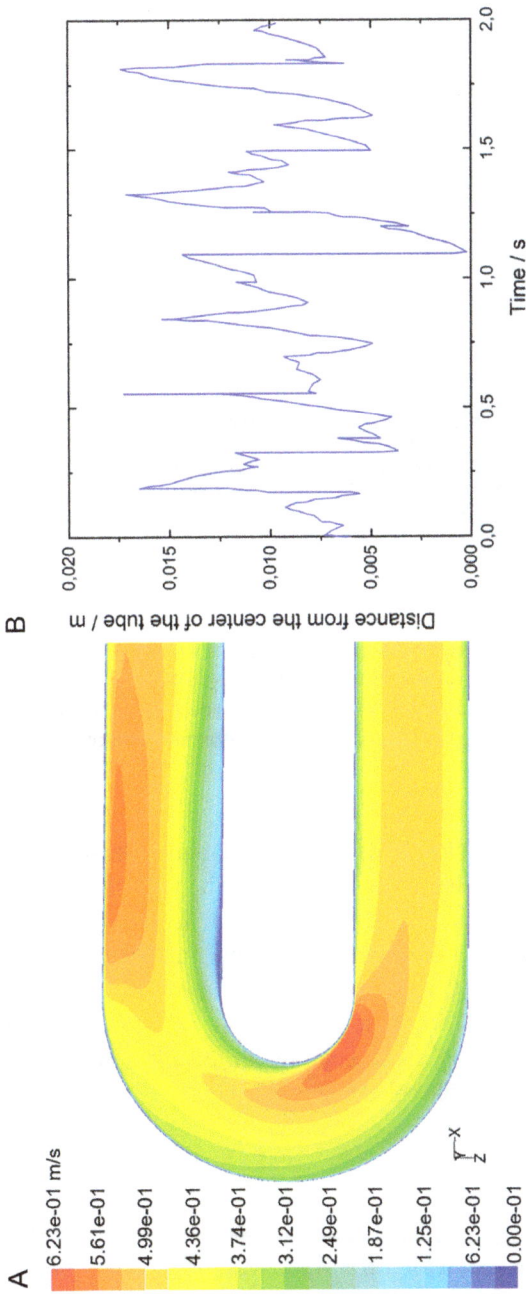

Fig. 5: CFD simulation of a tubular reactor; (A) velocity distribution at the bend — red areas indicate high velocity, which may not be good for the algae, the blue region indicates back flow, which is not good for mixing and is an unnecessary energy dispersion; (B) representation of particle tracking as the distance from the central axis of the tube; light/dark cycles are induced by the light field gradient from outside the tube to the central axis, after [13].

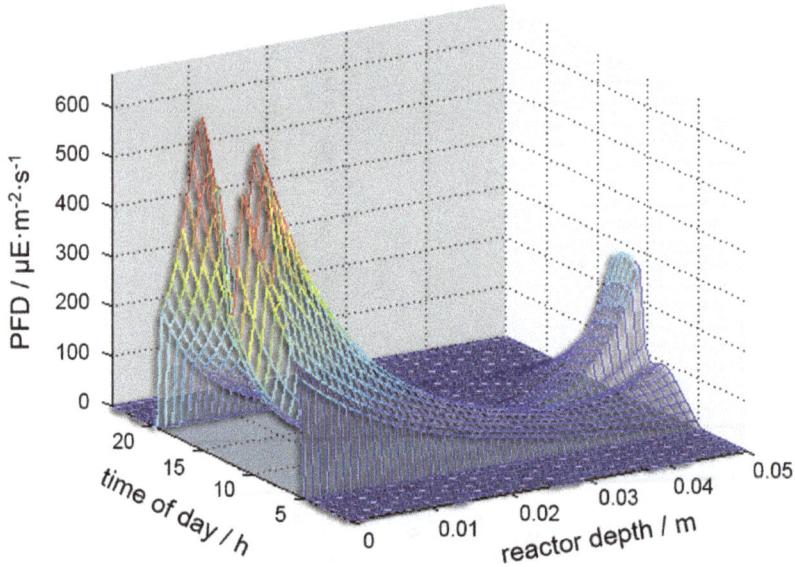

Fig. 6: Light simulation of a flat plate reactor located in Brisbane south of the southern tropic. The Sun rises in the morning shining stronger and stronger on the reactor surface (left). But simultaneously, the sunbeams become steeper. Around noon the light beams are nearly vertical to the ground and they only hit the top of the reactor; light on the surface (left) decreases. The Sun then moves to the other side (right), which is illuminated stronger. In the afternoon, the course is going backwards in the same manner.

the "Monte-Carlo method" [16] or along given grids called the "Euler–Euler method" [17]. In any case, the optical characteristics of the cells need to be known. This can be handled either by measurement of absorption and scattering in a spectrophotom-eter or by estimation via chlorophyll content. Measurements of complete scattering functions have hardly been done for any algae species but lead to the result that forward scattering is the most relevant. Without forward scattering, a few layers of algae cells would be completely non-transparent (e.g., biofilm), while a microalgal suspension is translucent up to a reasonable cell density.

7.3.3 Scale-down and scale-up

While the mentioned programs can predict light/dark cycles, they cannot predict cell reactions. On the other hand, it is impossible to make local measurements of growth rates on a large scale. Here, the concept of scale-down helps [18]. A small PBR with a low biomass concentration and therefore a low light gradient is inter-preted as a small volume element of a large reactor. Light fluctuations occurring natu-rally in the large reactor, while this volume element moves through it, are simulated by flickering illumination in the small model reactor. Here all cells thus have the same light history and the volume is large enough to take samples and to measure physiological parameters. This can be performed by small artificially illuminated reactors, for example, Fig. 12. The results then have to be computed for each volume element of a large reactor.

Scaling for production areas are visualized in Fig. 7. Even without complicated calculations, it is clear that the light axis of a production reactor cannot be changed, as light absorption is fixed. Also, height cannot be increased ad infinitum due to the optimum gas transfer along the height axis. Only the transverse axis can be enlarged, provided that the wholes for aeration and the input of medium or pH adjustment are also multiplied. To fill a large field, plate reactors have to be installed one beside the other according to the needs of light dilution. This principle of "numbering up" is shown in Fig. 7, which is a picture of a real production plant.

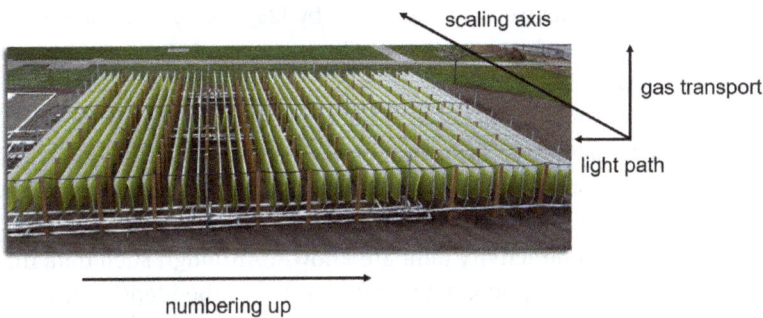

Fig. 7: Production system of the iSeaMC GmbH (Phytolutions Department) in the AUFWIND project at the Jülich National Research Centre, IGB2, Germany [19]. The picture visualizes the three different scaling axes of flat plate reactors.

The last element that counts is the area of transparent reactor surface per ground area with the given constraints. In contrast to a rule of thumb in chemical engineering, the volume is not decisive. Biomass production depends solely on the total light energy falling on the area. A higher volume (e.g., higher or thicker plates) only reduces the biomass concentration and increases the energy required for mixing [20].

7.4 State of the art

Hundreds of patents of closed PBRs have already been published and are still being published nowadays [21]. Many of them are based on good ideas but could not be developed to the point of widespread application or even become a standard. This is due to the different aspects like material, surface coating, mechanical stability, aeration and many others, which cannot be handled on top level from small research groups or start-up companies. Most of the algae biomass is still produced in raceway ponds. Another point leading to diversity is the different constraints with respect to location or product value. Nevertheless, some basic reactor types have been established in the past and can be regarded as commonly accepted standards [4, 22].

7.4.1 The flat plate reactor

The flat plate reactor is very simple and very common. It is basically a cuboid, where the small sides define the light path length. The width is a free design parameter and the only dimension that can be scaled up directly. With respect to gas transfer, the FPR represents a bubble column. The height is determined by the necessary bubble retention time. Early realizations were demonstrated by [23] and [6, 24]. Pictures of these "Green Wall Panels" are shown in Fig. 8. The transparent walls are usually made of PVC plastic bags. It can be seen that the hydrostatic pressure leads to the deformation of the weak material, which makes the application of supporting structures like wire-mesh fences necessary. Even acryl glass of several millimeter thickness suffers from visible deformations. This simple construction leads to quite long light path lengths ranging between 2 and 5 cm.

For flat plate reactors, productivity data are known. Although such data differ strongly between different reports, a general picture can be identified. Basic performance parameters are the volumetric productivity $P_V/g \cdot L^{-1} \cdot h^{-1}$, describing the efficiency of volume usage, and the areal productivity $P_A/g \cdot L^{-1} \cdot h^{-1}$, describing the efficiency of footprint area usage and sunlight, respectively. Further, energetic efficiency is a criterion showing the relation between the captured sunlight energy, the energy required for pumping and chemically bound energy in the microalgal biomass. Photo-conversion efficiency (PCE) describes the conversion of light energy to chemical energy bound in the biomass. Tables 1, 2, and 3. show typical data.

The PCE is typically in the range of 5%, being five times higher than the value, for example, for sugar cane or corn. Note that the mechanical energy effort is in the same order of magnitude as the produced chemical energy bound in the biomass.

Various variations have been proposed to overcome the technical problems. The waterbed reactor is legendary [25]. This consists of basically flat plates swimming in a water basin. The water offers support and cools the reactor during the day. The water body is big enough to equalize the system temperature during the day/night cycles. This basic idea has been further developed by Proviron [26]. Here the water body is realized in a water-filled plastic bag containing the flat plate reactors. Note that in both cases the height of the plates is much lower (e.g., 50 cm) than in the classic flate plate approach.

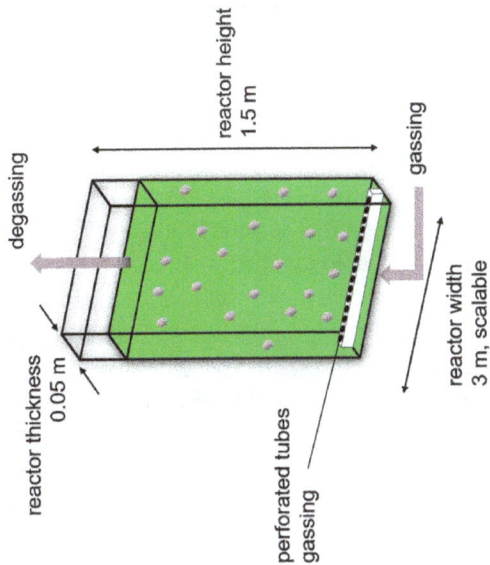

Fig. 8: (A) Schematic sketch of a typical flat plate reactor with aeration tube at the bottom; the gauges are typical values. (B) Picture of a set of flat plate reactors with mesh preventing the deflection of the plastic bag. Tubes and cables are for aeration, off-gas, control wires for mass flow, valves and pumps.

Tab. 1: Typical design aspects of a generic flat plate reactor.

Ground area A_G/m^2	Reactor thickness d_R/m	Reactor volume V_R/m^3	Reactor surface area A_R/m^2	Surface/ volume ratio, $SVR/m^2/m^3$	Surface/footprint ratio, $SFR/m^2/m^2$
100.0	0.01	1.0	500.0	500.0	5.0

Tab. 2: Typical operational parameters of a generic flat plate reactor.

Incident light intensity $I_0 /$ $W{\cdot}m^{-2}$	Daily light energy $E_{phot} /$ $MJ{\cdot}m^{-2}{\cdot}d^{-1}$	Daily mechanical energy $E_{mech} /$ $MJ{\cdot}m^{-2}{\cdot}d^{-1}$	Daily biomass energy $E_{biol} /$ $MJ{\cdot}m^{-2}{\cdot}d^{-1}$	Volumetric power input $P_V / W{\cdot}m^{-3}$	PCE / (MJ/MJ)
115.7	10.0	0.5	0.5	57.9	0.05

Tab. 3: Typical performance data of a generic flat plate reactor.

Heat of combustion biomass $H_x/$ $MJ{\cdot}kg^{-1}$	Areal biomass productivity $P_A/$ $kg{\cdot}m^{-2}{\cdot}day^{-1}$	Volumetric biomass productivity $P_V/$ $kg{\cdot}m^{-3}{\cdot}day^{-1}$	Daily total product/$kg{\cdot}day^{-1}$	Technical energetic efficiency E_{bio}/E_{mech}
20.0	0.0025	2.5	2.5	1.0

A second line of designs begins with the idea of hanging bags. These typically have a circular cross-section that prevents stretching with increasing hydrostatic pressure. They are fixed at the top to keep the bottom clear, for example, to attach the aeration and harvest the tube. This design requires specific kinds of plastics to carry the weight of the water column. Variations include tubes positioned in parallel either isolated or as part of a long plastic sheet, as well as V-shaped arrangements, which can be installed very easily from a coil of the appropriate plastic tube [19, 27].

The most sophisticated development has been presented by [28]. The flat plate is modified in several ways concerning fluid dynamics. Firstly, the rectangular cross section forming the main tube is amended by a second smaller tube. The main tube and the second tube form the riser and the downcomer of an airlift reactor. Therefore, this reactor is called a flat plate airlift (FPA) PBR. The aeration of the main tube leads to a circular medium flow in the axial direction. This solves the problem of low axial mixing times. Secondly, the riser has horizontal baffles inducing horizontal vortices. These lead to short mixing times along the light path, thus supporting the flashing light effect. Additionally, the baffles enhance light transport into the medium layer. The FPA shows the best-reported productivity and biomass concentration values among the flat plate reactors. A drawback is the necessary high aeration rate leading to a relatively high energy demand.

7.4.2 The tubular reactor

One way to achieve even higher surface areas is the employment of tubes. These have two radial axes to allow a short light path from all directions. Typical diameters are 3–5 cm. The tubes are horizontally oriented making the length axis of the tubes the length axis of the reactor. This can be in the range of several hundred meters, as is the case in the first large installations [29]. In general, the surface to volume ratio is better than in the case of FPRs. To take real advantage of light dilution, several tubes are mounted above each other to form a so-called fence as shown in Fig. 9. At the ends, the single tubes are connected by bends or are coupled directly to the pumping and aeration unit. The engineering basis, which is still valid today, was elaborated by [30]. In the early periods of PBR applications, thicker tubes were also built laying in one layer on the ground. As this is against the rule of short light path lengths, they were not successful.

The medium is pumped through the tubes with a typical velocity of $0.3 \, \text{m·s}^{-1}$. This is to enforce mixing, suppress sedimentation or fouling and lead the medium back to the gas exchange. Actually, one thing to be considered is that two axes are used for light transfer leaving only one axis for mass transfer. This cannot be scaled arbitrarily. While flowing through the tube, microalgae use up CO_2 and produce O_2 leading to axial gas concentration gradients. After a retention time of, for example, only 2 min, the partial pressure of oxygen especially can reach an inhibiting level and degassing is necessary. This limits the length of the tubes – or, more precisely, the length/velocity ratio. Degassing is often done using a normal aerated vessel. Fresh gas is injected after the pump forming big bubbles inside the tubes. A gas fraction ("gas hold up") of up to 20% is an advantage as it reduces the axial gas concentration gradient and supports mixing. Some authors propose pulsed aeration. This causes large air bubbles inducing a kind of wave turbulence in the fluid phase. A drawback of tubular reactors is the relatively high power needed for pumping, with wall friction being the origin of power losses in the range of $200 \, \text{W·m}^{-3}$.

Besides the energy demand, costs of the tubes are also quite high. There is still an ongoing debate about the pros and cons of glass compared to PMMA. A big point for glass is its light and chemical resistance leading to a long shelf life. One point for plastics is that it is cheap and can be extruded on-site reducing transport and construction costs considerably. Nevertheless, tubular reactors are so far the best of the closed PBRs for large-scale production, as they can be installed and operated in large volumes comparatively easily while performing with high productivities.

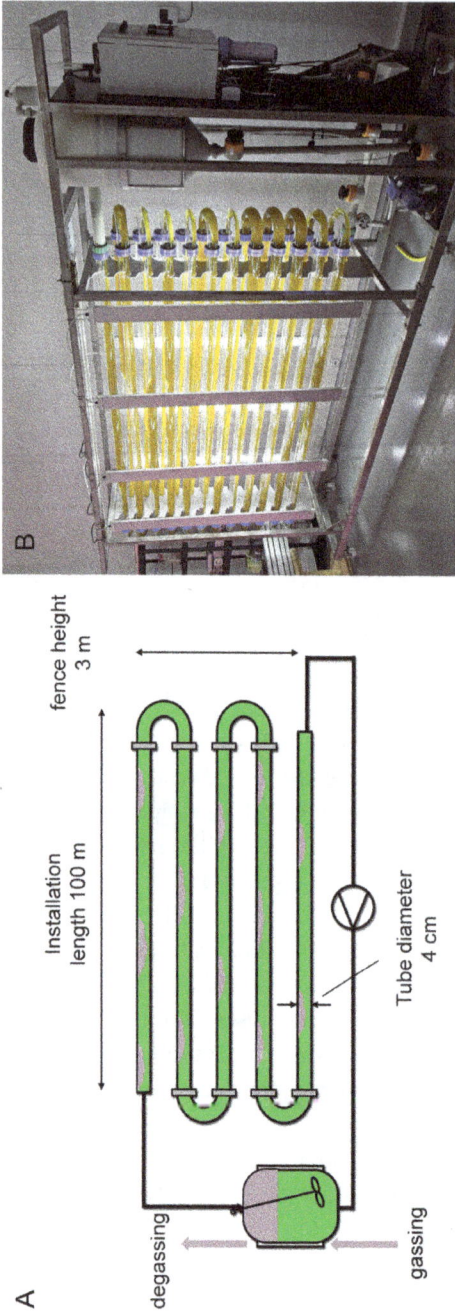

Fig. 9: Tubular reactor: (A) Sketch showing also the peripheral equipment like pump and gas exchange vessel. Remarkably, elongated bubbles supports soft mixing. (B) Pilot-scale test reactor, with gas exchange vessel on the right side (up), pump (down) and electrical equipment – all parts as inexpensive as possible. Elongated bubbles are evoked here by pulse-wise aeration and pumping [31].

7.4.3 Tubular airlift reactor

A hybrid between the flat panel and the tubular reactor is the multiple airlift design [32]. It first became known as the "hanging gardens" and became increasingly popular [33]. Tubes are mounted vertically up to heights of several meters. This is possible, as solid tubes can stand higher hydrostatic pressure. Unlike the bags in some panel designs, the tubes are in pairs in an airlift reactor, where one tube is more aerated than its neighbor leading to a circular flow. This increases axial mixing. Furthermore, the cell suspension can move slowly through the reactor construction.

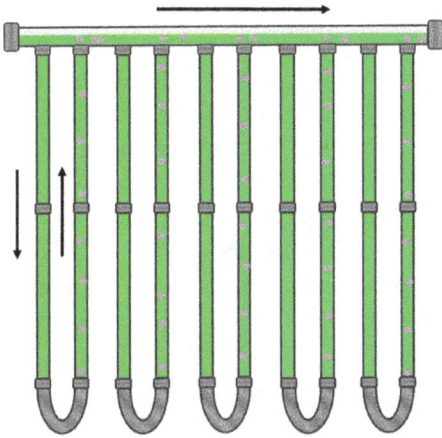

Fig. 10: Sketch of a tubular airlift: Vertical tubes operated in pairs like an airlift reactor induced by different aeration rates in consecutive tubes. As an overlay, a small horizontal flow, for example, from left to right can also be adjusted. This is mimicking a long continuous flow leading eventually to the outwash of contaminants.

7.5 Outlook: future perspectives

From the paragraphs above, it becomes clear that current PBR designs still have their limitations with respect to performance, energy demand, material effort or investment costs. Until now, no reactor design has fulfilled all the demands of the economically viable production of microalgae in the medium value range, for example, for regular food like proteins. In this chapter, some recent trends to tackle these problems are introduced.

7.5.1 Gas transfer by membranes

The supply and removal of dissolved gases by bubbles is an energetically expensive approach because of pressure losses along the supply chain, especially in the pores or holes of the aeration membrane or tube. Furthermore, the energy stored in the surface tension is lost when the bubbles burst at the surface. The surface tension can also be disadvantageous for sensitive algae strains. While the bubbles take over the mixing task, the two processes are not decoupled. An alternative is gas transfer by membranes, as also happens in analogy in microorganisms, plants and animals. Especially the geometry of at least one short dimension supports application of membranes, possibly as one of the transparent surfaces or below the reactor as described in Section 7.2.1 or in Fig. 11.

The membrane separates a gas phase and the liquid phase, that is, the medium. Carbon dioxide diffuses through the membrane into the cell suspension, while oxygen diffuses out into the gas phase. A two-tier reactor is available for lab use [34] (Patent DE 10 2013 015 969 B4; Fig. 11). CO_2 is provided via a lower gas volume, from where it diffuses through a membrane into the upper suspension volume. Oxygen leaves the algal suspension via the surface into the head space. Mixing is provided by a normal orbital shaker. Very good cultivation results showing high growth rates and high cell densities are reported [35].

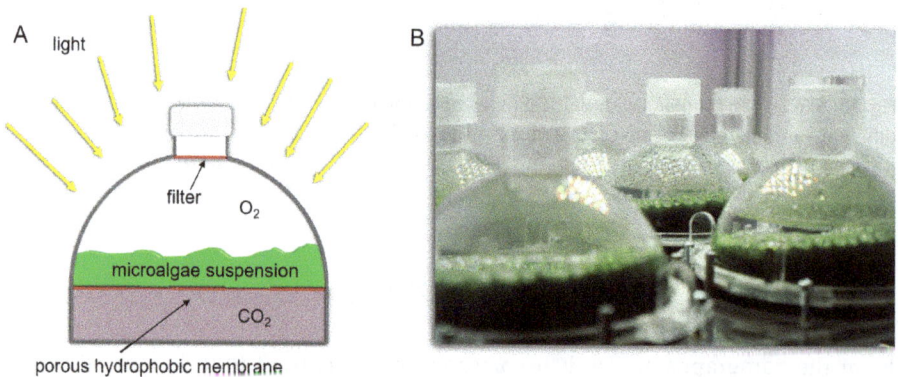

Fig. 11: Membrane lab reactor: (A) sketch, (B) picture in operation [36] (Patent DE 10 2013 015 969 B4). Carbon dioxide supplied from below through a membrane being impermeable for water. Excess oxygen removed through the headspace using a filter into the environment. Mechanical energy input for mixing provided by shaking.

Another good reason for avoiding air bubbles is the use of microalgae in life support systems, for example, during a flight to Mars [37]. The lack of gravity also means a lack of buoyancy; that is, the bubbles will not leave the system and even block the tubes. Here the usage of membranes is mandatory (Fig. 11). It must also be ensured that bubbles are not generated inside the suspension as can be observed in the summer, when algae in lakes take up CO_2 and produce O_2 during sunny days. Increasing the partial pressure then leads to bubble formation. When one mole of carbon dioxide is taken up by the algae and one mole of oxygen is produced, the total partial gas pressures rise, due to the lower solubility of oxygen. If the total partial gas pressure tends to exceed 100%, bubble formation occurs. Carbon dioxide needs hydrophilic membranes while oxygen prefers hydrophobic membranes for diffusion. Furthermore, there are hardly any membranes with high gas transfer coefficients but good selectivity for oxygen. Consequently, the solution lays in decomposition of the problem into two different membranes and two different gas compartments.

7.5.2 Artificial illumination

The ultimate goal of microalgal production is to deliver biomass through carbon-neutral and sustainable processes using sunlight as the sole energy source. For scientific purposes in the lab, artificial light is regularly employed up to pilot scale [38]. Here, LEDs are mainly the light source of choice. They can also be equipped with collimator lenses to produce coherent light comparable with sunlight, Fig. 12. The use of such "modeling reactors" for process development is further treated in Chapter 8. But commercial applications are envisaged as well. The assumed advantages are axenic and trouble-free operation throughout the day and the whole year, constant high quality of the product and compact production plants without the usage of large ground area. Also, the location is not prescribed by sunshine, as long as cheap electrical energy is provided. In combination with photovoltaics, additional advantages like higher spectral and intensity range in light usage, combined with an avoidance of saturation or dark volumes in the reactor become decisive. This is sensible to some extent, as microalgae not only have photosynthesis as a unique selling point, but also the formation of specific high-value compounds. In the following paragraphs, some design constraints for artificial illumination are outlined.

The first issue to be considered is the energy balance. Microalgae exhibit a content of chemical energy (heat of combustion) of about 20 $MJ \cdot kg^{-1}$ and up to 30 $MJ \cdot kg^{-1}$ for lipid-rich algae. Using optimal illumination conditions with respect to the light color and the uniformity of the light field, the energetic efficiency of the growth processes can be up to 0.1 MJ_{bio}/MJ_{light}. This means that 200 MJ of light energy must be made available. Assuming further highly efficient LEDs (50% efficiency of electrical energy to light energy conversion), 400 MJ corresponding to 110 kWh electrical energy is the expenditure for the production of 1 kg microalgal biomass. This makes sense for high-value compounds and mainly in countries with low prices and high availability of sustainable electrical energy. The approach to produce organic molecules by electrical energy and CO_2 has a parallel in the power-2-fuel approaches. Here, electrical energy is used for the production of hydrogen by electrolysis followed by

Fig. 12: Illuminated stirred tank reactor: Commercial glass reactor equipped with an illumination jacket which is closed in real operation. Collimator lenses on each individual LED to ensure parallel light in a radial direction. The quite homogeneous light distribution of this reactor is suited for modeling experiments; for details, see [18].

using H_2 and CO_2 for the synthesis of synthetic fuels ("e-fuels") in chemical reactors, which achieves an energetic efficiency of about 50%. This analogy means that the production of microalgal biomass with artificial light is not unreasonable in any case.

The second question to be answered is about light color: As all photons have the same photosynthetic effect, photons with low energy; that is, long wavelength (Planck's law) will lead to the highest energetic efficiency, which can be expected for red part in the absorbance band of chlorophyll. The use of white light is not required. However, similar to terrestrial plants microalgae also have sensors for light color, which regulate physiological processes like cell division. Several studies showed [39] that a combination of 90% red and 10% blue – resembling artificial light in greenhouses – leads to good results. Apparently, red light loses its advantages against white light at high biomass concentrations and high light intensities (Fig. 13). This is due to the high absorbance of red light at the saturation level in the first algae layer leading to photon losses. This effect can be reduced by using light at the trailing edge of the red absorbance band, for example, in the orange range, which leads to a greater reach of the photons.

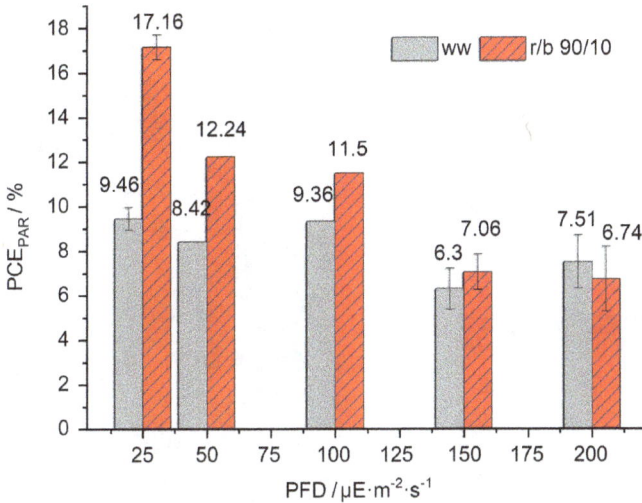

Fig. 13: Effect of colored light in different intensities from [39]. Under low light intensities, a mix of 90% red and 10% blue shows the highest values, close to the theoretical maximum. White light is less effective due to the lower energetic efficiency of its green and blue components. This advantage is increasingly lost for higher light intensities, as red light shows higher absorbance and therefore less penetration.

7.5.3 Horizontal design

An ideal reactor would be one with a low hydraulic pressure, low amount of plastics and easy installation. Eventually, such a horizontal design would look like green turf or a moss bed. It is already known that horizontal raceway ponds are not optimal, as sunlight hits the surface without light dilution. Several ideas have been published to overcome this problem. One is the use of swimming glass cones or fiber optics as light guides. Even collecting light outside and transporting it into a closed classic steel reactor has been tried out. With respect to closed PBRs, a swimming bag like an air mattress has been reported [40]. Light distribution into the depth of the horizontal design is done using inbuilt light-guiding baffle elements. These also lead to a directed flow supporting mixing between the upper and lower parts.

It has already been shown that tubes simply installed horizontally on the ground are not very effective. Realization of light dilution comparable to flat plates or tubes – however on a small scale – is the idea of the zig-zag reactor (Fig. 14) [41] (EP 2 388 310 B1). It has a corrugated surface leading to good light dilution. The medium body is a thin film of several millimeter thickness following the zig-zag line of the surface. This design can be produced simply by extrusion and is operated at low pressure. Membrane aeration from below overcomes the formation of an axial oxygen gradient. Furthermore, all photons impinging on the surface are

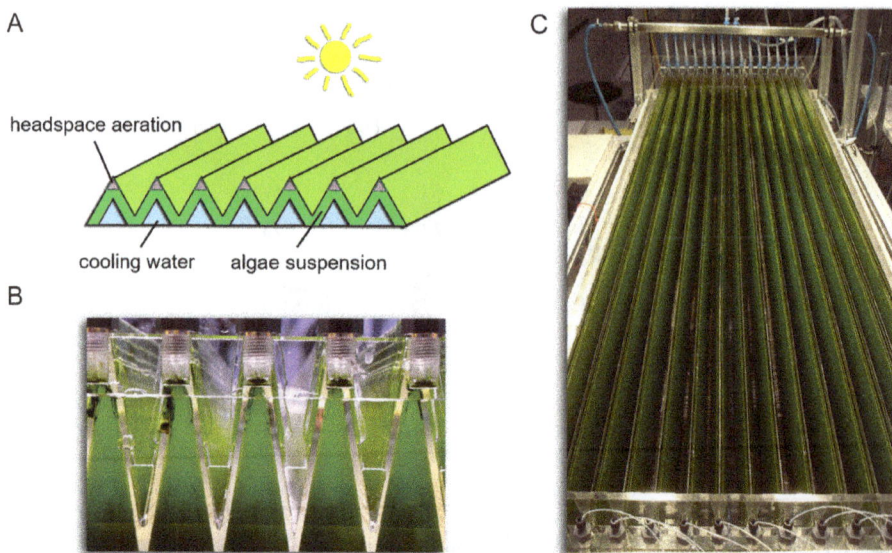

Fig. 14: Horizontal zig-zag reactor [41] (Patent EP 2 388 310 B1): Equivalent to many very low plate reactors clipped to each other to form jags, which prevents photons from being reflected back into the atmosphere. (A) Sketch; (B) front view of the jags; (C) view of the pilot reactor from diagonally above.

used directly or after being reflected, which at least prevents most of the photons to be reflected back into the atmosphere.

7.5.4 Biofilm reactors

Most microorganisms in nature do not live in suspension but in biofilms. This also holds for microalgae. There is evidence that the cells of some strains also have a more stable physiological state, for example, less shear stress. Using this property in technical systems would allow the running of a reactor with minimum mechanical energy, practically no mixing, and the application of membranes for aeration [42, 43]. For instance, for the extraction of the extracellular product Glycolate as a platform chemical according to [44], the cells have to be fixed on these membranes. This can be achieved either by natural biological adhesive mechanisms or by artificial molecular linkers. These could also be switchable, for example, for harvesting by pH shift. The whole system would only need a very small amount of auxiliary energy as material transport is mainly performed by diffusion. However, the problem is light transfer: A microalgae biofilm of four cell layers is nearly opaque and many of these films have to be implemented in a given reactor volume. Here again a solution of nature can be copied as we have already seen in the case of light dilution be leaves. Light distribution happens further in the palisade mesophyll by light conduction. This can be copied in artificial reactors as well. Considerable scientific efforts are also being undertaken to find technically feasible solutions [45, 46].

7.5.5 Persisting problems and economic feasibility

So far, most microalgal biomass, especially *Arthrospira*, is produced in raceway ponds despite their lower productivity. The reason is their simple scaling and low costs per square meter, as well as their low energy demand. Large installations for the production of high- or medium-value products are in most cases done using horizontal tubular reactors. They are often, especially in Europe, installed in greenhouses covering up to 1 ha ground area, though there are larger outdoor facilities in Asia. Nevertheless, microalgae are far from being "biomass 3.0" to supplement the biomass from terrestrial plants for food, feed or plastics in noteworthy amounts. The reasons are related to public perception, cost issues and energy aspects. PBR design contributes to the last two items at least, so it will be briefly discussed in the next paragraphs.

One of the most challenging and persistent problems in PBR operation is cooling. Temperatures of even a few degrees (e.g., 5 °C) over or below the optimum growth temperature can cut productivity by 50%. Irradiation from the Sun is absorbed by the microalgal suspension but only about 5% is converted into chemical energy. So nearly 100% of the solar energy is converted into heat, which somehow has to be removed from the reactor and disposed of through a heat sink. To give an example: On a nice summer day, the power of the Sun can be 500 $W \cdot m^{-2}$. Assuming a PBR of 100 L content on this square meter, the medium including the microalgal cells would heat up by 4.3 K per hour. Open ponds can evaporate 30% of the medium per day, which at least stabilizes the temperature. For closed PBRs, active cooling systems include cooling tubes inside the individual reactor modules or letting water rinse over the surface. Both measures require cooling energy and/or water and are not sustainable. In some regions of the world, reactors are built with the bottom in the water of a lagoon. However, cooling is expensive and requires additional costly equipment and auxiliary energy to provide cold water, even if it is somehow taken from the environment. Otherwise, productivity is reduced to unacceptably low values.

Besides fouling on the inner reactor surface [47], contamination with bacteria, other microalgae, protozoa, fungi or oligo-cellular predators can ruin the harvest of microalgae even in closed reactors or greenhouses. This cannot be completely prevented through reactor design. In the best case, only the probability and time of incubation can be reduced. The classic approach is to use extremophilic strains, for example high salinity, for *Haematococcus* cultivation [48]. Also, some producers use marine algae far from the coast to reduce the number of possible contaminants. Some help comes from proper sanitization procedures: In tubular reactors, brushes can be pumped through the tubes and disinfection media can be used, whereas thermal sterilization is practically impossible on a large scale. A new approach is the cultivation of symbiotic co-cultures. Altogether, this highlights the main problem of the robustness of microalgae cultures, which is still ongoing, while terrestrial plants have gone through 10,000 years of domestication. In any case, the current problems reduce the annual productivity by increasing standstill times.

Shelf-life studies of PBR installations yield a picture of the main cost issues of microalgal production. For extensive and detailed examples see [49, 50]. In this chapter, only technical measures such as energy and material are discussed as shown in Tab. 4.

Tab. 4: Possible shelf life calculation for a PBR (without ground, ground preparation, medium, cooling, downstream processing).

Amount plastics/ $t \cdot ha^{-1}$	Cost plastics and mounting/ $T\euro \cdot ha^{-1}$	Auxiliary energy/ $kW \cdot ha^{-1}$	Areal productivity/ $t \cdot ha^{-1}$	Shelf life/ year	Peripheral devices/ $T\euro \cdot ha^{-1}$
20	100	50	70	10	50

In any case, the electrical energy effort per produced amounts of algae and the investment costs are critical. For high-value products, the situation does not look too bad and leads to economically feasible solutions [51, 52]. Such products as Astaxanthin are typically produced in relatively small installations. Handwork for mounting the reactors and the use of local opportunities, for example, for cooling water or CO_2 supply can be feasible to allow for economically viable production.

Really (nowadays), low-value products like fuels are out of reach, simply because of the necessary expenditure of auxiliary energy. Even if cheap and sustainable energy is available, microalgal technology will be in competition with power2fuel technologies. The next sensible step should be medium-value products like high-quality food or feed or biochemicals, which are less critical. Here, not only the energy but the added value of the molecules counts.

Some shelf-life studies identify the reactor costs as the main obstacles for cheaper commercial microalgae production in the medium-value range [53]. Actually, current reactor costs are high due to a lot of handwork involved. Although the costs of the ground and ground preparation have only to be paid once, they cannot be neglected. Manufacturers try to reduce cost by using standard equipment, for example, from aquaculture, developing endless films that are easily installed, or offering the extrusion of endless PMMA-tubes directly on-site. Unfortunately, economy of scale does not help sufficiently because the effects are only weak under numbering up policy. Nevertheless, the community expects lower prices through the reduced costs of the equipment manufacturers, as soon as large installations increase. Specific parts could then be produced in larger series, with the costs for a reactor decreasing to depositing a box of plastic water bottles.

Even then, it will take half a year for the microalgae to bind the same amount of carbon as the plastics contains. On the other hand, PMMA can last for more than 10 years without serious changes in quality. Also, the auxiliary energy could be taken from sustainable electrical energy. For a microalgae production area, another idea is to use the infra-red part of the illuminating light. Applying transparent photovoltaics for this task could therefore both reduce the need for extra energy costs and heat load. No storage, transformation or conduction of electricity would be necessary. Of course, other costs for land, transport of products, algae nutrients, CO_2 and so on have to be accounted. But this example shows that algae technology can increase and has to increase in view of the growing need for food and feed in a world with a continuously growing population. Reactor technology has paved the way to bring the cost down to approximately the same level as for fruits or vegetables based on their dry weight. Also, more and more food companies offer specialties with microalgae (e.g., buggy power). Further ideas, especially on a large scale, will enable the production of cheap microalgal biomass of high quality. In the future, the lack of robustness can be solved only through a better "domestication" of the microalgae, a process that requires close teamwork between biology and technology.

References

[1]   Pulz O, Gross W. Valuable products from biotechnology of microalgae. Appl Microbiol Biotechnol 2004, 65(6), 635–48. [https://doi.org/10.1007/s00253-004-1647-x][PMID: 15300417].

[2]   Pulz O. Photobioreactors: Production systems for phototrophic microorganisms. Appl Microbiol Biotechnol 2001, 57(3), 287–93. [https://doi.org/10.1007/s002530100702] [PMID: 11759675].

[3]   Molina Grima E, Acién Fernández FG, García Camacho F, Chisti Y. Photobioreactors: Light regime, mass transfer, and scaleup. J Biotechnol 1999, 231–47.

[4]   Lehr F, Posten C. Closed photo-bioreactors as tools for biofuel production. Curr Opin Biotechnol 2009, 20(3), 280–5. [https://doi.org/10.1016/j.copbio.2009.04.004][PMID: 19501503].

[5]   Dillschneider R, Posten C. Closed Bioreactors as Tools for Microalgae Production 2012.

[6]   Rodolfi L, Chini ZG, Bassi N, Padovani G, Biondi N, Bonini G, Tredici MR. Microalgae for oil: Strain selection, induction of lipid synthesis and outdoor mass cultivation in a low-cost photobioreactor. Biotechnol Bioeng 2009, 102(1), 100–12. [https://doi.org/10.1002/bit.22033][PMID: 18683258].

[7]   Raes EJ, Isdepsky A, Muylaert K, Borowitzka MA, Moheimani NR. Comparison of growth of Tetraselmis in a tubular photobioreactor (Biocoil) and a raceway pond. J Appl Phycol 2014, 26(1), 247–55. [https://doi.org/10.1007/s10811-013-0077-5].

[8]   Schediwy K, Trautmann A, Steinweg C, Posten C. Microalgal kinetics – a guideline for photobioreactor design and process development. Eng Life Sci 2019, 19(12), 830–43. [https://doi.org/10.1002/elsc.201900107][PMID: 32624976].

[9]   Camacho Rubio F, Acién Fernández FG, Sánchez Pérez JA, García Camacho F, Molina Grima E. Prediction of dissolved oxygen and carbon dioxide concentration profiles in tubular photobioreactors for microalgal culture. Biotechnol. Bioeng 1999, 62(1), 71–86. [https://doi.org/10.1002/(SICI)1097-0290(19990105)62:1<71::AID-BIT9>3.0.CO;2-T].

[10] Acién Fernández FG, Camacho FG, Pérez JAS, Sevilla JMF, Grima EM. A model for light distribution and average solar irradiance inside outdoor tubular photobioreactors for the microalgal mass culture. Biotechnol Bioeng 1997, 55(5), 701–14. [https://doi.org/10.1002/(SICI)1097-0290(19970905)55:5<701::AID-BIT1>3.0.CO;2-F].

[11] Wolf J, Stephens E, Steinbusch S, et al. Multifactorial comparison of photobioreactor geometries in parallel microalgae cultivations. Algal Res 2016, 15, 187–201. [https://doi.org/10.1016/j.algal.2016.02.018].

[12] Schulze PSC, Brindley C, Fernández JM, et al. Flashing light does not improve photosynthetic performance and growth of green microalgae. Bioresour Technol Rep 2020, 9, 100367. [https://doi.org/10.1016/j.biteb.2019.100367].

[13] Perner-Nochta I, Posten C. Simulations of light intensity variation in photobioreactors. J Biotechnol 2007, 131(3), 276–85. [https://doi.org/10.1016/j.jbiotec.2007.05.024][PMID: 17681391].

[14] Pruvost J, Cornet JF, Le Borgne F, Goetz V, Legrand J. Theoretical investigation of microalgae culture in the light changing conditions of solar photobioreactor production and comparison with cyanobacteria. Algal Res 2015, 10, 87–99. [https://doi.org/10.1016/j.algal.2015.04.005].

[15] Kong B, Shanks JV, Vigil RD. Enhanced algal growth rate in a Taylor vortex reactor. Biotechnol Bioeng 2013, 110(8), 2140–9. [https://doi.org/10.1002/bit.24886][PMID: 23456851].

[16] Csögör Z, Herrenbauer M, Schmidt K, Posten C. Light distribution in a novel photobioreactor – modelling for optimization. J Appl Phycol 2001, 13(4), 325–33.

[17] Mink A, Thäter G, Nirschl H, Krause MJ. A 3D Lattice Boltzmann method for light simulation in participating media. J Comput Sci 2016, 17, 431–7. [https://doi.org/10.1016/j.jocs.2016.03.014].

[18] Jacobi A, Steinweg C, Sastre RR, Posten C. Advanced photobioreactor LED illumination system: Scale-down approach to study microalgal growth kinetics. Eng Life Sci 2012, 12(6), 621–30. [https://doi.org/10.1002/elsc.201200004].

[19] Thomsen C, Rill S, Thomsen L. Case study of a temperature-controlled outdoor PBR system in Bremen. In: Posten C, Walter C, eds. Microalgal Biotechnology: Integration and Economy. 2012, 73–7.

[20] Posten C. Design principles of photo-bioreactors for cultivation of microalgae. Eng Life Sci 2009, 9(3), 165–77. [https://doi.org/10.1002/elsc.200900003].

[21] Ramírez-Mérida LG, Zepka LQ, Jacob-Lopes E. Current status, future developments and recent patents on photobioreactor technology. Recent Pat Eng 2015, 9, 80–90.

[22] Morweiser M, Kruse O, Hankamer B, Posten C. Developments and perspectives of photobioreactors for biofuel production. Appl Microbiol Biotechnol 2010, 87(4), 1291–301. [https://doi.org/10.1007/s00253-010-2697-x][PMID: 20535467].

[23] Sierra E, Acién FG, Fernández JM, García JL, González C, Molina E. Characterization of a flat plate photobioreactor for the production of microalgae. Chem Eng J 2008, 138(1–3), 136–47. [https://doi.org/10.1016/j.cej.2007.06.004].

[24] Guccione A, Biondi N, Sampietro G, Rodolfi L, Bassi N, Tredici MR. Chlorella for protein and biofuels: From strain selection to outdoor cultivation in a Green Wall Panel photobioreactor. Biotechnol Biofuels 2014, 7, 84. [https://doi.org/10.1186/1754-6834-7-84][PMID: 24932216].

[25] Posten C, Rosello-Sastre R. Microalgae reactors. In: 2012-Ullmann's Encyclopedis of Industrial Chemistry. 145–56.

[26] https://algae.proviron.com/en/technology/; accessed 2021.

[27] https://www.phytolutions.de/anwendungen/algenanlage/; accessed 2021.

[28] Degen J, Uebele A, Retze A, Schmid-Staiger U, Trösch W. A novel airlift photobioreactor with baffles for improved light utilization through the flashing light effect. J Biotechnol 2001, 92 (2), 89–94. [https://doi.org/10.1016/S0168-1656(01)00350-9].

[29] Richmond A, Boussiba A, Vonshak A, Kopel R. A new tubular reactor for mass production of microalgae outdoors. J Appl Phycol 1993, 5, 327–32.

[30] Molina E, Fernández J, Acién FG, Chisti Y. Tubular photobioreactor design for algal cultures. J Biotechnol 2001, 92(2), 113–31. [https://doi.org/10.1016/S0168-1656(01)00353-4].

[31] https://lgem.nl/; accessed 2021.

[32] Wilson MH, Mohler DT, Groppo JG, et al. Capture and recycle of industrial CO2 emissions using microalgae. Appl Petrochem Res 2016, 6(3), 279–93. [https://doi.org/10.1007/s13203-016-0162-1].

[33] https://jongerius-ecoduna.at/; accessed 2021.

[34] Bähr L, Wüstenberg A, Ehwald R. Two-tier vessel for photoautotrophic high-density cultures. J Appl Phycol 2016, 28(2), 783–93. [https://doi.org/10.1007/s10811-015-0614-5].

[35] Schultze LKP, Simon M-V, Li T, Langenbach D, Podola B, Melkonian M. High light and carbon dioxide optimize surface productivity in a Twin-Layer biofilm photobioreactor. Algal Res 2015, 8, 37–44.

[36] Ehwald R, Prof. Dr, Bähr L. Inventors. Labor-Photobioreaktor: Deutsches Patent.

[37] Wagner I, Braun M, Slenzka K, Posten C. Photobioreactors in Life Support Systems. Adv Biochem Eng Biotechnol 2016, 153, 143–84. [https://doi.org/10.1007/10_2015_327] [PMID: 26206570].

[38] Rosello Sastre R, Csögör Z, Perner-Nochta I, Fleck-Schneider P, Posten C. Scale-down of microalgae cultivations in tubular photo-bioreactors–a conceptual approach. J Biotechnol 2007, 132(2), 127–33. [https://doi.org/10.1016/j.jbiotec.2007.04.022][PMID: 17561299].

[39] Wagner I, Steinweg C, Posten C. Mono- and dichromatic LED illumination leads to enhanced growth and energy conversion for high-efficiency cultivation of microalgae for application in space. Biotechnol J 2016, 11(8), 1060–71. [https://doi.org/10.1002/biot.201500357] [PMID: 27168092].

[40] Yan C, Zhang Q, Xue S, et al. A novel low-cost thin-film flat plate photobioreactor for microalgae cultivation. Biotechnol Bioproc E 2016, 21(1), 103–9. [https://doi.org/10.1007/s12257-015-0327-2].

[41] Posten C, Jacobi A, Steinweg C, Lehr F, Rosello R. Inventors. Photobioreactor: Europäisches Pat 13110189, 2010 May 21.

[42] Gross M, Jarboe D, Wen Z. Biofilm-based algal cultivation systems. Appl Microbiol Biotechnol 2015, 99(14), 5781–9. [https://doi.org/10.1007/s00253-015-6736-5][PMID: 26078112].

[43] Berner F, Heimann K, Sheehan M. Microalgal biofilms for biomass production. J Appl Phycol 2015, 27(5), 1793–804. [https://doi.org/10.1007/s10811-014-0489-x].

[44] Günther A, Jakob T, Goss R, et al. Methane production from glycolate excreting algae as a new concept in the production of biofuels. Bioresour Technol 2012, 121, 454–7. [https://doi.org/10.1016/j.biortech.2012.06.120][PMID: 22850169].

[45] Strieth D, Ulber R, Muffler K. Application of phototrophic biofilms: From fundamentals to processes. Bioprocess Biosyst Eng 2018, 41(3), 295–312. [https://doi.org/10.1007/s00449-017-1870-3][PMID: 29198024].

[46]   Jacob A, Bucharsky EC, GuenterSchell K. The application of transparent glass sponge for improvement of light distribution in photobioreactors. J Bioproces Biotechniq 2012, 02(01). [https://doi.org/10.4172/2155-9821.1000113].

[47]   Zeriouh O, Reinoso-Moreno JV, López-Rosales L, et al. Biofouling in photobioreactors for marine microalgae. Crit Rev Biotechnol 2017, 37(8), 1006–23. [https://doi.org/10.1080/07388551.2017.1299681][PMID: 28427282].

[48]   Mooij PR, Stouten GR, Van Loosdrecht MCM, Kleerebezem R. Ecology-based selective environments as solution to contamination in microalgal cultivation. Curr Opin Biotechnol 2015, 33, 46–51. [https://doi.org/10.1016/j.copbio.2014.11.001][PMID: 25445547].

[49]   Stephens E, Ross IL, King Z, et al. An economic and technical evaluation of microalgal biofuels. Nat Biotechnol 2010, 28(2), 126–8. [https://doi.org/10.1038/nbt0210-126][PMID: 20139944].

[50]   Acién FG, Molina E, Fernández-Sevilla JM, Barbosa M, Gouveia L, Sepúlveda C, Bazaes J, Arbib Z. Economics of microalgae production. In: Gonzalez-Fernandez C, Muñoz R, ed. Microalgae-Based Biofuels and Bioproducts: From Feedstock Cultivation to End-Products. Cambridge, Elsevier Science & Technology, 2017, 485–503.

[51]   Panis G, Carreon JR. Commercial astaxanthin production derived by green alga Haematococcus pluvialis: A microalgae process model and a techno-economic assessment all through production line. Algal Res 2016, 18, 175–90. [https://doi.org/10.1016/j.algal.2016.06.007].

[52]   Borowitzka MA. High-value products from microalgae – their development and commercialisation. J Appl Phycol 2013, 25(3), 743–56. [https://doi.org/10.1007/s10811-013-9983-9].

[53]   Acién FG, Fernández JM, Magán JJ, Molina E. Production cost of a real microalgae production plant and strategies to reduce it. Biotechnol Adv 2012, 30(6), 1344–53. [https://doi.org/10.1016/j.biotechadv.2012.02.005][PMID: 22361647].

Simon MoonGeun Jung, Jong-Hee Kwon

8 State-of-the-art cultivation process development for microalgae

Abstract: The application of microalgae to red, green and white biotechnology raises the need for cost-efficient cultivation processes. Here an engineered flat-panel photo-bioreactor system was introduced for continuous photo-cultivation of microalgae. Based on direct determination of the growth rate at constant cell densities and the continuous measurement of O_2 evolution, abiotic cultivation conditions (such as light, iron, pH and shear stress) and their effect on the photosynthetic productivity can be directly observed and optimized. In addition, an enrichment process in a continuous cultivation system was developed to screen a high-growth-rate microalga from a mixed culture of the pool.

To enhance cellular growth and lipid productivity of microalga, a light filtering through a solution of soluble colored additives was used to alter the light spectrum. A cost-efficient process devoid of several washing steps was developed, which is related to direct cultivation following the catalytic decomposition of peracetic acid.

Keywords: photobioreactor, microalgae, optimization, cultivation, sterilization

8.1 Introduction

Microalgae including cyanobacteria constitute attractive biomass for the commercial production of renewable energy sources and pharmacologically active compounds as well as human nutrients. In addition, microalgae contribute to the quality of the environment by fixing CO_2 for their autotrophic growth, thereby reducing greenhouse gas production, and removing pollutants from wastewater for their hetero/mixotrophic growth.

Most commercially produced microalgal biomasses involve growth in open ponds, which are inexpensive, easy to operate and rely on sunlight as a free source of energy [1]. However, biomass productivity in open ponds is not high, because of the limited ability to control environmental parameters, such as light, temperature and contaminants, leading to low photosynthetic efficiency [2]. Growing photosynthetic microorganisms in a photobioreactor (PBR) allows greater control of environmental parameters [3–5]. Because of the interactions among fluid dynamics, biochemical reactions and light transfer in PBRs are complex; however, photobiological processes in these vessels are difficult to optimize. Biological responses are induced by mutually dependent factors, including light quality and quantity, pH, and CO_2 [6], with complex strategies required for their adjustment (Fig. 1). High productivity under cost-efficient

https://doi.org/10.1515/9783110716979-008

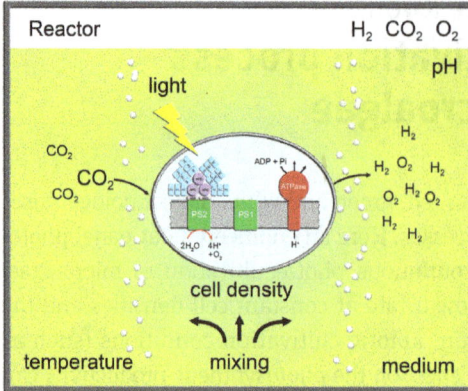

Fig. 1: Schematic presentation of physical and physicochemical parameters required for the cultivation of cyanobacteria. Major parameters affecting the yield of hydrogen and cell growth include light quantity and quality, pH, temperature, CO_2 concentration, substrate composition and concentration, and degree of mixing.

photosynthetic cultivation conditions requires an in-depth characterization of each bioprocess and its subsequent optimization.

This chapter describes the direct optimization of abiotic parameters, based on real-time visualization of growth and oxygen generation rates and a new chemical sterilization process consisting of the catalytic decomposition of peracetic acid. In addition, the light was effectively utilized to induce high lipid production, and an enrichment process in a continuous cultivation system was developed to screen a fast-growing strain. These findings provide the perspectives on new methods of technological improvement that will enable the commercialization of microalgae-based products.

8.2 Direct optimization of bioprocesses in a continuous flat-panel photobioreactor system

Continuous cultivation has several advantages over batch cultures, such as constant production rates, high photosynthetic productivity under defined cultivation conditions [6] and the elimination of downtime for cleaning and sterilization, which are necessary for repeated batch cultivation.

Continuous cultivation under turbidostatic process control can be attained by feedback control between a media pump and measurements of turbidity. This type of cultivation is characterized by the continuous addition of fresh media and the withdrawal of the same volume of biomass suspension. The main parameter characterizing continuous cultivation is the dilution rate D, defined as F/V_L, with F being the flow rate of fresh medium and V_L the working volume of the mass culture. The biomass balance under continuous cultivation conditions is expressed as $dX/dT = (\mu-D)\cdot X$, where X is the biomass in the reactor, T is the cultivation time and μ is the specific growth rate. At steady-state under continuous cultivation conditions, $dX/dT = 0$; that is, the specific growth rate is equivalent to the dilution rate. The specific growth rate can reflect cellular conditions. A decreasing specific growth rate indicates that cells are exposed to stress, resulting in a reduced growth rate. To focus on long-term changes, the effective growth rate under these conditions (μ_{eff}) was averaged over a period of 10 or 20 h, depending on cultivation conditions.

A strategy to optimize the continuous growth of cyanobacteria in PBRs was based on monitoring the effective bacterial growth rate, as well as photosynthetic O_2 generation. O_2 generation during cultivation was quantified using a BCP-O_2 sensor (BlueSens GmbH, Herten, Germany) and recorded continuously as an indicator of photosynthetic productivity. The sensor was inserted directly into the exhaust-gas stream of the fermenter, with the help of a flow adapter (Fig. 2). The stable determination of effective growth rate and photosynthetic O_2 generation at constant cell density facilitates both the detection and evaluation of changes in cellular metabolism during the cultivation process. Although changes in growth rate indicate metabolic changes in the cells, possibly induced by environmental changes, the altered O_2 evolution rate reflects differences in oxygen balance due to photosynthesis and oxygen consumption. Oxygen consumption may be caused by electron transfer from the respiratory chain or by the Mehler reaction [7], which indicates stress conditions, such as photoinhibitory zones within the reactor.

The high potential of this approach can be illustrated by showing the optimization of light intensity, iron supply, pH and shear stress due to the aeration flow rate.

8.2.1 Light intensity

Optimizing the growth of photosynthetic organisms requires the optimization of light supply. In addition to the three typical phases of bioreactor systems – the fluid medium as liquid phase, the cell as solid phase and the gas phase – light is sometimes regarded as the fourth phase [6]. Depending on the light supply and the driving force of photosynthesis, various volume elements of the PBR can be grouped into productive light zones with sufficient light intensities for photosynthetic metabolism and unproductive dark zones with insufficient light intensities. The depth of light penetration, which defines light and dark zones, depends on both the density of the light-absorbing microorganism in culture and the geometry of the reactor.

An oversupply of light, however, can have negative effects on the culture of photosynthetic microorganisms, including photoinhibition, defined as inactivation of the photosynthetic machinery [8]. Although photosynthetic organisms have developed many strategies to protect themselves from photodamage [9–11], most of these strategies reduce productivity and may have negative effects on the quality of bio-products

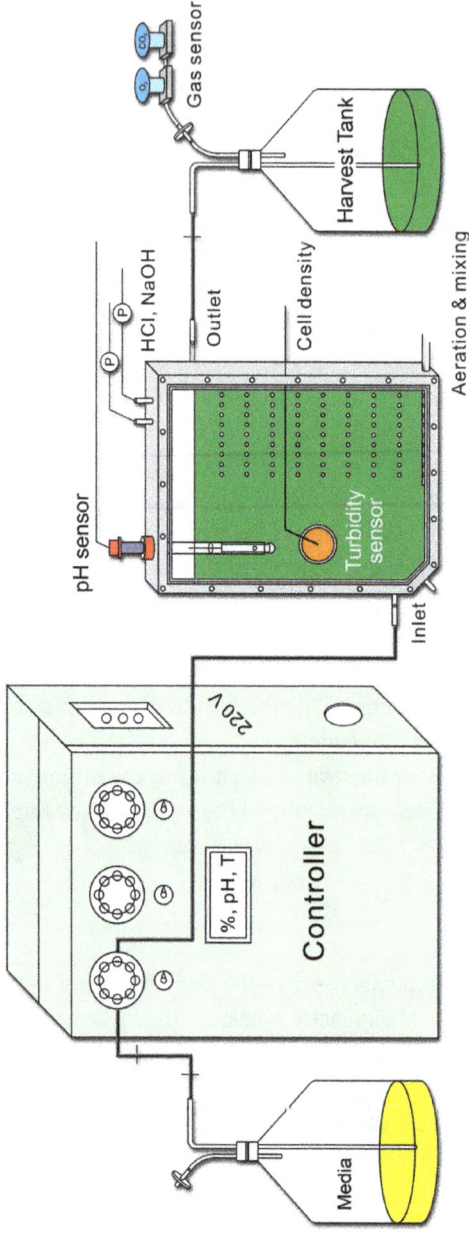

Fig. 2: Schematic view of the experimental set-up for the continuous cultivation of cyanobacteria in 2 L flat-panel PBR.

derived from microalgae [12]. Light intensity during cultivation of *Synechocystis* wild type was optimized based on changes in μ_{eff} and O_2 generation (Fig. 3). Exposure of this organism, grown under constant conditions, to stepwise increases in light intensity from 100 to 400 µmol photons $m^{-2}\,s^{-1}$, showed that growth rates μ_{eff} were slightly higher at 200 and 300 µmol photons $m^{-2}\,s^{-1}$ than at the initial light intensity, but that further increases in incident light intensity led to severe light stress and irreversible damage to the cyanobacterial culture (Fig. 3A). Overall, O_2 generation, which correlated strongly with growth rate, was dependent on light intensity between 100 and 300 µmol photons $m^{-2}\,s^{-1}$, increasing from 224 ± 10 µmol $O_2\,L^{-1}\,h^{-1}$ at 100 µmol photons $m^{-2}\,s^{-1}$ to 288 ± 9 µmol $O_2\,L^{-1}\,h^{-1}$ at 200 µmol photons $m^{-2}\,s^{-1}$ and then to

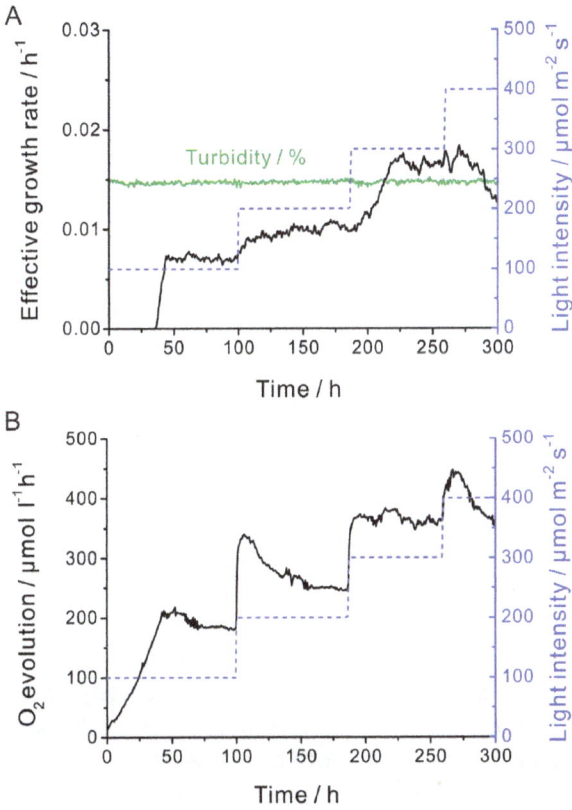

Fig. 3: Optimization of light intensity for *Synechocystis* PCC 6803 cells grown under continuous cultivation conditions. Effect of light intensities from 100 to 400 µmol photons $m^{-2}\,s^{-1}$ on: (A) Constant turbidity and changes in the effective growth rate corresponding to constant cell density. (B) O_2 evolution. Turbidity of cells, μ_{eff}, and O_2 evolution are shown as solid lines and light intensities as dashed lines [13].

396 ± 8 µmol O_2 L^{-1} h^{-1}, or 177% of the initial rate, at 300 µmol photons m^{-2} s^{-1}. This results may reflect larger rearrangements in the photosynthetic apparatus of the cells in response to light conditions (Fig. 3B).

8.2.2 Iron

The photosynthetic apparatus of cyanobacteria, including the PS1, PS2 and cytochrome b_6f enzyme complexes, requires iron for the formation of their active centers [14–16]. *Synechocystis* PCC 6803 has a high demand for iron, that is, up to 10 times that of *Escherichia coli* [17]. However, excessive iron in cells induces the formation of reactive oxygen species, which damage DNA and proteins, inhibiting many physiological processes [18]. Because of their limited access to bioavailable iron in their natural habitats, cyanobacteria have developed complex strategies for the transport and storage of iron, as well as to compensate for iron limitation [17–21].

Although varying the light intensity has immediate effects on growth and photosynthetic efficiency, other parameters, including iron supply, exert effects over a longer time scale. Because of the limited availability of iron in nature, evolutionary pressure has induced compensatory and storage strategies to overcome iron limitations over extended time periods. *Synechocystis* possesses storage facilities to accumulate iron for several cellular generations [22], compensating for the lack of iron without significantly affecting growth rates for extended periods of time. Besides monitoring iron supply externally by measuring growth rate, iron supply can be monitored internally by measuring the IsiA-protein, which is selectively expressed under conditions of iron deficiency [23–25] and is easy to monitor due to its strong fluorescence emission peak at 685 nm at 77ºK [26–28]. Using the same reactor setup as for optimization of light intensity, *Synechocystis* was exposed to stepwise increases in Fe^{3+} concentrations, from 6 to 120 µM (Fig. 4).

Cells were grown under turbidostatic cultivation conditions for 50 h, and Fe^{3+} was added to the medium at a steady-state concentration of 3 µM, based on ICP-OES analysis. Despite the constant iron concentration, IsiA fluorescence increased after 100 h, indicating the depletion of internal iron storage of the cyanobacterial cells (Fig. 4C). In parallel, the growth rate and O_2 generation rate decreased slightly. Both the increase in fluorescence and the reduction in growth rate were reversed rapidly in medium containing 30 µM Fe^{3+}, indicating that these parameters correlated with the availability of iron. Growth in a medium containing 60 µM $FeCl_3$ resulted in a five-fold increase in μ_{eff} and an increase in O_2 generation to 417 µmol O_2 L^{-1} h^{-1} (Fig. 4A, B), whereas higher iron concentrations resulted in cell coagulation rather than a higher growth rate.

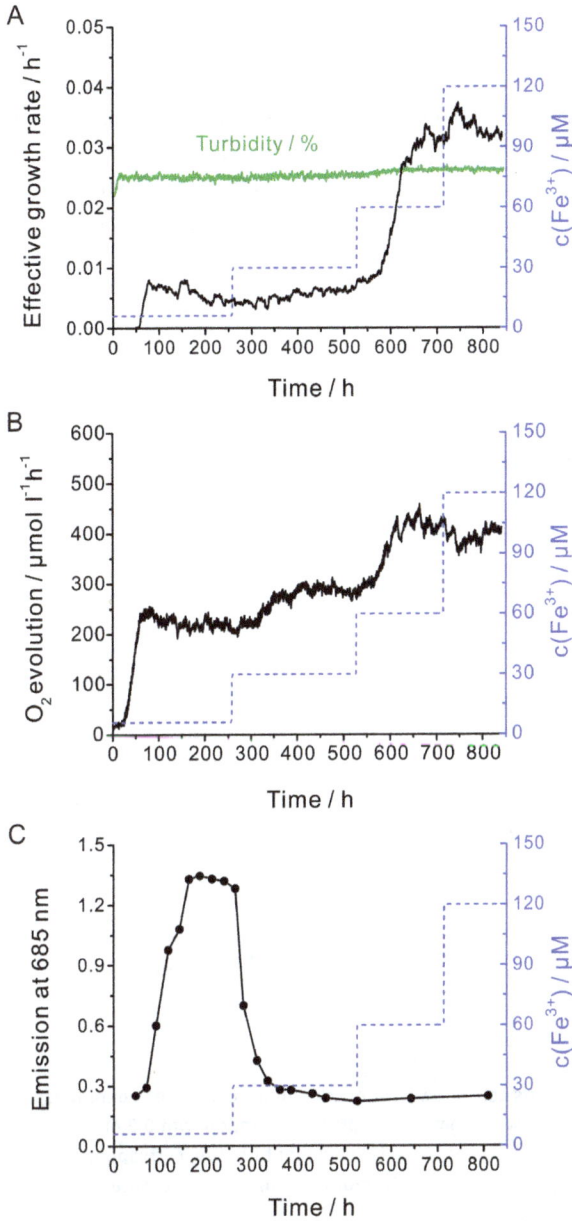

Fig. 4: Optimization of iron concentration for *Synechocystis* PCC 6803 cells grown under continuous cultivation conditions at 100 µmol photons m^{-2} s^{-1}. Effects of iron concentration ($FeCl_3$) on (A) cell growth rate, (B) O_2 generation at constant cell density and (C) 77°K Chl fluorescence emission peak at 685 nm. Turbidity, μ_{eff}, O_2 generation and emission intensity represented by solid lines, and iron concentration in the cultivation media by dashed lines [13].

8.2.3 pH

Cell growth and effective H_2 production are dependent on the pH of the growth medium.

To optimize pH, *Synechocystis* was grown under turbidostatic conditions in a 5 L flat-panel PBR, with the cells exposed to a stepwise increase in pH, from pH 7 to pH 10, and all other parameters kept constant. Although the highest effective growth rate μ_{eff} (0.01 h^{-1}) was observed at pH 7, μ_{eff} did not increase between pH 8 and pH 10 (Fig. 5).

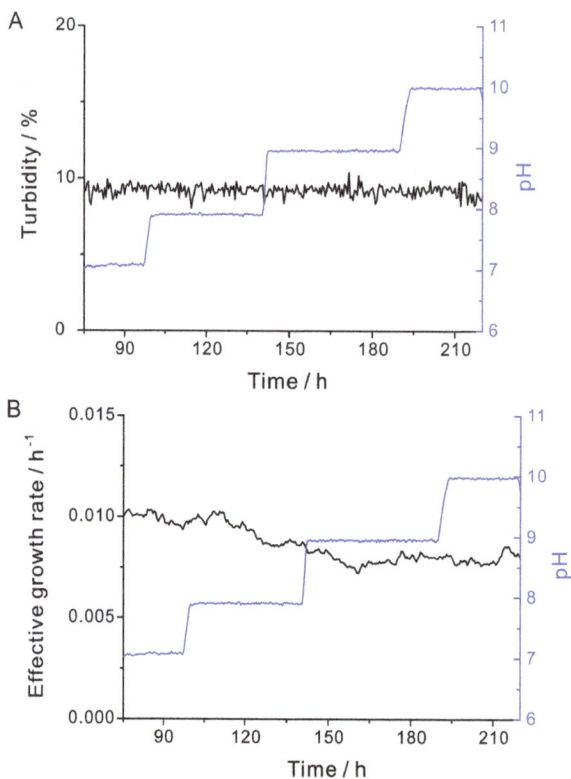

Fig. 5: Optimization of pH for *Synechocystis* cells grown under continuous cultivation conditions. External pH kept constant or adjusted stepwise from pH 7 to pH 10 by titration with 0.2 M hydrochloric acid and sodium hydroxide. Effects of pH on (A) constant turbidity, corresponding to constant cell density and (B) effective growth rate (μ_{eff}). Turbidity and μ_{eff} are represented by black lines and pH by blue lines.

8.2.4 Gas velocity (shear stress)

Cell growth is also affected by several environmental parameters, in particular fluid dynamics induced by pneumatic mixing, which restrict the hydrodynamic and static pressure of the reactor system. High superficial gas velocity is necessary to achieve a high degree of turbulence for sufficient mixing of the cell suspension with CO_2 gas, and various nutrients [29]. High photosynthetic efficiency has been demonstrated in fast gas and liquid circulations [30]. However, the increased hydrodynamic force caused by fast gas flow leads to shear stress, which may reduce cell growth rate or even induce cell death [31]. Shear stress-induced damage to microalgae was demonstrated in a bubble column with a gas sparger [29]. These findings suggested that the gas volume related to superficial gas velocity is an important operative parameter for sustainable cell growth, by providing a balance between mixing and shear stress.

The impact of shear stress on cell viability has been evaluated by direct measurement of effective growth rate in 5 L flat-panel PBR [13]. The fluctuation in cell turbidity following an increase in gas flow rate from 100 to 500 mL min^{-1} resulted in a rapid reduction in cellular growth rate (Fig. 6A), with 5 h required for this rate to approach zero. Significantly, cell density decreased at a gas flow rate of 500 mL min^{-1}. The subsequent reduction in the gas flow rate from stressful conditions resulted in an increase and stabilization of the effective growth rate, indicating that the retardation of cell growth was induced by shear stress resulting from the high gas flow rate. Although cell phenotypes, including cell density and color, recovered somewhat, the effective growth rate of these cells did not return to its initial, pre-stress value (Fig. 6A). Cells damaged by shear stress likely require more time to recover full biological stability.

To optimize gas flow rates, *Synechocystis* PCC 6803 cells were grown under continuous culture conditions at gas flow rate from 50 to 300 mL min^{-1}. Direct determination of the effective growth rate at constant cell density showed cell damage due to shear stress induced by the increased gas velocity at the sparger and/or bubble bursting at the surface. A gas flow rate of 200 mL min^{-1} resulted in a significant reduction in effective growth rate, indicating that shear stress induced cell damage (Fig. 6B). Reliable cell growth therefore requires optimization of gas volume and the development of an effective aeration system for each newly designed reactor setup, as well as scale-up of the reactor.

8.3 Enrichment as a screening method for high-growth-rate microalgal strains under continuous cultivation

Currently available microalgal biodiesel production systems have yet to achieve huge commercial success, as the production rates are insufficient to meet market demands or trading prices on the global biodiesel market [33]. The productivity of biodiesels is dependent on two major factors: growth rate and oil content [34]. The initial step in

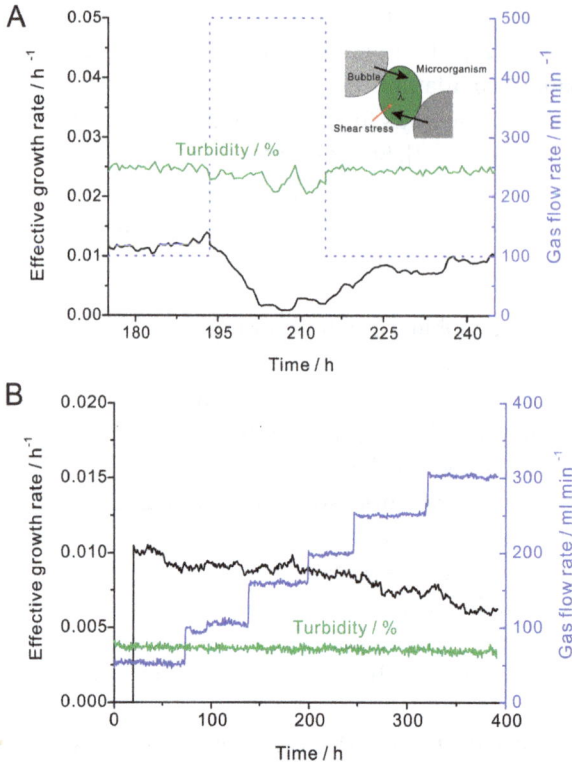

Fig. 6: Effect of shear stress on the growth of *Synechocystis* WT cells grown under continuous cultivation conditions. (A) Cells exposed to a rapid change in gas flow rate, from 100 to 500 mL min^{-1}. (B) Cells exposed to a stepwise increase in gas flow rate, from 50 to 300 mL min^{-1}, while keeping other parameters constant. μ_{eff} represented by black lines and gas flow rate by blue lines [32].

biodiesel production is to screen potential strains to determine their biomass accumulation and oil content. The effective utilization of fast-growing microalgae can be maximized by hydrothermal liquefaction (HTL) processes [35, 36]. HTL can convert microalgae with low lipid content into biocrude oil at high temperature and pressure [37]. Thus, regardless of whether fast-growing microalgae contain high lipid contents, the screening process to identify species in which lipids can be induced and converted to biocrude oil is important for effective biodiesel and biofuel production systems.

Although genetically modified oleaginous strains have been screened [38–42], a time-efficient process for screening fast-growing strains has not yet been developed. An enrichment method has been used to screen a high-growth-rate microalgal strain grown in an engineered 2 L flat-panel PBR under continuous cultivation conditions. A fast-growing microalgal strain was enriched from a mixed microalgal pool using the turbidostat cultivation system, which allows the time-dependent washout of relatively slow-growing microalgal strains. In one model, *Chlorella vulgaris*, *Chlorella protothecoides* and *Chlamydomonas reinhardtii* were grown in mixed culture, and enrichment of relatively faster growing species was validated.

Changes in cell populations in mixed culture were analyzed by flow cytometry (MoFlo XDP; Beckman Coulter, Fullerton, CA), with the excitation wavelength of the argon laser being 488 nm. Signals were measured in forward scatter channels (FSC) and side scatter channels (SSC) and analyzed using SUMMIT Software Version 5.2. The flow-cytometry sheath fluid consisted of Coulter Isoton II Diluent fluid (Beckman Coulter, La Brea, CA). One milliliter aliquots of the mixed culture sample were analyzed at the time of inoculation and at 45 and 90 h after the start of turbidostat cultivation.

Analysis of cell counts showed that the maximal growth rates of *C. vulgaris*, *C. protothecoides* and *C. reinhardtii* were 0.050 h^{-1}, 0.051 h^{-1}, and 0.056 h^{-1}, respectively. *C. reinhardtii* therefore had the highest maximal growth rate, suggesting that it would become dominant in mixed cultures under turbidostat conditions. Indeed, *C. reinhardtii* became the dominant species in mixed cultures, indicating that flow cytometry is an effective method to analyze cell composition and population in culture. Because FSC and SSC can be used to distinguish among various types of cells, these parameters have been used to characterize cellular morphology [43–45]. Two signals were observed at the time of inoculation (Fig. 7A), and each was compared with the signals of monocultures of *C. reinhardtii*, *C. protothecoides* and *C. vulgaris*. The upper right signal of the mixed culture was identified as *C. reinhardtii*, and the lower left signal was identified as a mixture of *C. protothecoides* and *C. vulgaris* (Fig. 7A). The stronger intensity spot, representing *Chlorella* rather than *Chlamydomonas*, originated from the overlapped signal of the two *Chlorella* species, indicating that the two had the same FSC and SSC signal positions. After 45 h, the intensity of the two signals had changed, indicating a change in microalgal populations during batch cultivation (Fig. 7B). The signal corresponding to *C. reinhardtii* (upper right) increased, whereas the signal corresponding to the two *Chlorella* species (lower left) decreased. After about 75 h, however, only one signal was observed, corresponding to *C. reinhardtii* (Fig. 7C). The FSC and SSC signals of both *C. vulgaris* and *C. protothecoides* were no longer observed after 90 h.

These findings suggest that the turbidostat system can be used to distinguish high-growth-rate microalgal species from others, as yet uncharacterized species. The difference in growth rates of the three species grown in the customized cultivation system resulted in the domination of the fastest growing species, *C. reinhardtii*. The final screening step, isolation of a single colony of one strain, can be completed by spreading the enriched culture on an agar plate. This selected strain can be used in the time-efficient production of microalgal biomasses, which can be further coupled with downstream processes such as lipid induction and hydrothermal liquefaction processes for efficient biodiesel production. Moreover, this pre-programmed cultivation system can be applied to screen for stress-tolerant strains under stressful conditions, including high exposure to light, high and low temperatures, and nutrient depletion conditions.

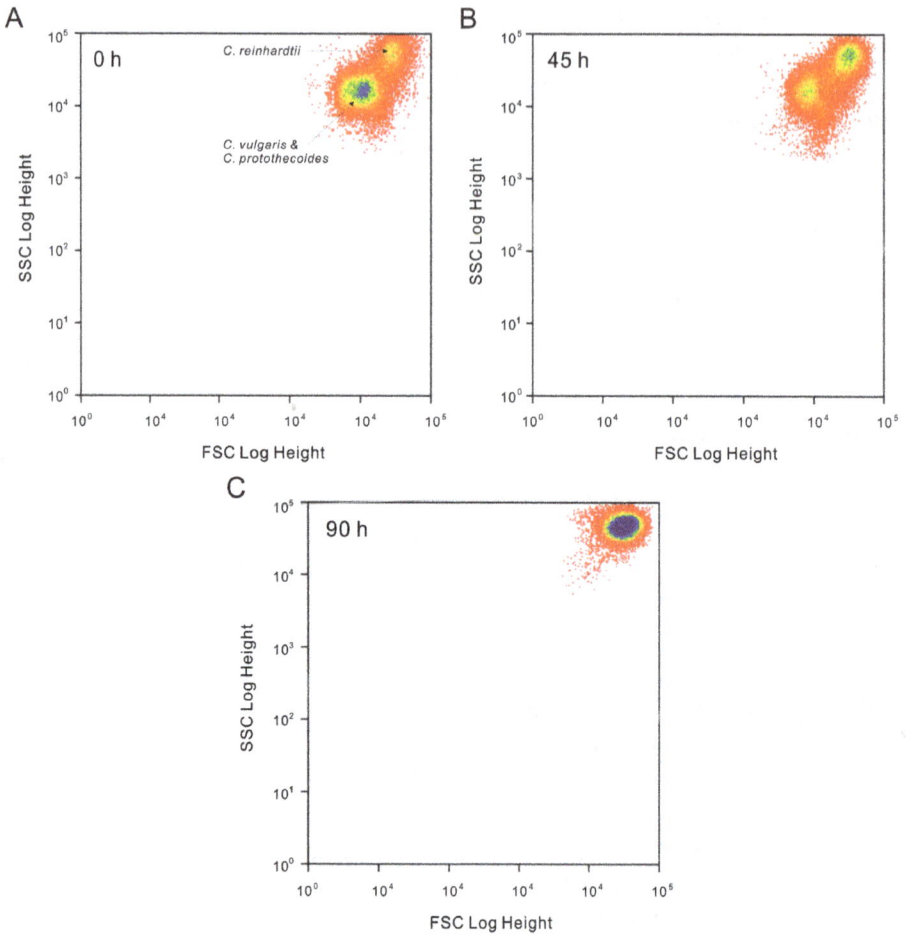

Fig. 7: Representative dot plots of forward and side scattering of microalgal cells, as assessed by FACS, at (A) the time of inoculation and at (B) 45 and (C) 90 h after the start of turbidostat cultivation [46].

8.4 Use of tar color additives as light filters to enhance growth and lipid production by the microalga *Nannochloropsis gaditana*

Photosynthetic organisms convert light energy into chemical energy and accumulate biomass and lipids [47]. Improved microalgal productivity requires the optimization of light conditions because light influences photosynthesis and subsequent metabolic steps. In addition to light intensity, the light spectrum also affects photosystem stoichiometry and cellular metabolism [11, 48]. Studies using monochromatic illumination indicate that light quality can improve cellular growth and lipid content at the same time [34, 49]. However, the optimal wavelength is dependent on the species and other factors [50–58]. Moreover, cost considerations make it difficult to scale up laboratory experiments that use monochromatic light from LEDs. It is also difficult to change the spectrum economically when sunlight is used as the light source.

The effects of wavelength on growth and lipid production of *Nannochloropsis gaditana* –an oleaginous microalga with high lipid productivity that can be cultivated on a large scale –were assessed using an inexpensive light filter, in which simple pigments were used to change the light spectrum [59, 60]. Various color additives (red, yellow, green and pink, Chunwoo Co. Ltd., Kyungbuk, Korea) were dissolved in the medium to make solutions filtering out specific areas of the spectrum from a white LED source. Wavelengths absorbed by the solutions were determined by UV/Vis spectrophotometry (JASCO-V-730) and their absorbance spectra were determined (Fig. 8). The red and yellow additives were mixed at a 4:1 ratio to filter out light below 600 nm.

The effects of filtered illumination on the growth and lipid production of *Nannochloropsis* were tested, similar to tests of the effects of monochromatic illumination with blue, red and green LEDs. The major pigments of *N. gaditana* are chlorophyll a, β-carotene and violaxanthin [62], with this species absorbing blue (400–500 nm) and red (600–700 nm) light. Culture under red light illumination resulted in a higher accumulation of biomass and lipids than under blue light, as well as in better biodiesel quality [34]. The presented study used a mixture of red and yellow pigments as simple and inexpensive light filter. As the red additive had an absorption peak around 506 nm (Fig. 8A) and did not effectively absorb light below 450 nm, it was mixed with a yellow additive at a ratio of 4:1 (red:yellow) to achieve more uniform absorbance below 506 nm (Fig. 8B). The half-maximal absorbance of the mixture was at about 550 nm (Fig. 8B). A 0.02% solution of this mixture, used as a color filter, allowed the passage of 50% of incident light intensity.

N. gaditana cultures were grown under photoautotrophic conditions with white or filtered light. Compared with cells grown under white light, cells grown under filtered light had a significantly higher specific growth rate (0.55 h^{-1} vs. 0.38 h^{-1}) (Fig. 8C) and produced 46.7% more biomass (0.107 g L^{-1} day^{-1} vs. 0.073 g L^{-1} day^{-1}) after 9 days. Also, the maximum daily biomass production was higher under filtered light (0.156 g L^{-1} day^{-1} on day 4) than under white light (0.082 g L^{-1} day^{-1} on day 8). The yield of FAME was also significantly higher under filtered than under white light (23.52% vs. 19.18%, $p < 0.05$). These results suggest that the use of a simple light filter can enhance lipid productivity in *N. gaditana*. Further studies are needed to determine the optimal light spectrum required for FAME production by *N. gaditana* and other microalgal species. Moreover, this method may be applicable in outdoor mass cultivation systems as a cost-effective replacement for artificial light sources.

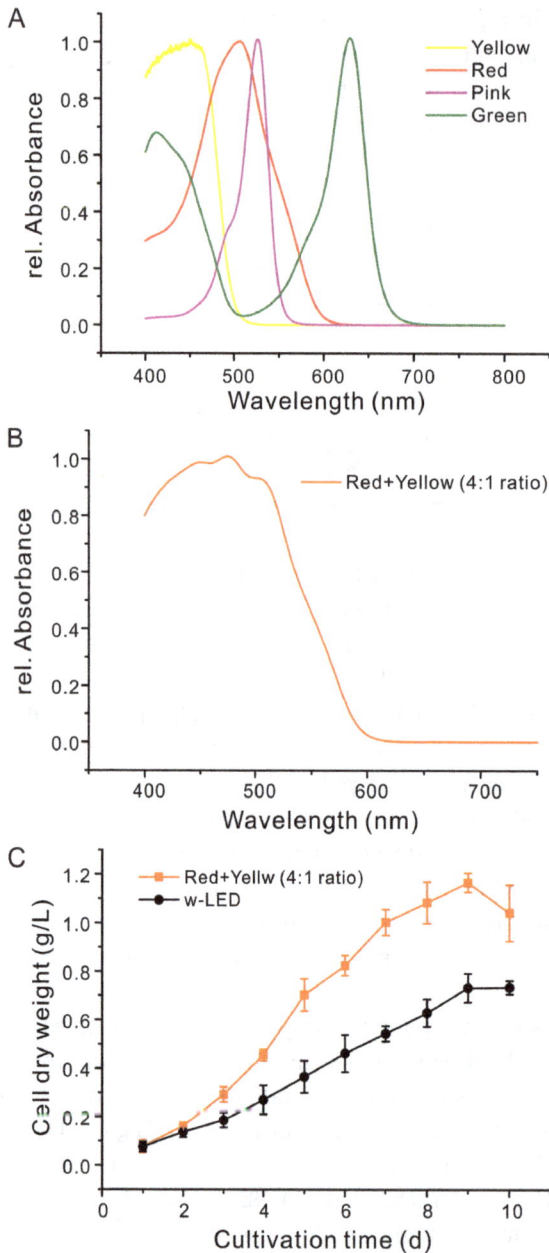

Fig. 8: (A) Absorption spectra of yellow, red, pink and green pigments, with the red and yellow pigments showing maximal absorption at 506 nm (red line) and 455 nm (yellow line), respectively. (B) Absorption spectrum of a 4:1 mixture of the red and yellow pigments, showing broad absorption in the blue region and a half-maximal absorbance at about 550 nm. (C) Growth of *N. gaditana* under white (circles) and filtered (squares) light, with each point representing the mean ± standard deviation of three biological replicates [61].

8.5 Development of a new chemical sterilization process

Sterility within the reactor is indispensable for reliable cultivation and can be achieved by maintaining a temperature of 121 °C for 20 min or longer. Although heating is the most reliable method to eliminate the risk of contamination, it is not appropriate for heat-sensitive materials, including plastics and electronic sensors as well as low-priced products requiring a low cost of investment. While chemical sterilization, which is widely used in medicine, avoids the problem of heat damage, it has some disadvantages, including the release of hazardous materials and the need for an additional rinse with sterile water after treatment. As the preparation of sterile water by autoclaving requires high amounts of energy and is expensive, there is a need for energy- and cost-efficient, as well as eco-friendly sterilization processes for the economical preparation of microalgae-based biofuels and biorefinery processes.

8.5.1 Cost-efficient cultivation of *Spirulina platensis* by chemisorption of CO_2 into medium containing NaOH

Sodium hydroxide (NaOH) is widely utilized for cleaning and sanitization (Healthcare, 2006). Incubation of culture medium with NaOH for 1 h successfully reduced bacterial populations. The antimicrobial activity of aqueous solutions of 0.2 M NaOH was investigated by measuring colony-forming units (CFU) on LB and TYG agar plates. Medium contaminated by bacteria, including strains of *Tristrella, Pseudomonas, Stappia, Labrenzia, and Microbacterium*, formed $1.57 \pm 0.11 \times 10^5$ CFU mL^{-1} on LB agar and $1.24 \pm 0.14 \times 10^5$ CFU mL^{-1} on TYG agar plates. However, incubation of contaminated medium with 0.2 M NaOH solution for 1 h yielded no colonies on either LB or TYG plates.

High alkalinity induced by NaOH could be reduced to pH 8 or pH 9 by chemisorption of CO_2 gas prior to cultivation. Among the chemicals used for CO_2 absorption, NaOH has the advantage of high load capacity and fast reaction time due to the high binding energy of chemical sorbents [63]. Inorganic bicarbonate for the cultivation of microalgae can be supplied by chemical absorption of gaseous CO_2, resulting in soluble inorganic carbon compounds such as $NaHCO_3$ and Na_2CO_3 [63–65]. This approach to sanitization can be applied to the cultivation of other species of microalgae and cyanobacteria, which require large amounts of bicarbonate for growth.

Spirulina platensis is a blue-green alga with many beneficial properties, including its centuries-long use as a food source in Africa, the Americas and Asia. *S. platensis* contains high-quality proteins, pigments, carbohydrates, lipids and many other essential nutrients, including vitamins and anti-oxidants [66]. It is generally cultivated in Zarrouk medium, which contains sodium bicarbonate (76 wt. %) as a major component. The potential ability of NaOH to chemisorb CO_2 was investigated, for chemical sterilization and to determine whether $NaHCO_3$ in standard Zarrouk medium could be replaced by NaOH for the cultivation of *S. platensis* (Fig. 9).

Bicarbonate (16.8 g L^{-1}) in standard Zarrouk medium was found to be completely replaceable by chemical CO_2 absorption using 0.2 M NaOH, with 1 L medium absorbing 154.2 mmol CO_2 gas in the form of $NaHCO_3/Na_2CO_3$ during this process. Incorporation of this process into *Spirulina* cultivation reduced the total cost of preparing Zarrouk medium by 34.3% without reducing biomass and pigment production. The advantages of this CO_2 convergence process using NaOH for *S. platensis* cultivation include the rapid chemical absorption of CO_2 gas, the reduction of cultivation costs by replacing $NaHCO_3$ with NaOH in standard Zarrouk medium, the biological conversion of fixed CO_2 into valuable compounds, and the reduction of sterilization costs.

8.5.2 A simple method for decomposition of peracetic acid in a microalgal cultivation system

Several chemicals have been tested for their ability to act as liquid sterilizing agents, including glutaraldehyde, hydrogen peroxide, sodium hypochlorite, alcohol and peracetic acid (PAA). PAA is widely used in food processing and handling, as a sanitizer for surfaces with food contact, and as a disinfectant for fruits, vegetables, meat and eggs [68]. Due to its high oxidizing potential, even at very low concentrations, PAA is considered as ideal antimicrobial agent. It has been shown to disinfect by oxidizing the outer membranes of vegetative bacterial cells, endospores, yeast cells and mold spores *via* formation of hydroxyl- and acetyl-hydroxyl radicals. These reactions can lead to cell lysis and death of all types of microorganisms [69].

Using chemical components, a resource-friendly process has been developed for the direct degradation of PAA. This process allows the direct cultivation of microorganisms without subsequent costly and time-consuming washing processes (Fig. 10). The decomposition of PAA and hydrogen peroxide (H_2O_2), which is formed by the reaction of PAA and water, was also catalyzed by ferric iron (III), *via* a process based on the Fenton reaction [70]. In addition to ferric iron, pH and organic compounds also play important roles in the effective decomposition of PAA reagents [71, 72]. The direct decomposition of PAA within PBRs has several advantages, including low energy and material costs, low contamination risk during the sterilization process, reduced requirement for water, applicability to reactors made of heat-sensitive materials, avoidance of harmful end compounds and production of acetic acid as a secondary carbon source enhancing cell growth. *Chlorella vulgaris*, a green freshwater alga, is widely cultivated in mass culture systems due to its high growth rate and high lipid productivity [73], while *Synechocystis* PCC 6803 is one of the best characterized

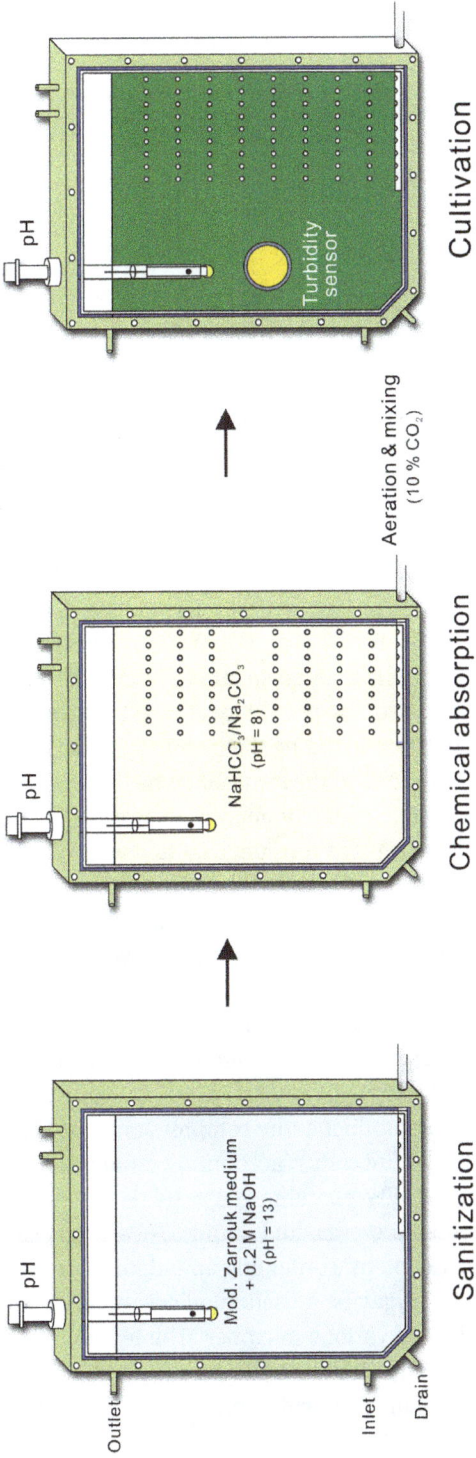

Fig. 9: Overall process of chemisorption of CO_2 by NaOH solution from sanitization to cultivation [67].

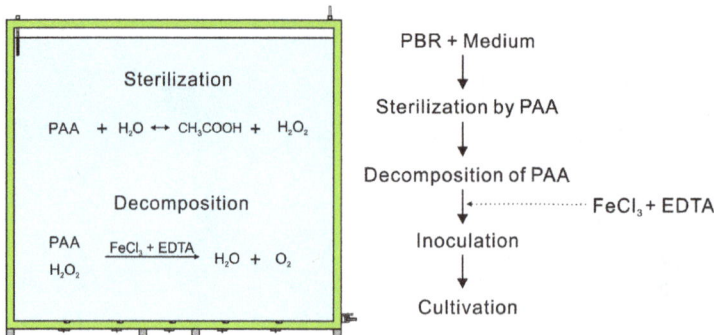

Fig. 10: Process flow diagram from sterilization to direct cultivation. In the absence of additional costly washing steps, sterilization by and decomposition of PAA were found to reduce the PBR-driven maintenance costs [70].

cyanobacterial strains with primary interest for biofuel production [74]. *Aurantiochytrium* sp. is a marine microalga that can produce significant quantities of omega-3 fatty acids, such as docosahexaenoic acid (DHA) [75, 76]. The fact that degradation of PAA enhanced the growth of *C. vulgaris*, *Synechocystis* PCC6803 and *Aurantiochytrium* sp. in TAP, BG11 and M7 media, respectively, suggests the high potential of this approach for a wide application in bioprocess engineering. Also, the possibility of direct cultivation following the total decomposition of PAA was confirmed by the healthy growth of all three species, with the additional carbon source due to the sterilization method having a positive growth effect. Additionally, gas chromatography analysis showed a higher amount of omega-3 fatty acids (DHA) in cells grown in a medium sterilized with PAA at all glucose levels tested [77].

Although culture medium is usually sterilized by autoclaving, this leads to the formation of hydroxymethylfurfural (HMF) via a Maillard reaction between amino acids and reducing sugars [78]. HMF has been shown to inhibit the growth and metabolism of microorganisms but may be carcinogenic to humans [79]. For this reason, sterilization by treatment with PAA has an especially great potential when wastewater is used as a microbial growth medium.

Soybean curd wastewater (SCWW) is a by-product of the compression of protein coagulants during the production of soybean curd (tofu), accounting for about 50% of the total amount of added water [80]. SCWW contains 2–3% solids, of which about 16% are nitrogen compounds, such as proteins, and about 53% are saccharides [81–83]. As SCWW has a high polymer content, containing abundant inorganic salts and lacking toxic compounds [84–86], it may be a useful and cost-efficient alternative substrate for the industrial growth of various microbes. The fermentative potential of SCWW was examined by assessing its ability to act as a nutrient source for the cultivation of *Aurantiochytrium* [87]. Maillard reaction products in the SCWW medium, induced rapidly by heat or slowly by long-term incubation, depressed cell growth. Sterilization of SCWW medium by PAA-treatment can allow SCWW to successfully replace commercial nutrients (glucose, peptone and yeast extract), resulting in a fourfold higher production of omega-3 fatty acids than the commercial nutrients [87].

These results indicate that the use of an appropriate sterilization process is critically important for high biomass enrichment when using organic wastewater as a source of microbial nutrients. However, although chemical sterilization followed by direct degradation of PAA seems to be a cost-efficient, eco-friendly process, this process also increases organic compounds which may result in microbial regrowth due to the remaining acetic acid. This may limit its application to closed systems such as PBRs.

References

[1]   Borowitzka MA. Commercial Production of Microalgae: Ponds, Tanks, and Fermenters. Progress in Industrial Microbiology. Vol. 35, Elsevier, 1999, 313–21.
[2]   Brennan L, Owende P. Biofuels from microalgae –a review of technologies for production, processing, and extractions of biofuels and co-products. Renewable Sustainable Energy Rev 2010, 14(2), 557–77.
[3]   Moheimani NR, Isdepsky A, Lisec J, Raes E, Borowitzka MA. Coccolithophorid algae culture in closed photobioreactors. Biotechnol Bioeng 2011, 108(9), 2078–87.
[4]   Pulz O, Scheibenbogen K. Photobioreactors: Design and Performance with Respect to Light Energy Input. Bioprocess and Algae Reactor Technology, Apoptosis. Springer, 1998, 123–52.
[5]   Singh R, Sharma S. Development of suitable photobioreactor for algae production–A review. Renewable Sustainable Energy Rev 2012, 16(4), 2347–53.
[6]   Posten C. Design principles of photo-bioreactors for cultivation of microalgae. Eng Life Sci 2009, 9(3), 165–77.
[7]   Mehler AH. Studies on reactions of illuminated chloroplasts. II. Stimulation and inhibition of the reaction with molecular oxygen. Arch Biochem Biophys 1951, 34(2), 339–51.
[8]   Aro E-M, Virgin I, Andersson B. Photoinhibition of photosystem II. Inactivation, protein damage and turnover. Biochimica et Biophysica Acta (BBA)-Bioenergetics 1993, 1143(2), 113–34.
[9]   Fujita Y, Murakami A, Aizawa K, Ohki K. Short-term and Long-Term Adaptation of the Photosynthetic Apparatus: Homeostatic Properties of Thylakoids. The Molecular Biology of Cyanobacteria. Springer, 1994, 677–92.
[10] Joshua S, Mullineaux CW. Phycobilisome diffusion is required for light-state transitions in cyanobacteria. Plant Physiol 2004, 135(4), 2112–9.
[11] Singh AK, Bhattacharyya-Pakrasi M, Elvitigala T, Ghosh B, Aurora R, Pakrasi HB. A systems-level analysis of the effects of light quality on the metabolism of a cyanobacterium. Plant Physiol 2009, 151(3), 1596–608.
[12] Chisti Y. Biodiesel from microalgae. Biotechnol Adv 2007, 25(3), 294–306.
[13] Kwon J-H, Rögner M, Rexroth S. Direct approach for bioprocess optimization in a continuous flat-bed photobioreactor system. J Biotechnol 2012, 162(1), 156–62.

[14] McDermott AE, Yachandra VK, Guiles R, Britt R, Dexheimer S, Sauer K, et al. Low-potential iron-sulfur centers in photosystem I: An X-ray absorption spectroscopy study. Biochemistry 1988, 27(11), 4013–20.

[15] Molik S, Karnauchov I, Weidlich C, Herrmann RG, Klösgen RB. The Rieske Fe/S protein of the cytochromeb 6/f complex in chloroplasts missing link in the evolution of protein transport pathways in chloroplasts? J Biol Chem 2001, 276(46), 42761–6.

[16] Yu L, Bryant DA, Golbeck JH. Evidence for a mixed-ligand [4Fe-4S] cluster in the C14D mutant of PsaC. Altered reduction potentials and EPR spectral properties of the FA and FB clusters on rebinding to the P700-FX core. Biochemistry 1995, 34(24), 7861–8.

[17] Badarau A, Firbank SJ, Waldron KJ, Yanagisawa S, Robinson NJ, Banfield MJ, et al. FutA2 is a ferric binding protein from *Synechocystis* PCC 6803. J Biol Chem 2008, 283(18), 12520–7.

[18] Shcolnick S, Summerfield TC, Reytman L, Sherman LA, Keren N. The mechanism of iron homeostasis in the unicellular cyanobacterium *Synechocystis* sp. PCC 6803 and its relationship to oxidative stress. Plant Physiol 2009, 150(4), 2045–56.

[19] Foster JS, Havemann SA, Singh AK, Sherman LA. Role of mrgA in peroxide and light stress in the cyanobacterium *Synechocystis* sp. PCC 6803. FEMS Microbiol Lett 2009, 293(2), 298–304.

[20] Katoh H, Hagino N, Ogawa T. Iron-binding activity of FutA1 subunit of an ABC-type iron transporter in the cyanobacterium *Synechocystis* sp. strain PCC 6803. Plant Cell Physiol 2001, 42(8), 823–7.

[21] Lewin A, Moore GR, Le Brun NE. Formation of protein-coated iron minerals. Dalton Trans 2005, 22, 3597–610.

[22] Shcolnick S, Shaked Y, Keren N. A role for mrgA, a DPS family protein, in the internal transport of Fe in the cyanobacterium *Synechocystis* sp. PCC6803. Biochimica et Biophysica Acta (BBA)-Bioenergetics 2007, 1767(6), 814–9.

[23] Falk S, Samson G, Bruce D, Huner NP, Laudenbach DE. Functional analysis of the iron-stress induced CP 43′ polypeptide of PS II in the cyanobacterium *Synechococcus* sp. PCC 7942. Photosynth Res 1995, 45(1), 51–60.

[24] Laudenbach D, Reith M, Straus N. Isolation, sequence analysis, and transcriptional studies of the flavodoxin gene from *Anacystis nidulans* R2. J Bacteriol 1988, 170(1), 258–65.

[25] Straus NA. Iron Deprivation: Physiology and Gene Regulation. The Molecular Biology of Cyanobacteria. Springer, 1994, 731–50.

[26] Burnap RL, Troyan T, Sherman LA. The highly abundant chlorophyll-protein complex of iron-deficient *Synechococcus* sp. PCC7942 (CP43 [prime]) is encoded by the isiA gene. Plant Physiol 1993, 103(3), 893–902.

[27] Park YI, Sandström S, Gustafsson P, Öquist G. Expression of the isiA gene is essential for the survival of the cyanobacterium *Synechococcus* sp. PCC 7942 by protecting photosystem II from excess light under iron limitation. Mol Microbiol 1999, 32(1), 123–9.

[28] Rakhimberdieva MG, Vavilin DV, Vermaas WF, Elanskaya IV, Karapetyan NV. Phycobilin/chlorophyll excitation equilibration upon carotenoid-induced non-photochemical fluorescence quenching in phycobilisomes of the cyanobacterium *Synechocystis* sp. PCC 6803. Biochimica et Biophysica Acta (BBA)-Bioenergetics 2007, 1767(6), 757–65.

[29] Barbosa MJ, Albrecht M, Wijffels RH. Hydrodynamic stress and lethal events in sparged microalgae cultures. Biotechnol Bioeng 2003, 83(1), 112–20.

[30] Janssen M, Tramper J, Mur LR, Wijffels RH. Enclosed outdoor photobioreactors: Light regime, photosynthetic efficiency, scale-up, and future prospects. Biotechnol Bioeng 2003, 81(2), 193–210.

[31] Contreras A, García F, Molina E, Merchuk J. Interaction between CO2-mass transfer, light availability, and hydrodynamic stress in the growth of *Phaeodactylum tricornutum* in a concentric tube airlift photobioreactor. Biotechnol Bioeng 1998, 60(3), 317–25.

[32] Sung M-G, Shin W-S, Kim W, Kwon J-H, Yang J-W. Effect of shear stress on the growth of continuous culture of *Synechocystis* PCC 6803 in a flat-panel photobioreactor. Korean J Chem Eng 2014, 31(7), 1233–6.

[33] Mata TM, Martins AA, Caetano NS. Microalgae for biodiesel production and other applications: A review. Renewable Sustainable Energy Rev 2010, 14(1), 217–32.

[34] Kim CW, Sung M-G, Nam K, Moon M, Kwon J-H, Yang J-W. Effect of monochromatic illumination on lipid accumulation of *Nannochloropsis gaditana* under continuous cultivation. Bioresour Technol 2014, 159, 30–5.

[35] Barreiro DL, Prins W, Ronsse F, Brilman W. Hydrothermal liquefaction (HTL) of microalgae for biofuel production: State of the art review and future prospects. Biomass Bioenergy 2013, 53, 113–27.

[36] Biller P, Ross A. Potential yields and properties of oil from the hydrothermal liquefaction of microalgae with different biochemical content. Bioresour Technol 2011, 102(1), 215–25.

[37] Yu G, Zhang Y, Schideman L, Funk T, Wang Z. Hydrothermal liquefaction of low lipid content microalgae into bio-crude oil. Trans ASABE 2011, 54(1), 239–46.

[38] Anandarajah K, Mahendraperumal G, Sommerfeld M, Hu Q. Characterization of microalga *Nannochloropsis* sp. mutants for improved production of biofuels. Appl Energy 2012, 96, 371–7.

[39] Choi J-I, Yoon M, Joe M, Park H, Lee SG, Han SJ, et al. Development of microalga *Scenedesmus dimorphus* mutant with higher lipid content by radiation breeding. Bioprocess Biosyst Eng 2014, 37(12), 2437–44.

[40] Courchesne NMD, Parisien A, Wang B, Lan CQ. Enhancement of lipid production using biochemical, genetic and transcription factor engineering approaches. J Biotechnol 2009, 141 (1–2), 31–41.

[41] Doan TTY, Obbard JP. Enhanced intracellular lipid in *Nannochloropsis* sp. via random mutagenesis and flow cytometric cell sorting. Algal Res 2012, 1(1), 17–21.

[42] Vigeolas H, Duby F, Kaymak E, Niessen G, Motte P, Franck F, et al. Isolation and partial characterization of mutants with elevated lipid content in *Chlorella sorokiniana* and *Scenedesmus obliquus*. J Biotechnol 2012, 162(1), 3–12.

[43] Crispin JC, Martinez A, Alcocer-Varela J. Quantification of regulatory T cells in patients with systemic lupus erythematosus. J Autoimmun 2003, 21(3), 273–6.

[44] Hyka P, Lickova S, Přibyl P, Melzoch K, Kovar K. Flow cytometry for the development of biotechnological processes with microalgae. Biotechnol Adv 2013, 31(1), 2–16.

[45] Karpagam R, Preeti R, Ashokkumar B, Varalakshmi P. Enhancement of lipid production and fatty acid profiling in *Chlamydomonas reinhardtii*, CC1010 for biodiesel production. Ecotoxicol Environ Saf 2015, 121, 253–7.

[46] Shin W-S, Lee H, Sung M-G, Hwang K-T, Jung SM, Kwon J-H. Enrichment as a screening method for a high-growth-rate microalgal strain under continuous cultivation system. Biotechnol Bioprocess Eng 2016, 21(2), 268–73.

[47] Wilhelm C, Jakob T. From photons to biomass and biofuels: Evaluation of different strategies for the improvement of algal biotechnology based on comparative energy balances. Appl Microbiol Biotechnol 2011, 92(5), 909–19.

[48] Ooms MD, Dinh CT, Sargent EH, Sinton D. Photon management for augmented photosynthesis. Nat Commun 2016, 7(1), 1–13.

[49] Ra C-H, Kang C-H, Jung J-H, Jeong G-T, Kim S-K. Effects of light-emitting diodes (LEDs) on the accumulation of lipid content using a two-phase culture process with three microalgae. Bioresour Technol 2016, 212, 254–61.

[50] Bland E, Angenent LT. Pigment-targeted light wavelength and intensity promotes efficient photoautotrophic growth of Cyanobacteria. Bioresour Technol 2016, 216, 579–86.

[51] Hultberg M, Jönsson HL, Bergstrand K-J, Carlsson AS. Impact of light quality on biomass production and fatty acid content in the microalga *Chlorella vulgaris*. Bioresour Technol 2014, 159, 465–7.

[52] Khalili A, Najafpour GD, Amini G, Samkhaniyani F. Influence of nutrients and LED light intensities on biomass production of microalgae *Chlorella vulgaris*. Biotechnol Bioprocess Eng 2015, 20(2), 284–90.

[53] Kumar MS, Hwang J-H, Abou-Shanab RA, Kabra AN, Ji M-K, Jeon B-H. Influence of CO_2 and light spectra on the enhancement of microalgal growth and lipid content. J Renewable Sustainable Energy 2014, 6(6), 063107.

[54] Mohsenpour SF, Richards B, Willoughby N. Spectral conversion of light for enhanced microalgae growth rates and photosynthetic pigment production. Bioresour Technol 2012, 125, 75–81.

[55] Mohsenpour SF, Willoughby N. Effect of CO_2 aeration on cultivation of microalgae in luminescent photobioreactors. Biomass Bioenergy 2016, 85, 168–77.

[56] Okumura C, Saffreena N, Rahman MA, Hasegawa H, Miki O, Takimoto A. Economic efficiency of different light wavelengths and intensities using LEDs for the cultivation of green microalga *Botryococcus braunii* (NIES-836) for biofuel production. Environ Prog Sustain Energy 2015, 34(1), 269–75.

[57] Teo CL, Atta M, Bukhari A, Taisir M, Yusuf AM, Idris A. Enhancing growth and lipid production of marine microalgae for biodiesel production via the use of different LED wavelengths. Bioresour Technol 2014, 162, 38–44.

[58] Vadiveloo A, Moheimani NR, Kosterink NR, Cosgrove JJ, Parlevliet D, Gonzalez-Garcia C, et al. Photosynthetic performance of two *Nannochloropsis* spp. under different filtered light spectra. Algal Res 2016, 19, 168–77.

[59] Moazami N, Ashori A, Ranjbar R, Tangestani M, Eghtesadi R, Nejad AS. Large-scale biodiesel production using microalgae biomass of *Nannochloropsis*. Biomass Bioenergy 2012, 39, 449–53.

[60] Simionato D, Sforza E, Carpinelli EC, Bertucco A, Giacometti GM, Morosinotto T. Acclimation of *Nannochloropsis gaditana* to different illumination regimes: Effects on lipids accumulation. Bioresour Technol 2011, 102(10), 6026–32.

[61] Shin W-S, Jung SM, Cho C-H, Woo D-W, Kim W, Kwon J-H, et al. Use of tar color additives as a light filter to enhance growth and lipid production by the microalga *Nannochloropsis gaditana*. Environ Eng Res 2018, 23(2), 205–9.

[62] Lubián LM, Montero O, Moreno-Garrido I, Huertas IE, Sobrino C, González-del Valle M, et al. *Nannochloropsis* (Eustigmatophyceae) as source of commercially valuable pigments. J Appl Phycol 2000, 12(3–5), 249–55.

[63] Peng Y, Zhao B, Li L. Advance in post-combustion CO_2 capture with alkaline solution: A brief review. Energy Procedia 2012, 14, 1515–22.

[64] González-López C, Acién Fernández F, Fernández-Sevilla J, Sánchez Fernández J, Molina Grima E. Development of a process for efficient use of CO_2 from flue gases in the production of photosynthetic microorganisms. Biotechnol Bioeng 2012, 109(7), 1637–50.

[65] Moran R. Formulae for determination of chlorophyllous pigments extracted with N, N-dimethylformamide. Plant Physiol 1982, 69(6), 1376–81.

[66] Belay A, Ota Y, Miyakawa K, Shimamatsu H. Current knowledge on potential health benefits of *Spirulina*. J Appl Phycol 1993, 5(2), 235–41.

[67] Jung J-Y, Yang J-W, Kim K, Hwang K-T, Jung SM, Kwon J-H. Cost-efficient cultivation of *Spirulina platensis* by chemical absorption of CO_2 into medium containing NaOH. Korean J Chem Eng 2015, 32(11), 2285–9.

[68] Baldry M. The bactericidal, fungicidal and sporicidal properties of hydrogen peroxide and peracetic acid. J Appl Bacteriol 1983, 54(3), 417–23.

[69] Pedersen L-F, Meinelt T, Straus DL. Peracetic acid degradation in freshwater aquaculture systems and possible practical implications. Aquacult Eng 2013, 53, 65–71.

[70] Sung M-G, Lee H, Nam K, Rexroth S, Rögner M, Kwon J-H, et al. A simple method for decomposition of peracetic acid in a microalgal cultivation system. Bioprocess Biosyst Eng 2015, 38(3), 517–22.

[71] Chi Z, Pyle D, Wen Z, Frear C, Chen S. A laboratory study of producing docosahexaenoic acid from biodiesel-waste glycerol by microalgal fermentation. Process Biochem 2007, 42(11), 1537–45.

[72] Floch J. A simple method for the isolation and purification of total lipids from animal tissues. J biol Chem 1957, 226, 497–509.

[73] Lardon L, Hélias A, Sialve B, Steyer J-P, Bernard O. Life-Cycle Assessment of Biodiesel Production from Microalgae. ACS Publications, 2009.

[74] Parmar A, Singh NK, Pandey A, Gnansounou E, Madamwar D. Cyanobacteria and microalgae: A positive prospect for biofuels. Bioresour Technol 2011, 102(22), 10163–72.

[75] Gao M, Song X, Feng Y, Li W, Cui Q. Isolation and characterization of *Aurantiochytrium* species: High docosahexaenoic acid (DHA) production by the newly isolated microalga, *Aurantiochytrium* sp. SD116. J Oleo Sci 2013, 62(3), 143–51.

[76] Kim K, Kim EJ, Ryu B-G, Park S, Choi Y-E, Yang J-W. A novel fed-batch process based on the biology of *Aurantiochytrium* sp. KRS101 for the production of biodiesel and docosahexaenoic acid. Bioresour Technol 2013, 135, 269–74.

[77] Cho C-H, Shin W-S, Woo D-W, Kwon J-H. Growth medium sterilization using decomposition of peracetic acid for more cost-efficient production of omega-3 fatty acids by *Aurantiochytrium*. Bioprocess Biosyst Eng 2018, 41(6), 803–9.

[78] Bhat H, Qazi G, Chopra C. Effect of 5-hydroxymethylfurfural on production of citric acid by *Aspergillus niger*. Indian J Exp Biol 1984.

[79] Modig T, Lidén G, Taherzadeh MJ. Inhibition effects of furfural on alcohol dehydrogenase, aldehyde dehydrogenase and pyruvate dehydrogenase. Biochem J 2002, 363(3), 769–76.

[80] Kim H, Eom K, Kim J, Kim W. Drying of isoflavone and oligosaccharides retentates separated by membrane filtration from tofu sunmul. Food Eng Prog 2005.

[81] Belén F, Benedetti S, Sánchez J, Hernández E, Auleda J, Prudêncio E, et al. Behavior of functional compounds during freeze concentration of tofu whey. J Food Eng 2013, 116(3), 681–8.

[82] Belén F, Sánchez J, Hernández E, Auleda J, Raventós M. One option for the management of wastewater from tofu production: Freeze concentration in a falling-film system. J Food Eng 2012, 110(3), 364–73.

[83] Chua J-Y, Liu S-Q. Soy whey: More than just wastewater from tofu and soy protein isolate industry. Trends Food Sci Technol 2019, 91, 24–32.

[84] Mahdiana A, Siregar AS, Januar CS, Prayogo NA, editors. The effect in the wastewater treatment at soybean curd of contact time modification of artificial wetland using SSF by using schoenoplectus corymbosus to improve water quality. E3S Web of Conferences; 2018: EDP Sciences.

[85] Wang S-K, Wang X, Miao J, Tian Y-T. Tofu whey wastewater is a promising basal medium for microalgae culture. Bioresour Technol 2018, 253, 79–84.

[86] Wang Y, Serventi L. Sustainability of dairy and soy processing: A review on wastewater recycling. J Clean Prod 2019, 237, 117821.

[87] Lee G-I, Shin W-S, Jung SM, Kim W, Lee C, Kwon J-H. Effects of soybean curd wastewater on growth and DHA production in *Aurantiochytrium* sp. LWT 2020, 134, 110245.

Felix Melcher*, Michael Paper*, Thomas B. Brück

9 Photosynthetic conversion of CO_2 into bioenergy and materials using microalgae

Abstract: Fuel production for mobility is highly dependent on finite fossil resources, such as oil. The energetic use of these fossil resources has led to excessive CO_2 emissions, which result in the climate change effects that endanger the global ecosystem today. In order to mitigate CO_2-emissions, fast-growing phototropic organisms, such as microalgae have the capacity for efficient CO_2 fixation. The resulting biomass can be converted into advanced biofuels and CO_2 negative advanced materials, such as carbon fiber that can be applied in lightweight building applications. Due to their efficient photosynthetic mechanisms, microalgae have several advantages over land plants like higher growth rates or land efficiency. Nevertheless, only a small percentage of today's commercial biofuel production is based on algae biomass because of high infrastructure development and process costs. Intensive research and advances in technology might facilitate the use of microalgae as commercially viable energy source in the future. This chapter provides an overview over current application areas of microalgae for the production of sustainable biofuels and biomaterials. The last section focuses especially on potential applications of algae biomass as starting material for jet fuels in the aviation industry and the production of carbon fibers.

9.1 Goal

The goal of this chapter to give is a short overview of methods for the production of microalgae-based materials and bioenergy. Later on, challenges and opportunities for the aviation industry will be pointed out using the benefits of microalgae.[*]

9.2 Basic background

The global demand for energy has been steadily increasing for the past decades due to economic growth around the world and an increased global population. So far technological advances have not been able to reduce total CO_2 emission sufficiently to effectively combat the impact on climate change. The vast majority of nations declared to fight climate change in the Paris Agreement of 2015 by stopping the rising atmospheric emissions of greenhouse gases rapidly and effectively. Thus, alternative

---

* Both authors contributed equally to this manuscript

https://doi.org/10.1515/9783110716979-009

energy sources that are sustainable and do not depend on fossil fuels are greatly gaining in importance.

Biofuels have the potential to replace fossil fuels in many applications but their production often requires vast areas of land that could be used for food production [1]. To be beneficial, biofuels must be produced without impacting arable land or tropical rainforests and provide significant greenhouse-gas emission savings compared to fossil fuels [2]. The higher costs of biofuels, which is mainly due to the high expenses in biomass production and processing, are the main obstacles for its broader commercialization [3, 4].

9.2.1 Potential and advantages of algae cultivation

Microalgae have a vast range of industrial applications such as wastewater treatment, high-value food, cosmetic and pharmaceutical products, pigments, feedstock for aquacultures or bioplastic production [5–8]. Currently microalgae are being promoted as an ideal third-generation biofuel feedstock because of their rapid growth rate, high photosynthetic efficiency and CO_2 fixation ability [9]. Biomass production with microalgae has several advantages in comparison to conventional cultivation of land plants. As unicellular organisms, microalgae lack non-photogenic structures (i.e., stems or roots). Compared to land plants they also need fewer structural polysaccharides like cellulose [10]. Additionally, microalgae are more photosynthetically efficient and have a higher productivity than terrestrial plants. As a consequence, algae have a greater capacity to generate and store carbon resources [11]. Many land plants that can be used for the production of biofuels go through a cycle of growth and maturation. This limits the total amount of possible harvests to only a few per year.

On the other hand, microalgae cultures can be harvested frequently throughout the year. Areas that are not favorable for terrestrial plants can be used nonetheless to grow microalgae. This can circumvent the problem of competition for areas that can be deployed for food production [12].Taking all of these factors into account, the productivity per space is significantly higher in microalgae cultures compared to land-based crops (Tab. 1).

Furthermore, microalgae are especially interesting for the production of renewable biofuels as many known species contain lipids and/or different sugars as storage compounds. As we know, each biochemical fraction – lipids, carbohydrates and proteins – can be used for the production of biofuels through different processes. Thus, biofuels that can potentially be produced from algal biomass include biodiesel from lipids, bioethanol or biobutanol from sugars, hydrogen, methane, long-chain hydrocarbons and crude oils from the pyrolysis of algal biomass [13–16].

Tab. 1: Comparison of selected sources of biodiesel [12].

Crop	Oil yield (L ha^{-1})
Corn	172
Soybean	446
Canola	1190
Jatropha	1892
Coconut	2689
Palm	5950
Microalgae[1]	58,700
Microalgae[2]	136,900

[1] 30% oil in biomass (by weight).
[2] 70% oil in biomass (by weight).

9.2.2 Strain selection

A critical step in the development of a commercially viable process is the selection of a suitable microalgae strain. An estimated 35,000 microalgae species have been described to date, even though the actual total number is most likely much higher [17]. Only few species of microalgae have achieved commercial importance. These include *Arthrospira* (*Spirulina*), *Chlorella*, *Haematococcus*, *Microchloropsis* and *Dunaliella* [18, 19]. The extensive biodiversity of microalgae species presents both an opportunity and a challenge in the selection process.

As biofuels are low-value-added products, low production costs are key in order to compete with conventional fossil fuels. Intracellular lipid or carbohydrate content is one of the characteristics that play a central role for the economic viability of microalgae-based production processes [20]. In addition, biomass productivity, harvesting and downstream processing have to be taken into consideration.

For the production of bioenergy, especially biofuels, it is desirable to use algae strains that combine a rapid growth rate and a high lipid content (Tab. 2). However, strains with lower oil content grow generally faster than strains which exhibit high production of oil. Thus, total lipid production with a fast-growing algae strain might be superior to a slowly growing strain with high lipid content due to total production of biomass [23].

Nevertheless, total lipid production is only one of many aspects that should to be considered. Other relevant factors for strain selection can be: lipid content, distribution of free fatty acids and triglycerides, growth rate, CO$_2$ tolerance and uptake, nutrient demand, tolerance toward changes in environmental conditions, and the

Tab. 2: Lipid content of different microalgae [21, 22].

Microalgae species	Lipid content % dry weight biomass	Microalgae species	Lipid content % dry weight biomass
Ankistrodesmus species	28–40	Scenedesmus dimorphus	16–40
Anabaena cylindrica	4–7	Euglena gracilis	14–20
Botryococcus braunii	25–86	Haemotococcus pluvialis	25
Chaetoceros mueller	33	Isochrysis galbana	21.2
Chlorella emersonii	25–63	Nannochloropsis sp.	20–56
Chlorella minutissima	57	Nitschia closterium	27.8
Chlorella prototothecoides	14–57	Prostanthera incisa	62
Chlorella sorokiniana	22	Prymnesium parvum	22–39
Chlorella vulgaris	14–56	Pyrrosia laevis	69.1
Crypthecodinium cohnii	20–51	Arthrospira plantensis	16.6
Dunaliella primolecta	23	Tetraselmis suecia	15–23
Dunaliella salina	28.1	Thalassiosira pseudonana	20
Dunaliella tertiolecta	36–42	Zitzschia sp.	45–47

competition with other microalgae or bacteria [24, 25]. Tolerance toward salinity of microalgae is oftentimes also an important characteristic for strain selection. Saline water such as seawater or wastewater is a preferable source for biofuel production with microalgae in a sustainable commercial scale as it does not compete with freshwater that could be used for land-based agriculture [26]. Algal cultures growing in high-salinity media tend to grow under more alkaline conditions which in turn facilitate the effective conversion of gaseous CO_2 to water soluble bicarbonate (CO_3^{2-}). The latter can directly be assimilated by microalgae and converted into biomass. High salinity furthermore inhibits the growth of terrestrial culture contaminates, such as bacteria and filamentous fungi, thereby significantly enhancing process robustness even in open pond cultivation systems. This results in an increased long-term stability of algae cultures that are not cultivated under sterile conditions.

9.3 Methods involved

This subchapter provides a brief overview of microalgae utilization for the production of bioenergy, encompassing biomass production, harvesting, biomass processing and the subsequent conversion into different biofuels (Fig. 1).

The basis for the production of renewable biofuels is the transformation of solar energy into biomass with microalgae. Compared to terrestrial plants harvested, microalgae biomass possesses higher moisture content and has to go through different dewatering process steps before being transformed into biofuels. Currently both the cultivation of microalgae and the reduction of the moisture content in harvested biomass are the most important cost factors that limit the commercial viability of algal-derived biofuels [4].

Depending on the composition, algal biomass can be converted into different biofuels. While lipids are predominantly converted into biodiesel or hydrocarbons, carbohydrates and proteins can be used for the production of biogas, hydrogen or bioethanol. Nevertheless, there are numerous processes that can be applied in order to convert algal biomass into bioenergy. This subchapter chiefly focuses on the production of biofuels designated for the mobility sector.

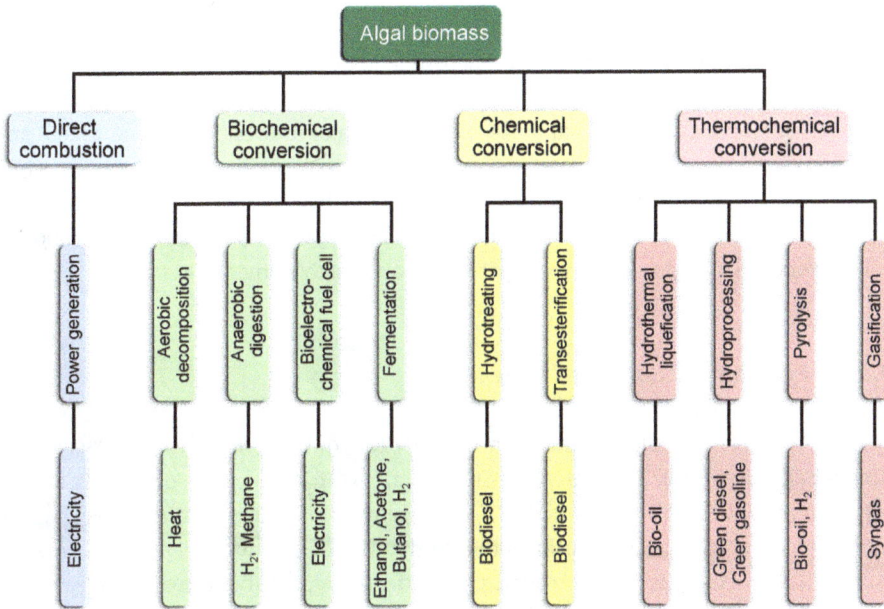

Fig. 1: Overview of conversion pathways for algal biomass to energy (adapted from [24]).

9.3.1 Biomass production – cultivation strategies

The main purpose of algae cultivation is the production of biomass that can be further processed to biofuels or biomaterials. The production of microalgal biomass can be carried out in open ponds or in closed systems like photobioreactors. Even though both systems have different advantages and drawbacks, open ponds are most widely used for the commercial production of microalgae today [27].

9.3.1.1 Open pond

Biofuels are a low-value product. Hence, the cultivation of microalgae for the production of biofuels on a commercial-scale demands the ability to produce large quantities of biomass in a cheap process. Oftentimes this requires culture volumes from 10,000 to greater than 1,000,000 L [28]. Therefore, the vast majority of algal biomass production is currently located outdoors in open pond systems [19]. Open ponds can have various shapes and sizes depending on the location, algae strain and produced product. A frequently used design for the cultivation of microalgae are raceway ponds with a paddle wheel to improve circulation [29]. One important disadvantage of open systems compared to photobioreactors is that cultivation parameters are more difficult to control. Besides relatively poor mixing, a long light path and low photosynthetic efficiency can lead to lower biomass concentrations, volumetric productivity and CO_2 uptake [30]. As there is no artificial lighting and temperature control, maximum biomass productivity is usually attained in subtropical and tropical areas [22]. Other major drawbacks of open ponds are the high loss of water through evaporation, higher space requirements and the exposure to biological contaminants [31, 32]. Even though productivity and growth rates are significantly lower compared to photobioreactors, classical open pond systems are economically favorable so far due to cheaper operational and establishment costs.

9.3.1.2 Photobioreactor/closed systems

Closed systems allow algae cultivation with better control over relevant process parameters like aeration, light intensity or mixing. So far, a multitude of different reactor designs have been developed [33–36]. Many photobioreactor systems can be sterilized which theoretically allows the cultivation of axenic algal cultures. As the cultivation process can be controlled more effectively, algal cultures usually reach much higher growth rates and cell densities compared to open ponds. To achieve a more cost-effective production of biomass closed systems are especially interesting for the cultivation of algae stains that can grow mixotrophically or heterotrophically [37]. Even though there are promising advances and developments in the cultivation of microalgae in closed systems for the production of biofuels [38, 39], they have not caught on commercially yet, primarily due to higher investment and energy costs. Current closed systems are oftentimes solely commercially viable for high-value products but not for mass production of cheap energy carriers [40].

9.3.2 Downstream processing

In order to be converted into biofuels, algal biomass has to be further processed. The major drawback of algal biomass is its high moisture content after harvest making it unsuitable for the direct conversion into biofuels in many processes. The energy-efficient removal of moisture is one major challenge to improve the commercial competitiveness for microalgae-based biofuels. Algal biomass can later on be separated into lipids and so-called residue with various processes (Fig. 2). The oil, on the one hand, can then be converted into biodiesel or hydrocarbon-fuels, while the residue (carbohydrates and proteins), on the other hand, can be utilized for the production of bioethanol or biogas. The following part gives an overview of various critical processing steps that are undertaken when converting algal-biomass into biofuels.

9.3.2.1 Harvest, dewatering and drying

Algal cultures are usually harvested after they have reached a sufficient biomass or product concentration. There are several techniques for the recovery of algal biomass. Depending on the process design and required product properties different procedures can be applied. Harvesting methods that are used in industrial processes include chemical, mechanical and biological operations (Tab. 3). Freshly harvested algal biomass typically still contains a high percentage of water. In most cases, this biomass is not suited for conversion to biofuel products until it has undergone some degree of dewatering and drying [30]. Dewatering further decreases the moisture content of harvested biomass by draining or mechanical means. The removal of water oftentimes is very energy intensive and creates significant operational costs, thereby rendering algae-based fuels less economically attractive [41, 42]. Certain properties of microalgae such as large cell size and the capability to autoflocculate can simplify the dewatering process.

After harvesting the moisture content in dewatered algae, biomass has to be further decreased in many cases, depending on the requirements for subsequent processes. Fast dehydration or drying of the harvested biomass can also extend the shelf-life and quality of the final product. Drying methods that have been used for microalgae include spray drying, drum drying, rotary drying, freeze-drying and solar drying [44]. Many methods like spray drying or freeze-drying are suitable only for high-value products and cannot be applied in the commercial production of biofuels, as a consequence of their energy or time-consuming nature [45].

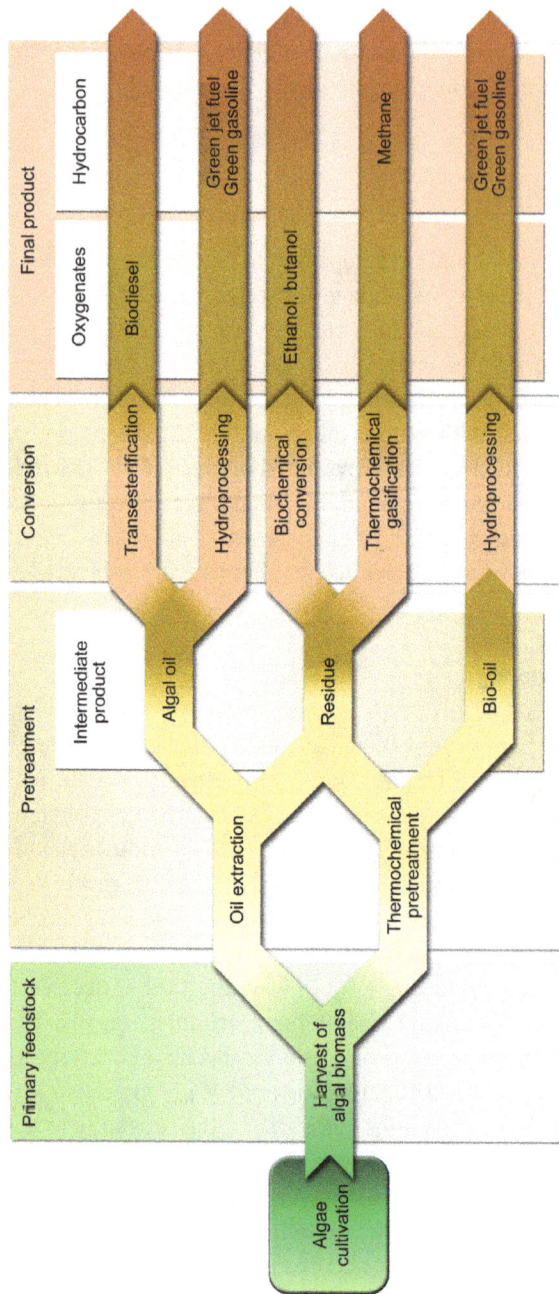

Fig. 2: Pathways for the downstream processing of biomass can follow a number of routes. Chemical makeup and quality of intermediate products (residue, algal oil, bio-oil) will play a role in determining which processes may be used, as well as how great the environmental externalities may be of a specific pathway; adapted from [30].

Tab. 3: Overview of common harvesting methods for microalgae cultures (adapted from [43]).

Method	Advantages	Disadvantages	Proportion of dry biomass after harvesting (%)
Flocculation + sedimentation of microalgal flocs	Cell recovery over 90%, wide range of flocculants available, variable price, can be low-cost	Removal of flocculants, chemical contamination, fragile flocs and/or longer settling times	3–8
Flotation	Cell recovery 50–90%, can be more rapid than sedimentation, possibility to combine with gaseous transfer	Algal species specific, high capital and operational cost, flocculants usually required	3–6
Sedimentation	Cell recovery 10–90%, low cost, potential for use as a first stage to reduce energy input and cost of subsequent stages	Algal species specific, best suited to dense (heavy) non-motile cells, separation can be slow or unreliable, low final concentration	0.5–3
Filtration	Cell recovery 70–90%, wide variety of filter and membrane types available, reliable, can handle delicate cells	Highly dependent on algal species, best suited to large algal cells, clogging or fouling an issue, high capital and operational costs	5–27
Centrifugation	Cell recovery over 90%, can handle most algal types with rapid efficient cell harvesting, reliable	High capital and operational costs, energy intensive	12–22

9.3.2.2 Oil extraction

As many algae species accumulate lipids inside their cells, a prominent way to produce algal-oil is the direct extraction from algal biomass. Depending on the strain and environmental conditions, lipids can make up more than 50% of total dry matter [46]. Compared to oil that is produced by land plants, algal oil contains a higher proportion of unsaturated fatty acids [47].

Oil extraction from algal biomass yields algal-oil and a residue primarily consisting of carbohydrates, proteins and minerals. The methods of oil extraction from algal biomass can be classified into mechanical and chemical methods. Chemical methods are generally solvent assisted extractions, for instance, Soxhlet extraction or supercritical fluid extraction, while mechanical methods include sonication, grinding or mechanical expulsion [30]. Since many algae species have multilayered cell walls, a combination of mechanical methods with chemical methods is deployed in the extraction process for the disruption of algal cell walls and an enhancement of the overall lipid yield. However, the cost- and energy-intensive mechanical methods for cell disruption and lipids extraction could render the process less attractive and limit its commercial viability [48].

9.3.2.3 Thermochemical pretreatment

Algal biomass can be pretreated thermochemically via pyrolysis or hydrothermal liquefication. Among other byproducts, these methods yield intermediary bio-oil which in turn can be further upgraded by hydroprocessing.

Pyrolysis

Pyrolysis is a thermochemical pretreatment process in which biomass is degraded with high temperatures in the absence of oxygen. Typically, temperatures between 500 and 600 °C are applied [30]. This method usually requires the feedstock to have a moisture content of around 10% which complicates the application to algal biomass. However, numerous studies were carried out on algae, including *Nannochloropsis* sp., *Dunaniella* sp. *or Chlorella* sp. [49–51]. The pyrolysis of algae yields bio-oil, gaseous products and biochar. Pyrolytic bio-oils have a relatively high oxygen content which causes a lower heating value, a low vapor pressure, low thermal stability and a high nitrogen content [52]. In order to avoid those undesirable aspects of bio-oil and to use it as transportation fuel, the oxygen and nitrogen content has to be reduced in additional process steps.

Hydrothermal liquefication

Hydrothermal liquefication (HTL) of algal biomass is conducted at temperatures between 200 °C and 400 °C in a pressurized water environment [53]. HTL is a versatile method as it can be applied to untreated, wet algal biomass. Therefore, energy intensive downstream process steps like the reduction of moisture in harvested algal biomass can be minimized. Even though the lipid fraction is typically the main target of oil production, HTL can convert both protein and carbohydrates into a bio-oil [54]. During liquefication biomass is decomposed into smaller, unstable molecules that can repolymerize into oily compounds. The main products of HTL are bio-oil, solid residue and an aqueous phase [52]. HTL-derived bio-oil has a lower oxygen content and a better overall quality than bio-oil from pyrolysis [55]. However, bio-oil derived from HTL still needs subsequent processing to remove oxygen and nitrogen in order to be utilized as a transportation fuel.

9.3.2.4 Pathways for oil conversion to biofuels

Algal oil is chemically distinct from bio-oil derived from thermochemical treatment and must therefore be converted to biofuel under different conditions [30]. Process parameters hereby can vary, for example, in temperature, pressure and catalyst type. Apart from bioethanol and biodiesel, biofuels are currently predominantly developed and produced as "drop-in biofuels." Drop-in biofuels can be defined as liquid biohydrocarbons that are functionally equivalent to petroleum fuels and are fully compatible with existing petroleum infrastructure [56]. This enables their application in modern engines and avoids the necessity to develop new technologies. The following part emphasizes on the most relevant biofuels for the mobility sector in Europe – biodiesel and hydrocarbons.

Biodiesel
Following the extraction of biomass algal oil or bio-oil, collected oils can be used for the production of biodiesel. Biodiesel has been defined as the monoalkyl esters of long-chain fatty acids derived from renewable feedstocks, such as vegetable or algal oil [57]. Despite the existence of different methods to produce biodiesel, transesterification is most commonly used to chemically convert raw plant or algal oil into biodiesel, which in turn can be used in conventional engines [58]. For the transesterification reaction, oil or fat and a short-chain alcohol (usually methanol) are used as reagents in the presence of a catalyst [59]. Different methods, for example, acid-catalyzed, alkali-catalyzed or enzymatic transesterification, can be applied [60]. In the course of the conversion, triacylglycerides (TAGs) are separated into fatty acid methyl esters and glycerol (Fig. 3). Glycerol has to be removed continuously during the process to enable the conversion toward the desired products. It can later be used as feedstock for mixotrophic or heterotrophic cultivation of certain algae species or alternatively for the production of carbon fibers (see 9.4.4 Biomaterials: Carbon fiber).

Hydrocarbons
Hydrocarbons include fuels such as gasoline or jet fuel which do not contain oxygen but exclusively hydrogen and carbon. Even though some algae strains like *Botryococcus braunii* are capable of producing hydrocarbons [16], the majority of microalgae primarily form oxygen molecule-containing lipids as TAGs. Bioderived hydrocarbon fuels are fabricated via thermochemical conversion from algal or bio-oil in which the chemical structure of lipids is either modified or oxygen-containing groups are removed in a process called hydroprocessing (Fig. 4) [24]. These biofuels are often referred to as green fuels. Mechanisms for the reduction of oxygen content in algae-based fuels

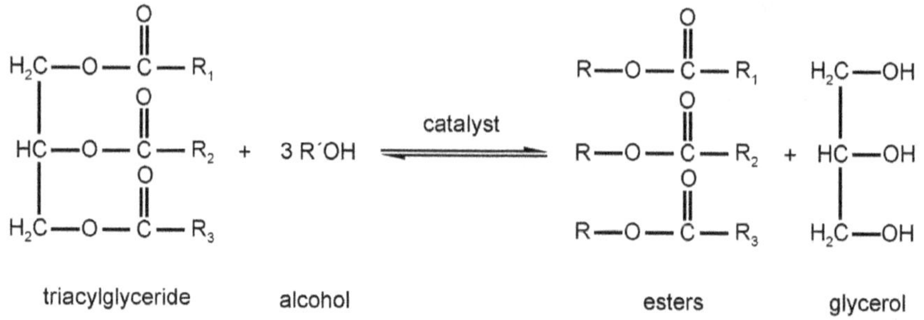

Fig. 3: Reaction scheme for the catalytic transesterification of triacylglycerides (adapted from [61]).

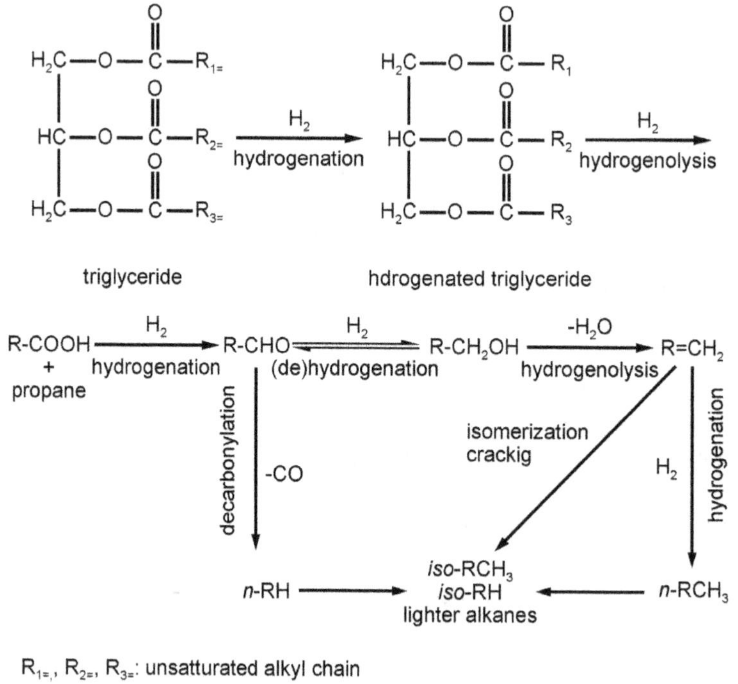

$R_{1=}$, $R_{2=}$, $R_{3=}$: unsatturated alkyl chain

Fig. 4: Proposed reactions pathway for a metal-catalyzed transformation of bio-oil to hydrocarbon fuels that can be used in transportation industry; adapted from [63].

include hydrodeoxygenation, decarboxylation/decarbonylation and catalytic cracking [62]. Hydrodeoxygenation eliminates oxygen through the formation of water, while decarboxylation and decarbonylation remove oxygen by the formation of CO_2 and CO, respectively. These three mechanisms might happen simultaneously in a single hydroprocessing reaction.

9.4 State of the art

9.4.1 Algae-based alternatives in the mobility sector

The mobility sector is a key player in the emission of greenhouse gases all over the world. For the automotive industry, different technologies to reduce the production of greenhouse gases have already been established. One viable option is to use electric power instead of fuels. Contrarily, a main disadvantage of this technology is the high demand for rare earth elements and metals, for instance, lithium. The latter is often produced under dubious conditions regarding human rights and the environment. The generation of electricity is often based on fossil sources related to a high release of greenhouse gases. Another possibility would be biofuels derived from sustainable sources like ethanol or biodiesel. Especially drop in biofuels as biodiesel can represent a short-term solution to reduce emissions in the automotive sector. There are already different generations of biofuels. The first one was based on biomolecules like sugar or proteins, but these molecules are in competition to the food industry. Therefore, the second generation is based on waste stream of the agriculture and food industry, while the third generation is based on microalgae or cyanobacteria. In order to increase the productivity and reduce the production costs, methods of genetic engineering can be applied to photosynthetic organisms, for example, to optimize the CO$_2$ uptake and the carbon flux. The biofuels based on these genetically modified organisms are known as the fourth-generation biofuels [64, 65].

9.4.2 Jet fuel

While there are different new technologies in the automotive industry using electricity, the aviation industry would be based on liquid fuels for the time being. Although several studies regarding electric flight have been conducted, especially for long distance flights this does not represent a viable alternative. Currently an annual increase of about 5% for the international aviation industry is predicted. This translates into an increase of about 400 million tons in jet fuels per year in 2020. This demand is mainly complied with fossil sources. To face the increasing demand of jet fuel while reducing the emission of greenhouse gases, the EU founded the "Flightpath 2050" project [66]. One major point of this project is the reduction of the carbon dioxide emissions within the aviation industry by 75% in comparison to the year 2005. Strikingly, this value surpasses even the ambitious value of 25% mentioned by the Air Transport Action Group. The International Air Transport Association (IATA) showed in their report "Vision 2050" that a reduction of the carbon dioxide emissions of 50% cannot be reached using solely technical improvements or an increase in the efficiency.

One of the most sought-after solutions mentioned by the IATA is the combination of technical improvement and the usage of biogenic jet fuels. These biofuels should be a drop-in, energy rich and liquid jet fuel derived from biological sources with a high efficiency in carbon dioxide fixing. Consequently, the importance of such a biobased alternative is much more profound if the value of 75% should be met. One of the main challenges in this field is the economic and ecological production of biobased jet fuels in the scale of the market's demands.

The guideline ASTM D7566 for the usage of jet fuel mixtures containing synthetic hydrocarbons, and also biomass-based HEFA fuels, was continuously updated by the IATA. The latest version is from 2019 [67]. In Annex A2 and A4 it is mentioned that mixtures with HEFA fuels have to be assessed and if the evaluations are positive, they can be equally deployed as petroleum-based fuels authorized according to ASTM D1655 [68]. As a result, it is allowed to test 1:1 mixtures of HEFA- and conventional fuels. In this regard, a consortium of the National Aeronautics and Space Administration, the Deutsches Luft- und Raumfahrt Zentrum, the University of Vienna, the Ludwig-Maximillian University, the California state University and the Gutenberg-University Mainz published a comprehensive study for the behavior of mixtures containing biofuels in the normal flight operation [69]. In this study it could be illustrated how the lower amount of sulfur in the biobased fuels had a reductive effect on the emission of fine particles (>10 nm) and carbon monoxide under all boost types (low, medium and high). In comparison to the conventional JetA fuel, the amount of fine particles could be halved in about 50% of the experiments. Contrarily, the amounts of emitted carbon monoxide in the process of combustion could not be decreased extensively, as the amount was between 80% and 100% compared to conventional JetA fuel.

This promising study, illustrating the positive environmental effects under normal flight operations in addition to the regulatory guidelines ASTM D7566-19d, which allow for the utilization of 1:1 mixtures of sustainable, bio-based HEFA fuels and conventional jet fuels, paved the way for future research in this field and ultimately production of biobased jet fuels.

Microalgae-based jet fuels

One suitable host for the production of biofuels, especially jet fuel, are microalgae. Microalgae are very efficient in carbon capture, grow about ten times faster than land crops and produce lipids up to 70% of their cell dry weight. In addition, microalgae can be cultivated all over the world in open pond systems on non-arable lands. Furthermore, most of the strains exploited for the production of oil are able to grow in waste-, brackish or sea water [70–72]. For a cost efficient production of algae-based jet fuels, it is pivotal to choose a rapidly growing algae strain with a high capacity for oil production.

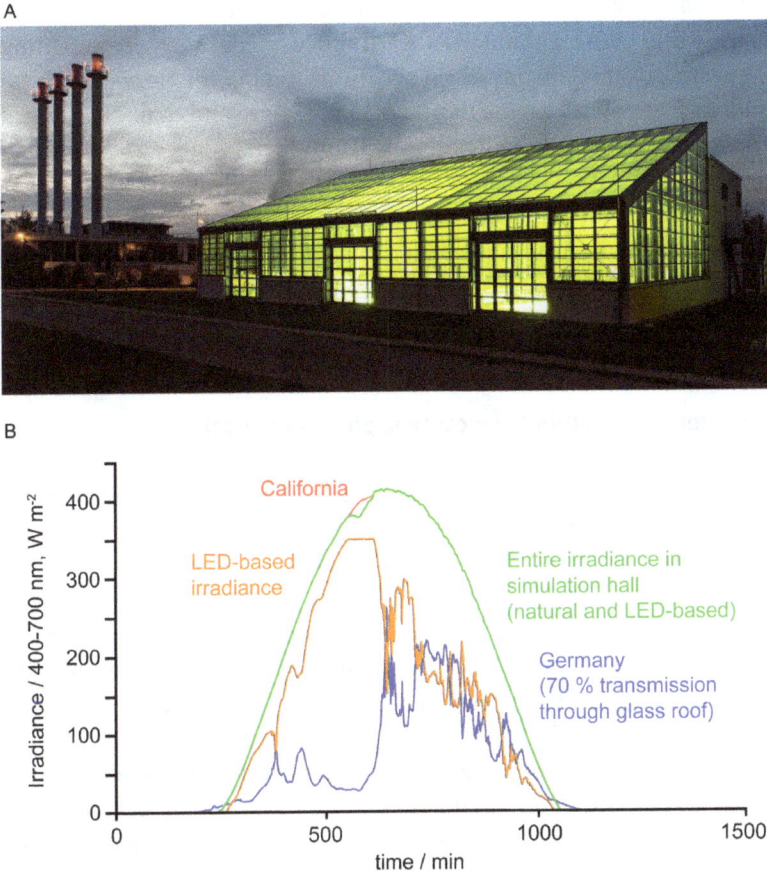

Fig. 5: TUM-AlgaeTec Center (Technical University of Munich, Germany) and an example for the simulated light spectrum of California. In red the light spectrum of California, in blue the light spectrum of Germany, in orange the emitted light by a combination of the eight different LED types, in green the resulted spectrum of the irradiance.

In this regard, the saline alga *Microchloropsis salina* represents one promising candidate, with an intracellular lipid production of up to 50% in relation to its cell dry weight. This microalga exhibits high potential for various technologies in the area of bioenergy and biomaterial. Hereby, a study has showcased how *M. salina* can be exploited for the biogas production in a comparable way to corn silage or corn-crob-mix [73]. Furthermore, this microalga is cultivated in the TUM-AlgaeTec Center (Technical University of Munich, Germany) (Fig. 5). It embodies one of the world's largest research centers for the mass cultivation of microalgae. The cultivation can be performed in three independent cultivation halls, 200 m^2 each, mainly built of highly UV-transparent specialty glass. Both the light spectrum and intensity of the day light are measured in real time by a sensor on the roof of the building.

Additionally, the day light can be supplemented by one of the largest high-voltage LED plants in the world with eight different, partly customized, LED types emitting distinct light spectra to simulate any given spectrum. Taking into account the facilities for irradiance coupled with the thorough temperature and humidity control comprising the three independent cultivation halls, the TUM-AlgaeTec Center represents the first building with a realistic climate control worldwide [74]. The condition simulation necessary for an efficient algae mass cultivation in the pilot scale enables an accelerated commercialization of different microalgae-based processes.

9.4.3 Technical improvements

9.4.3.1 Optimization of open thin-layer cascade photobioreactors

For the efficient production of microalgal biomass, low-cost open photobioreactor productions systems that achieve high productivities and product concentrations are required [75, 76]. Therefore, thin-layer cascade photobioreactors with a surface of 8 square meters were installed and optimized in several different studies concerning the production of biofuels. Hereby, the carbon dioxide conversion efficiency could be increased to about 84% by optimization of the mechanical carbon dioxide supply, a process-related control of the alkalinity ($\leq$15 mM) combined with the usage of urea as nitrogen source instead of potassium nitrate [77]. This efficiency is quite high in comparison to raceway ponds (13–20%) or 70% for other thin-layer photobioreactors [78]. Lipid accumulation is caused by external stressors, for example nutrient starvation like nitrogen limitation. During nutrient starvation the cells start to accumulate lipids but stop building up additional biomass. It could be shown in a batch process with different initial urea concentrations of 0.6 or 1.2 g L^{-1} that the alga reached a lipid amount of about 46% or 42% of its dry weight, which translates into 5.7 or 6.6 g L^{-1} of lipids. The initial urea is consumed over the cultivation time which leads to a nitrogen limitation. Accordingly, the lower initial urea concentration leads to a higher amount of intracellular lipids, with the drawback of a lower cell density observed during the cultivation [77]. This yield of lipids is one of the highest of its kind for a phototrophic fermentation, using open bioreactors. Remarkably, it is also profoundly higher when compared to the published yield of 1.7 g L^{-1} that could be reached employing the freshwater algae *Chlorella vulgaris*, for example [79]. In addition, the lipid productivity of 0.34 g L^{-1} day^{-1} per 7 days and the productivity of 0.19 g L^{-1} day^{-1} for the whole cultivation time of 35 days is rather high in comparison to existing literature.

9.4.3.2 Establishment of a continuous cultivation with high lipid yields

In an industrial scale, a continuous cultivation procedure is oftentimes used instead of a batch process, that is because it is not only easier to keep the conditions constant in an optimal range over a longer time but also to avoid down times of the bioreactors which in turn leads to a higher space-time yield. For the efficient production of lipids using microalgae, a two staged process is necessary (Fig. 6). Since the lipid accumulation is induced by nutrient starvation, this comes with negative implications on the growth of the culture. In order to realize such a process, thin-layer cascade photobioreactors were coupled in a cascade. In the first photobioreactor, the microalga is cultivated without any limitations to rapidly build up biomass. Therefore, nutrients were added continuously, being directly consumed by the alga due to the high biomass productivity of the culture. The expiration of the first bioreactor is transferred to the second one without any addition of nutrients, which induces the lipid accumulation. During this process, a biomass concentration of 10–11 g L^{-1} dry weight with about 3.0–3.5 g L^{-1} lipid yield could be kept continuously for a time span of 15 days. This converts into a space-time yield of about 0.2 g L^{-1} day^{-1} [77]. In comparison to other continuous processes described in the literature, this space-time yield is much higher [80–82].

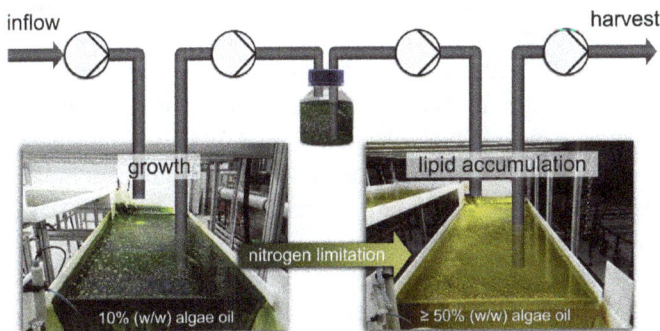

Fig. 6: Flowchart of the two-staged reactor cascade for the continuous production of algae oil induced by nitrogen limitation in the second stage.

9.4.3.3 Wastewater recycling to increase the cultivation efficiency and decrease cultivation costs

Another improvement of the cultivation of microalgae for the production of oil for biofuel production in a sustainable and economical manner is the wastewater recycling. One critical cost factor in the production of microalgal biomass are fertilizers for the nutrient supply [83]. Beside high productivity of the photobioreactors, recycling of the used process water in order to recover residual nutrients can prove helpful in cost reduction of medium preparation and wastewater treatment [71, 84]. The usage of urea as nitrogen source instead of potassium nitrate also has a beneficial effect on wastewater recycling. While the growth rate of *M. salina* is decreasing over time using potassium nitrate, it was stable over several batch processes using the recycled wastewater [85]. With the aid of this specific wastewater recycling process, all the nutrients in the medium responsible for good growth can be recovered without a loss of these compounds. Missing nutrients have to be replenished during the cultivation. This renders the whole process more sustainable, increasing the efficiency and lowering the cost of the cultivation.

9.4.4 Biomaterials: Carbon fiber – a high-performance material

Most of the biofuels are derived from fatty acids, but the storage lipids of most microalgae are triglycerides. For the production of biofuels, these triglycerides have to be split in free fatty acids and glycerol. Subsequently, the free fatty acids can be converted into different types of biofuel. Accordingly, the glycerol can also be used for the production of biofuels or alternatively as a feed stock for other microorganisms. Another possible application is the conversion of the purified algae-derived glycerol into acrylonitrile. In a following step this can be polymerized into polyacrylonitrile (PAN) and spanned to PAN fibers, which represents a universal precursor for carbon fibers (Fig. 7) [86]. Carbon fibers are a high-performance material that is mostly used in carbon fiber composites (CFC) blended with different matrix materials. Due to the high strength of the resulting material and its low weight in comparison with other materials, carbon fiber composites embody a viable alternative to steel, aluminum, and concrete in technical applications, such as aircraft, automotive, and building construction [87, 88]. The resulting weight reduction of air planes and cars comes with beneficial effects on the fuel demand and can thereby assist in reducing the carbon dioxide emissions over the life time of these vehicles [88, 89].

To reach the goals of the Paris Agreement of 2015, the mere substitution of fossil energy supplies by renewables will not be sufficient for a timely compliance with these goals. To that end, carbon dioxide sinks with lasting greenhouse gas extraction and permanent storage effects have to be installed in an industrially relevant scale [90]. At present, carbon fibers are mostly produced out of fossil-based oleochemicals with high carbon dioxide emissions [91, 92]. One possible realization of such a sink is the production of carbon fibers out of oil rich algal biomass [72].

Microalgal-based carbon fibers represent a two-dimensional opportunity to reduce the carbon dioxide emissions within the aviation industry. On the one hand, the usage of lightweight material reduces the weight of the air planes and the coupled demand of jet fuels. On the other hand, the usage of carbon dioxide–based biofuels

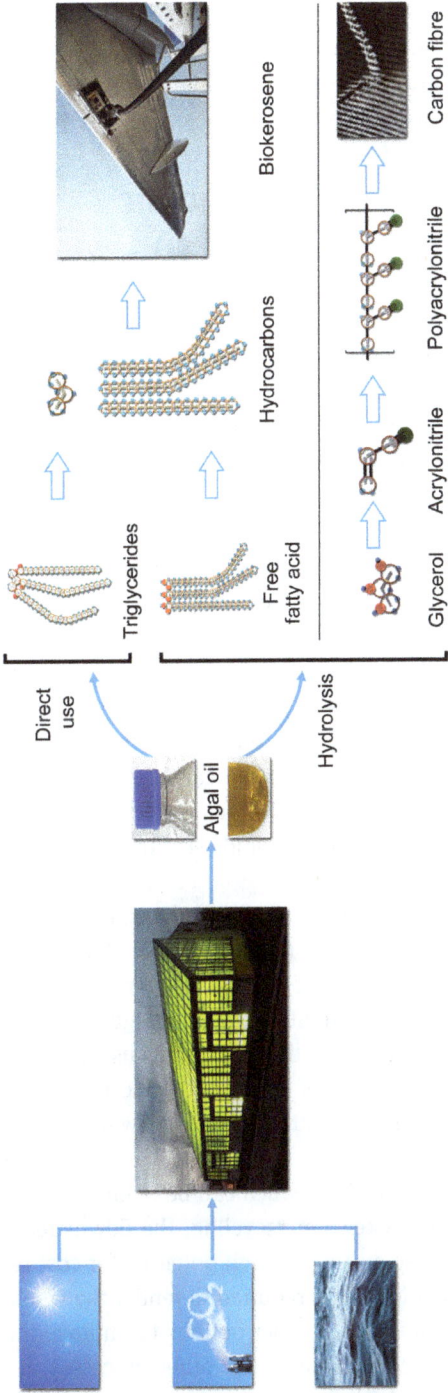

Fig. 7: Conversion routes of algal oil to biokerosene and carbon fiber.

can reduce the carbon dioxide emissions from fossil sources. Jet fuel is a low-cost product needed in high amounts; the carbon fiber is a high-value product and can increase the economic feasibility of the production of microalgae biomass.

9.5 Outlook: future perspective and economic feasibility

Reaching the goals of the Paris Agreement of 2015 is a severe challenge. To meet these goals, it is necessary to find renewable and sustainable alternatives for the different fields of modern life and technology. Microalgae and their advantages over land crops can do their part to fight climate change. Especially in the last few years, the increasing interest in the research of microalgae came with a lot of ideas, solutions and new techniques.

Innovative technologies for the cultivation of microalgae are necessary. This is especially true for the production of biofuels and other mass productions, as there exists a high demand which can be met only with efficient open pond cultivation systems in an industrial scale. In that regard, the generation of algal biomass in raceway ponds is cheap but comes with a low efficiency. On the opposite, closed photobioreactors with a high efficiency are expensive and not feasible for low-value products [4]. A viable alternative for the efficient production of algal biomass and products thereof could be open thin-layer cascade photobioreactors, as the employed ones in the TUM AlgaeTec Center. This type of photobioreactor combines a high microalgae cultivation efficiency while being cheap and easy to install with the optimizations shown in the pilot scale.

For a cost-efficient, sustainable production of, for example, biofuels, the knowledge has to be transferred from the pilot scale into the industrial scale. Areas with suitable environmental conditions for the production of microalgal biomass can be found nearly all over the world. Even if the environmental impact of algae farms in the industrial scale and economic factors for the production of biofuels using microalgae are taken into account, there is a suitable area of about 1,420,000 km^2 using fresh, brackish and seawater microalgae and 132,000 km^2 if using seawater microalgae exclusively [12].

In addition to the cost scenarios for the cultivation, which can be reduced by the optimization of the cultivation system or by wastewater recycling, the downstream processing is a major cost driver. This field also subsists a high potential for cost reduction, but new, efficient methods and techniques are required to render the whole process more sustainable and economically feasible. Further options to improve the economic viability of microalgae involve the combination with other processes, for example, the usage of waste stream from the dairy industry or sewage for the cultivation. In this scenario, outgas of power plants or devices utilized in the production process can be exploited as carbon dioxide sources for the cultivation and hence directly be converted into biomass.

Furthermore, to date the price of biofuels is too high in comparison to fossil fuels. To achieve the goals set by the Paris Agreement in 2015, it is of great importance to support these technologies. This could be done by implementing a carbon tax for fossil-based fuels in order to render them more expensive. In addition, fossil resources will eventually dwindle and it will become increasingly expensive to find and use reservoirs of fossil sources. Therefore, the price of fossil resources is bound to spiral upward over time.

9.5.1 Two-step process for the production of microalgae-based biogenic oils

The accumulation of lipids in microalgae is often induced by external stress like nutrient starvation. Therefore, the lipid accumulation comes along with decrease in the growth rate. To circumvent this issue, the process can be split in two steps (Fig. 8). In the first step, the carbon dioxide is fixed by the microalgae under non-limiting conditions translating into a high growth rate. In the second step, the microalgae biomass will be hydrolyzed and used as feedstock for other oleaginous organisms. These oleaginous organisms (e.g., some yeasts) are able to constitute oil in high amounts, up to 80% of their dry weight. The oil of these organisms can be used in the same manner as oil directly derived from microalgae for the conversion into biofuels and materials. Such a process combines the high carbon capture efficiency of the microalgae with the high efficiency of lipid accumulation of other oleaginous organisms. Aside from that, the combination of this process with other techniques is conceivable, for example, the heterologous expression of a decarboxylase in an oleaginous yeast, in order to produce hydrocarbons directly out of the fatty acids [93]. Additionally, the generation of other biopolymers is possible as already published for plant-derived oils [94–96]. In this method, the fatty acids get epoxidized and polymerized with a vulcanization agent like bisphenol A.

9.5.2 Sustainable solar-based carbonization of polyacrylonitrile (PAN) to carbon fiber

Major drawbacks of the conventional carbon fiber production include the extremely energy and cost intensive conversion of the PAN fiber into carbon fibers, causing significant carbon dioxide emissions. During the production process the PAN is spun into PAN fibers and thermochemically converted into carbon fibers. Therefore, the PAN fiber is firstly stabilized (oxidized) under air atmosphere at temperatures of up to 300 °C and afterward carbonized (denitrogenated) under nitrogen inter gas

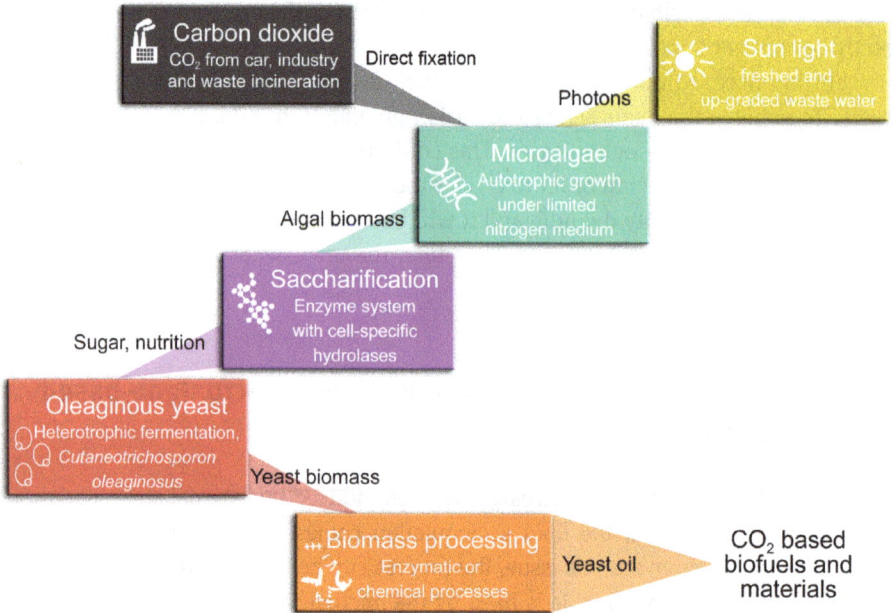

Fig. 8: Conversion routes of algal biomass to CO_2-based yeast oil and biofuels and materials derived thereof.

atmosphere in excess of 1,500 °C. For this process a typical energy demand amounts to about 340 MJ/kg without precursor production [97]. To be a suitable carbon capture and utilization, alternative new production processes with minimal energy and cost input have to be developed. Utilization of concentrated solar power for the carbonization process might provide an alternative to the conventional process. A study of Fraunhofer-Institute ILT at RWTH Aachen University was able to show that it is possible to carbonize a PAN fiber using a laser beam as a concentrated light source [98]. For the quality of resulting carbon fiber, sufficiently high temperatures are important but required temperatures should be achievable using sufficient collector aperture area and focusing quality. To that end, temperatures up to 3,200 °C for the 1 MW solar furnace at Odeillo in southern France are reported [99]. This reduction in carbon dioxide emissions and energy consumption may enable a diversified application of suitable high-quality carbon fibers in the construction, vehicle and other producing industries, leading to a permanent carbon dioxide sink [100].

References

[1]   Brennan L, Owende P. Biofuels from microalgae – a review of technologies for production, processing, and extractions of biofuels and co-products. Renewable Sustainable Energy Rev 2010, 14(2), 557–77.

[2]   Zittelli GC, Rodolfi L, Bassi N, Biondi N, Tredici MR. Photobioreactors for Microalgal Biofuel Production. Algae for Biofuels and Energy. Springer, 2013, 115–31.

[3]   Wright L. Worldwide commercial development of bioenergy with a focus on energy crop-based projects. Biomass Bioenergy 2006, 30(8–9), 706–14.

[4]   Slade R, Bauen A. Micro-algae cultivation for biofuels: Cost, energy balance, environmental impacts and future prospects. Biomass Bioenergy 2013, 53, 29–38.

[5]   Woertz I, Feffer A, Lundquist T, Nelson Y. Algae grown on dairy and municipal wastewater for simultaneous nutrient removal and lipid production for biofuel feedstock. J Environ Eng 2009, 135(11), 1115–22.

[6]   Begum H, Yusoff FM, Banerjee S, Khatoon H, Shariff M. Availability and utilization of pigments from microalgae. Crit Rev Food Sci Nutr 2016, 56(13), 2209–22.

[7]   Schwartz RE, Hirsch CF, Sesin DF, Flor JE, Chartrain M, Fromtling RE, et al. Pharmaceuticals from cultured algae. J Ind Microbiol 1990, 5(2–3), 113–23.

[8]   Maheswari NU, Ahilandeswari K. Production of bioplastic using Spirulina platensis and comparison with commercial plastic. Res Environ Life Sci 2011, 4(3), 133–6.

[9]   Dragone G, Fernandes BD, Vicente AA, Teixeira JA. Third generation biofuels from microalgae. 2010.

[10]   Lyon SR, Ahmadzadeh H, Murry MA. Algae-based wastewater treatment for biofuel production: Processes, species, and extraction methods. Biomass and Biofuels from Microalgae. Springer, 2015, 95–115.

[11]   Vassilev SV, Vassileva CG. Composition, properties and challenges of algae biomass for biofuel application: An overview. Fuel 2016, 181, 1–33.

[12]   Correa DF, Beyer HL, Possingham HP, Thomas-Hall SR, Schenk PM. Global mapping of cost-effective microalgal biofuel production areas with minimal environmental impact. GCB Bioenergy 2019, 11(8), 914–29

[13]   Torzillo G, Scoma A, Faraloni C, Ena A, Johanningmeier U. Increased hydrogen photoproduction by means of a sulfur-deprived Chlamydomonas reinhardtii D1 protein mutant. Int J Hydrogen Energy 2009, 34(10), 4529–36.

[14]   Demirbaş A. Oily products from mosses and algae via pyrolysis. Energy Sources Part A 2006, 28(10), 933–40.

[15]   Daroch M, Geng S, Wang G. Recent advances in liquid biofuel production from algal feedstocks. Appl Energ 2013, 102, 1371–81.

[16]   Metzger P, Largeau C. Botryococcus braunii: A rich source for hydrocarbons and related ether lipids. Appl Microbiol Biotechnol 2005, 66(5), 486–96.

[17]   Borowitzka MA. Species and strain selection. In: Borowitzka MA, Moheimani NR, ed. Algae for Biofuels and Energy. Dordrecht, Springer Netherlands, 2013, 77–89

[18]   Sathasivam R, Radhakrishnan R, Hashem A, Abd_Allah EF. Microalgae metabolites: A rich source for food and medicine. Saudi J Biol Sci 2019, 26(4), 709–22.

[19]   Benemann J. Microalgae for biofuels and animal feeds. Energies 2013, 6(11), 5869–86.

[20]   Sydney EB, Sydney ACN, De Carvalho JC, Soccol CR. Microalgal Strain Selection for Biofuel Production. Biofuels from Algae, Elsevier, 2019, 51–66.

[21]   Becker EW. Microalgae: Biotechnology and microbiology. Cambridge University Press, 1994.

[22] Verma NM, Mehrotra S, Shukla A, Mishra BN. Prospective of biodiesel production utilizing microalgae as the cell factories: A comprehensive discussion. Afr J Biotechnol 2010, 9(10), 1402–11.

[23] Griffiths MJ, Harrison ST. Lipid productivity as a key characteristic for choosing algal species for biodiesel production. J Appl Phycol 2009, 21(5), 493–507.

[24] Ghasemi Y, Rasoul-Amini S, Naseri A, Montazeri-Najafabady N, Mobasher M, Dabbagh F. Microalgae biofuel potentials. Appl Biochem Microbiol 2012, 48(2), 126–44.

[25] Grobbelaar JU. Physiological and technological considerations for optimising mass algal cultures. J Appl Phycol 2000, 12(3–5), 201–6.

[26] Borowitzka MA, Moheimani NR. Sustainable biofuels from algae. Mitigation Adapt Strategies Global Change 2013, 18(1), 13–25.

[27] Costa JAV, Freitas BCB, Santos TD, Mitchell BG, Morais MG. Open pond systems for microalgal culture. Biofuels from Algae. Elsevier, 2019, 199–223.

[28] Borowitzka MA. Culturing microalgae in outdoor ponds. Algal Culturing Techniques. Academic Press, 2005, 205–18.

[29] Sompech K, Chisti Y, Srinophakun T. Design of raceway ponds for producing microalgae. Biofuels 2012, 3(4), 387–97.

[30] Ryan C, Green TB, Hartley A, Browning B, Garvin C. Cultivation Clean Energy: The Promise of Algae Biofuels. 2009

[31] Wang H, Zhang W, Chen L, Wang J, Liu T. The contamination and control of biological pollutants in mass cultivation of microalgae. Bioresour Technol 2013, 128, 745–50.

[32] Day JG, Gong Y, Hu Q. Microzooplanktonic grazers–A potentially devastating threat to the commercial success of microalgal mass culture. Algal Res 2017, 27, 356–65.

[33] Delente JJ, Behrens PW, Hoeksema SD. Closed photobioreactor and method of use. Google Patents, 1992.

[34] Pulz O. Photobioreactors: Production systems for phototrophic microorganisms. Appl Microbiol Biotechnol 2001, 57(3), 287–93.

[35] Suh IS, Lee SB. A light distribution model for an internally radiating photobioreactor. Biotechnol Bioeng 2003, 82(2), 180–9.

[36] Bergmann P, Ripplinger P, Beyer L, Trösch W. Disposable flat panel airlift photobioreactors. Chem Ing Tech 2013, 85(1-2), 202–5.

[37] Behrens PW. Photobioreactors and fermentors: The light and dark sides of growing algae. Algal Culturing Techniques. 2005, 189–204.

[38] Li X, Xu H, Wu Q. Large-scale biodiesel production from microalga Chlorella protothecoides through heterotrophic cultivation in bioreactors. Biotechnol Bioeng 2007, 98(4), 764–71.

[39] Olivieri G, Salatino P, Marzocchella A. Advances in photobioreactors for intensive microalgal production: Configurations, operating strategies and applications. J Chem Technol Biotechnol 2014, 89(2), 178–95.

[40] Morweiser M, Kruse O, Hankamer B, Posten C. Developments and perspectives of photobioreactors for biofuel production. Appl Microbiol Biotechnol 2010, 87(4), 1291–301.

[41] Uduman N, Qi Y, Danquah MK, Forde GM, Hoadley A. Dewatering of microalgal cultures: A major bottleneck to algae-based fuels. J Renewable Sustainable Energy 2010, 2(1), 012701.

[42] Dismukes GC, Carrieri D, Bennette N, Ananyev GM, Posewitz MC. Aquatic phototrophs: Efficient alternatives to land-based crops for biofuels. Curr Opin Biotechnol 2008, 19(3), 235–40.

[43] Branyikova I, Prochazkova G, Potocar T, Jezkova Z, Branyik T. Harvesting of microalgae by flocculation. Fermentation 2018, 4(4), 93.

[44] Show K-Y, Lee D-J, Chang J-S. Algal biomass dehydration. Bioresour Technol 2013, 135, 720–9.

[45]   Grima EM, Belarbi E-H, Fernández FA, Medina AR, Chisti Y. Recovery of microalgal biomass and metabolites: Process options and economics. Biotechnol Adv 2003, 20(7–8), 491–515.

[46]   Choudhary AR, Karmakar R, Kundu K, Dahake V. "Algal" biodiesel: Future prospects and problems. Water Eenrgy Int 2011, 68(11), 44–51.

[47]   Huber GW, Iborra S, Corma A. Synthesis of transportation fuels from biomass: Chemistry, catalysts, and engineering. Chem Rev 2006, 106(9), 4044–98.

[48]   Zhou Y, Hu C. Catalytic thermochemical conversion of algae and upgrading of algal oil for the production of high-grade liquid fuel: A review. Catalysts 2020, 10(2), 145.

[49]   Sanchez-Silva L, López-González D, Garcia-Minguillan A, Valverde J. Pyrolysis, combustion and gasification characteristics of Nannochloropsis gaditana microalgae. Bioresour Technol 2013, 130, 321–31.

[50]   Grierson S, Strezov V, Ellem G, Mcgregor R, Herbertson J. Thermal characterisation of microalgae under slow pyrolysis conditions. J Anal Appl Pyrolysis 2009, 85(1–2), 118–23.

[51]   Aysu T, Sanna A. Nannochloropsis algae pyrolysis with ceria-based catalysts for production of high-quality bio-oils. Bioresour Technol 2015, 194, 108–16.

[52]   Saber M, Nakhshiniev B, Yoshikawa K. A review of production and upgrading of algal bio-oil. Renewable Sustainable Energy Rev 2016, 58, 918–30.

[53]   Galadima A, Muraza O. Hydrothermal liquefaction of algae and bio-oil upgrading into liquid fuels: Role of heterogeneous catalysts. Renewable Sustainable Energy Rev 2018, 81, 1037–48.

[54]   Guo Y, Yeh T, Song W, Xu D, Wang S. A review of bio-oil production from hydrothermal liquefaction of algae. Renewable Sustainable Energy Rev 2015, 48, 776–90.

[55]   Jena U, Das K. Comparative evaluation of thermochemical liquefaction and pyrolysis for bio-oil production from microalgae. Energy & fuels 2011, 25(11), 5472–82.

[56]   Karatzos S, McMillan JD, Saddler JN. The potential and challenges of drop-in biofuels. Report for IEA Bioenergy Task. 2014;, 39.

[57]   Dincer K. Lower emissions from biodiesel combustion. Energy Sources Part A 2008, 30(10), 963–8.

[58]   Ma F, Hanna MA. Biodiesel production: A review. Bioresour Technol 1999, 70(1), 1–15.

[59]   El-Sheekh M, Abomohra A. Biodiesel Production from Microalgae. Industrial Microbiology: Microbes in Action. Garg N, Aeron A, eds. New York, Nova Science Publishers, 2016, 355–66.

[60]   Otera J. Transesterification. Chem Rev 1993, 93(4), 1449–70.

[61]   Pereira CO, Portilho MF, Henriques CA, Zotin FM. SnSO4 as catalyst for simultaneous transesterification and esterification of acid soybean oil. J Braz Chem Soc 2014, 25(12), 2409–16.

[62]   Yang C, Li R, Cui C, Liu S, Qiu Q, Ding Y, et al. Catalytic hydroprocessing of microalgae-derived biofuels: A review. Green Chem 2016, 18(13), 3684–99.

[63]   Peng B, Yao Y, Zhao C, Lercher JA. Towards quantitative conversion of microalgae oil to diesel-range alkanes with bifunctional catalysts. Angew Chem Int Ed 2012, 51(9), 2072–5.

[64]   Mat Aron NS, Khoo KS, Chew KW, Show PL, Chen WH, Nguyen THP. Sustainability of the four generations of biofuels–A review. Int J Energy Res 2020, 44(12), 9266–82.

[65]   Lü J, Sheahan C, Fu P. Metabolic engineering of algae for fourth generation biofuels production. Energy Environ Sci 2011, 4(7), 2451–66.

[66]   (Transport) ECD-GfRalafMa. Flightpath 2050 – Europe's Vision for Aviation. Publication Office of the European Union (Luxembourg). 2011.

[67]   International A. ASTM D7566-19b: Standard Specification for Aviation Turbine Fuel Containing Synthesized Hydrocarbons. 2019.

[68]   International A. ASTM D1655-19a: Standard Specification for Aviation Turbine Fuels. 2019.

[69]   Moore RH, Thornhill KL, Weinzierl B, Sauer D, D'Ascoli E, Kim J, et al. Biofuel blending reduces particle emissions from aircraft engines at cruise conditions. Nature 2017, 543 (7645), 411–5.

[70]   Slocombe SP, Zhang Q, Ross M, Anderson A, Thomas NJ, Lapresa Á, et al. Unlocking nature's treasure-chest: Screening for oleaginous algae. Sci Rep 2015, 5, 9844.

[71]   Zhu L. Microalgal culture strategies for biofuel production: A review. Biofuels Bioprod Biorefin 2015, 9(6), 801–14.

[72]   Rosello-Sastre R. Products from microalgae: An overview. In: Posten C, Walter C, eds. Microalgal Biotechnology: Integration and Economy. De Gruyter, 2012.

[73]   Schwede S. Untersuchung und Optimierung des Biogasbildungspotentials der marinen Mikroalge\(\textit{Nannochloropsis salina}\). 2014.

[74]   Apel A, Pfaffinger C, Basedahl N, Mittwollen N, Göbel J, Sauter J, et al. Open thin-layer cascade reactors for saline microalgae production evaluated in a physically simulated Mediterranean summer climate. Algal Res 2017, 25, 381–90.

[75]   Apel AC, Weuster-Botz D. Engineering solutions for open microalgae mass cultivation and realistic indoor simulation of outdoor environments. Bioprocess Biosyst Eng 2015, 38(6), 995–1008.

[76]   Grivalský T, Ranglová K, Da Câmara Manoel JA, Lakatos GE, Lhotský R, Masojídek J. Development of thin-layer cascades for microalgae cultivation: Milestones. Folia microbiologica. 2019, 1–12.

[77]   Schädler T, Cerbon DC, De Oliveira L, Garbe D, Brück T, Weuster-Botz D. Production of lipids with Microchloropsis salina in open thin-layer cascade photobioreactors. Bioresour Technol 2019, 289, 121682.

[78]   Doucha J, Lívanský K. Productivity, CO 2/O 2 exchange and hydraulics in outdoor open high density microalgal (Chlorella sp.) photobioreactors operated in a Middle and Southern European climate. J Appl Phycol 2006, 18(6), 811–26.

[79]   Přibyl P, Cepák V, Zachleder V. Production of lipids in 10 strains of Chlorella and Parachlorella, and enhanced lipid productivity in Chlorella vulgaris. Appl Microbiol Biotechnol 2012, 94(2), 549–61.

[80]   San Pedro A, González-López C, Acién F, Molina-Grima E. Outdoor pilot-scale production of Nannochloropsis gaditana: Influence of culture parameters and lipid production rates in tubular photobioreactors. Bioresour Technol 2014, 169, 667–76.

[81]   San Pedro A, González-López C, Acién F, Molina-Grima E. Outdoor pilot production of Nannochloropsis gaditana: Influence of culture parameters and lipid production rates in raceway ponds. Algal Res 2015, 8, 205–13.

[82]   Sung M-G, Lee B, Kim CW, Nam K, Chang YK. Enhancement of lipid productivity by adopting multi-stage continuous cultivation strategy in Nannochloropsis gaditana. Bioresour Technol 2017, 229, 20–5

[83]   Laurens LM, Chen-Glasser M, McMillan JD. A perspective on renewable bioenergy from photosynthetic algae as feedstock for biofuels and bioproducts. Algal Res 2017, 24, 261–4.

[84]   Farooq W, Suh WI, Park MS, Yang J-W. Water use and its recycling in microalgae cultivation for biofuel application. Bioresour Technol 2015, 184, 73–81.

[85]   Schädler T, Neumann-Cip A-C, Wieland K, Glöckler D, Haisch C, Brück T, et al. High-density microalgae cultivation in open thin-layer cascade photobioreactors with water recycling. Appl Sci 2020, 10(11), 3883.

[86]   Arnold U, Brück T, De Palmenaer A, Kuse K. Carbon capture and sustainable utilization by algal polyacrylonitrile fiber production: Process design, techno-economic analysis, and climate related aspects. Ind Eng Chem Res 2018, 57(23), 7922–33.

[87] Eickenbusch H, Krauss O. Kohlenstofffaserverstärkte Kunststoffe im Fahrzeugbau–Ressourceneffizienz und Technologien. 3. überarbeitete Auflage. Berlin, VDI Zentrum Ressourceneffizienz GmbH (VDI ZRE), 2016, Accessed February 8, 2021, at http://www.ressource-deutschland.de/fileadmin/user_upload/downloads/kurzanalysen/2014-Kurzanalyse-03-VDI-ZRE-CFK.pdf.

[88] Lässig R, Eisenhut M, Mathias A, Schulte RT, Peters F, Kühmann T, et al. Serienproduktion von hochfesten Faserverbundbauteilen. Perspektiven für den deutschen Maschinen-und Anlagenbau, Studie Roland Berger. 2012.

[89] Yuan Z, Eden MR, Gani R. Toward the development and deployment of large-scale carbon dioxide capture and conversion processes. Ind Eng Chem Res 2016, 55(12), 3383–419

[90] Metz B. Adjusting the EU's climate targets to meet the Paris Agreement. 2016.

[91] Lisienko V, Chesnokov YN, Lapteva A, Noskov VY, ed. Types of greenhouse gas emissions in the production of cast iron and steel. IOP Conference Series: Materials Science and Engineering; 2016: IOP Publishing.

[92] Ma F, Sha A, Yang P, Huang Y. The greenhouse gas emission from Portland cement concrete pavement construction in China. Int J Environ Res Public Health 2016, 13(7), 632.

[93] Bruder S, Moldenhauer EJ, Lemke RD, Ledesma-Amaro R, Kabisch J. Drop-in biofuel production using fatty acid photodecarboxylase from Chlorella variabilis in the oleaginous yeast Yarrowia lipolytica. Biotechnol Biofuels 2019, 12(1), 202.

[94] Lligadas G, Ronda JC, Galia M, Cadiz V. Renewable polymeric materials from vegetable oils: A perspective. Mater Today 2013, 16(9), 337–43.

[95] Saurabh T, Patnaik M, Bhagt S, Renge V. Epoxidation of vegetable oils: A review. Int J Adv Eng Technol 2011, 2(4), 491–501.

[96] Wang R, Schuman TP. Vegetable oil-derived epoxy monomers and polymer blends: A comparative study with review. Express Polym Lett 2013, 7(3), 272–92.

[97] Liddell H, Brueske S, Carpenter A, Cresko J, ed. Manufacturing energy intensity and opportunity analysis for fiber-reinforced polymer composites and other lightweight materials. Proceedings of the American Society for Composites: Thirty-First Technical Conference; 2016.

[98] Lott P, Stollenberg J. Laserbasierte effiziente Herstellung von Carbonfasern. eports Fraunhofer-Institut für Lasertechnik ILT. Aachen, 2013.

[99] PROMES-CNRS. PROMES Scientific Report 2013–2015. Research Unit of CNRS, Institute for Engineering and System Sciences (INSIS), University of Perpignan (UPVD): Perpignan. 2016.

[100] Arnold U, De Palmenaer A, Brück T, Kuse K. Energy-efficient carbon fiber production with concentrated solar power: Process design and techno-economic analysis. Ind Eng Chem Res 2018, 57(23), 7934–45.

Part 3: **Emerging technologies**

Lauri Nikkanen, Michal Hubacek, Yagut Allahverdiyeva

10 Photosynthetic microorganisms as biocatalysts

Abstract: Photosynthetic cell factories based on living algae or cyanobacteria have vast potential for carbon-neutral production of desired compounds, and can thus become integral components of sustainable bioeconomies. Immobilization of the photosynthetic cells in solid-state matrices provides significant advantages over conventional suspension cultures such as higher cell densities, higher volumetric productivity, enhanced light distribution through elimination of self-shading, as well as lower energy and water consumption through lower volumes and unnecessity of stirring. State-of-the-art matrices of, for example, nanocellulose provide possibilities for highly tailored porous and hierarchical architectures mimicking the anatomy of the plant leaf, where light distribution, water use and gas exchange can be optimized. In addition to immobilization, the yields of photosynthesis-based biocatalyst platforms may be improved by redirecting photosynthetically produced reducing power toward the desired reactions via bioengineering of the electron transport pathways in algae and cyanobacteria. In their natural environments where light, temperature and nutrient availability can fluctuate dramatically, photosynthetic organisms protect the integrity of the photosynthetic apparatus by several alternative electron transport (AET) pathways that can in effect function as release valves for excessive reducing power. AET pathways include the Mehler-like reaction catalyzed by flavodiiron proteins, cyclic electron transport around photosystem 1, as well as terminal oxidases in the thylakoid membrane. Modification or removal of AET pathways has potential to enhance the channeling of electrons toward the desired bioproduction reaction(s). Other possible approaches include modification of the photosynthetic electron transport chain itself, as well as the introduction of heterologous electron sinks in the cell or chloroplast. Optimization of production conditions may also have a substantial impact on the yield of bioproduction. For example, the production of bio-hydrogen in algae has recently been enhanced by applying a pulse-illumination protocol that prevents the activation of competitive electron sinks. Ultimately, biotechnologically optimized photosynthetic whole-cell biotransformation platforms can be combined with tailored solid-state immobilization matrices for efficient and sustainable solar-driven production.

Acknowledgment: This work was funded by the NordForsk Nordic Center of Excellence "NordAqua" (no. 82845) and the Academy of Finland (project no. 315119 and no. 322754 to Y.A.).

https://doi.org/10.1515/9783110716979-010

10.1 Introduction

Fossil fuels are currently exploited to fulfill 80% of the needs of the energy and chemical industries globally [1]. However, any adequate response to the imminent societal and environmental challenges posed by climate change and resource scarcity necessitates a paradigm shift away from fossil fuels toward sustainable, carbon-neutral energy and chemicals production. As a key part of such a paradigm shift, it is imperative that we effectively harness the power of the process that has sustained most ecosystems on the planet for more than 3 billion years; oxygenic photosynthesis performed by cyanobacteria, algae and plants. The unique photosynthetic machinery utilizes the abundantly available energy of sunlight to split water and produce reducing power that is then used to fuel the cell metabolism and fix atmospheric CO_2 into energy-rich carbon compounds. Splitting water releases oxygen (O_2) as a by-product, which has led to oxygenation of the atmosphere, enabling complex life on the planet.

10.2 Photosynthetic microorganisms in blue biorefinery

Photosynthetically produced biomass serves as feedstock for bio-refineries. For example, algal and cyanobacterial biomass is regarded as a third-generation feedstock for the chemical and petrochemical industries [2, 3]. The advantages of microalgae and cyanobacteria based bioproduction are (i) fast growth (short doubling time), (ii) the small impact that cultivation has on agricultural land, (iii) possibility of integration with wastewater treatment and utilization of waste streams as nutrient stock and (iv) utilization of high concentrations of CO_2 and even flue gas [4] (Fig. 1).

Large investments into R&D in algae biorefineries in recent years have led to an increase in the variety of natural algal products, mainly in food markets. The high potential of microalgae as food is due to their balanced biochemical composition and high nutritional value. There is a firm belief that algae-based proteins will become market leaders before insect-based protein production is accepted by society. Microalgae and cyanobacteria are also a good source of long-chain polyunsaturated fatty acids such as docosahexaenoic and eicosapentaenoic acids. Pigments such as β-carotene, phycocyanin, astaxanthin and fucoxanthin are commonly produced and used in cosmetics, food colorants and food supplements. Importantly, cyanobacteria and algae are rich in natural bioactive compounds, for example, anticancer, antifungal, antiviral drugs. Cyanobacteria- and algae-based biostimulants and biofertilizers are utilized in agro-industry for their stimulating effect on crops' growth and development, and improvements to soil quality and fertility, respectively [5–14]. Furthermore, some algal metabolites possess a pesticidal activity and could potentially replace synthetic pesticides.

The diversity of cyanobacteria and microalgae is immense. While being an active research area, only a handful of algae and cyanobacteria strains are studied in research labs and utilized in different industrial applications. The abundance and diversity of microalgal and cyanobacterial strains requires comprehensive investigation and search for robust and superior strains for efficient bioproduction. Metabolic products of cyanobacteria and algae are still a largely unexplored natural resource, and the role of these metabolites in cyanobacterial metabolism has remained unknown. Furthermore, strains native to a particular area are advantageous in their innate adaptation to the local environmental conditions, a vital consideration in achieving maximal biomass accumulation with minimal energy input. High-throughput screening methods, omics technologies as well as various bioinformatics tools and computational strategies have strongly advanced during the past few years and opened wider possibilities to identify and characterize natural bioactive compounds and superior production strains. Further studies in these areas are necessary in order to discover novel bioactive compounds and improve bioproduction yield.

Biofuels and bioplastics from microalgae and cyanobacteria have the potential to mitigate the negative environmental impacts and depleting reserves of fossil fuels [15, 16]. Bioplastic production from microalgae is a new opportunity to be explored and further improved. However, due to the high cultivation (including energy-demanding mixing and harvesting) and production costs, biofuels (and other low-value products) derived from algal biomass are not economically competitive in the current market (Tab. 1 mentions some challenges of algal bioproduction). There are currently more than 150 SMEs (e.g., Ecoduna, MIAL, AstaREAL, AlgaEnergy, Cyano BioTech, Swedish Algae Factory) in Europe, many of which have turned their focus away from biofuels to high-value compounds in order to make biomass cultivation profitable. An alternative approach to increase profitability is to deploy a synthetic biology approach and multiproduct value chain in which multiple products (e.g., pigments + oils, pigments + peptides or oils + pigments + peptides) are extracted in a cascade. This increases the biorefinery cost significantly but can be mitigated by the revenue from multiple extracted products, allowing full valorization of biomass [17, 18].

This chapter focuses on the advanced bioproduction approach where photosynthetic microbes are utilized as biocatalysts to overcome the general insufficiency of biomass-based biorefineries.

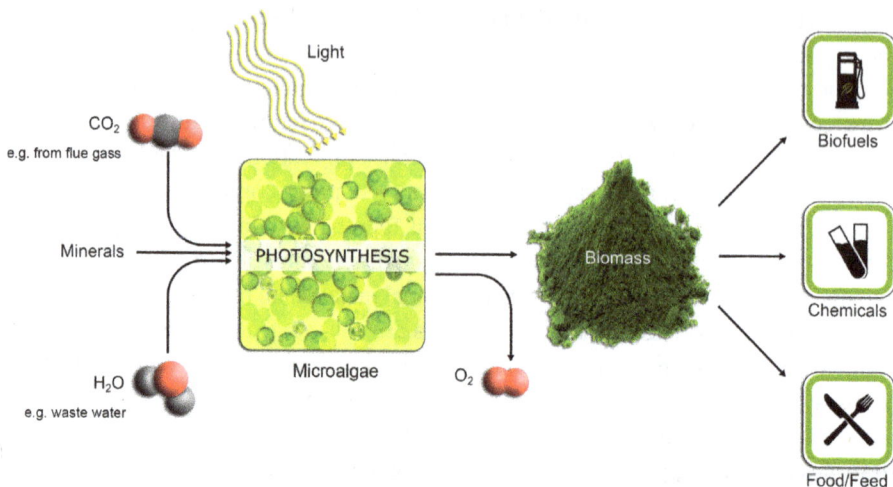

Fig. 1: Cyanobacteria and microalgae as a feedstock for blue biorefineries. Photosynthetic microbes utilize carbon dioxide (or flue gas), water (or wastewater), light and some minerals to accumulate biomass. The resulting biomass is further processed to extract desired products or used as feed or food.

10.3 Direct solar chemicals – without excessive biomass production

It is clear that the long-term sustainability of a bioeconomy cannot be guaranteed only by exploiting and refining biomass. Development of novel solar-based advanced biotechnologies relying on the efficient utilization of natural photosynthesis is necessary for the establishment of truly sustainable production of biofuel and biochemicals [19]. To this end, photosynthetic microbes can be engineered as long-living, self-regenerating biocatalysts for efficient conversion of solar energy and CO_2 into excreted targeted compounds.

10.3.1 Synthetic biology – targeted products

Development of synthetic biology toolkits and application of metabolic engineering has enabled the introduction of exogenous production pathways into photosynthetic microbes [20–22], allowing the cells to function as living cell factories producing desired solar fuels and chemicals. Currently, there are dozens of proof of concept for the production of chemicals in photosynthetic microbes. Some production systems, such as lactic acid production at Photanol B. P., are even functioning at Technology

Readiness Levels (TRL) 6–7 (https://photanol.com). In order to achieve high product titers, long-term microbial productivity is required.

Therefore, instead of using algal biomass as a feedstock for biorefineries, employing photosynthetic cells as long-lived biocatalysts is a promising strategy for production platforms. In a continuous production system, long-lived photosynthetic microbes convert light energy into desired products, excrete it into the cultivation medium or gaseous head space (if volatile), from where they can be collected and subjected to downstream processing when necessary.

Advantage of photosynthetic microbial biocatalysts are (i) unlike heterotrophs, this group of photosynthetic microorganisms does not require organic compounds; (ii) unlike chemical catalysts, the photosynthetic microorganisms are capable of synthesizing complex organic compounds with high efficiency and selectivity, combined with the ability of self-replication (recovery) and long-term operation under physiological conditions.

Many factors influence photosynthetic productivity of microalgae, but the most critical variable is the availability of light. The average amount of photons received by each cell is affected by light intensity, cell layer thickness (light path), biomass density, turbulence and rate of mixing. In line with this, thin-layer cultivation systems (<0.5 cm) allow enhanced photosynthetic efficiency [23]. Since in traditional mass culturing a significant fraction of the absorbed photon energy is wasted in biomass accumulation [24], uncoupling growth from production would be a promising approach for efficient biocatalyst production.

10.4 Immobilization – a tool for engineering solid-state biocatalysts

Entrapment of the photosynthetic microbes in thin-layer polymer matrices offers an easy and efficient technology for a transition of cell factories from suspension state to solid-state. Polymer scaffold surrounding photosynthetic cells restricts cell division and excess biomass accumulation, enabling direct conversion of light energy and CO_2 into desired chemicals instead of biomass. Moreover, it provides shelter and protection of cell integrity from harsh environmental factors such as substrate toxicity and allows control of pH, thus enabling the operation of photosynthetic cells as long-lived biocatalysts. Immobilized cells have been reported to maintain full functionality for more than 3–5 months [25, 26]. Besides the robustness and long-term functionality of a biofilm, the high density of the photosynthetic biocatalysts enables increased volumetric productivity and enhanced light-to-product conversion efficiency. Entrapment of cells in a polymer matrix allows simple separation of the production cells and extracted product, thus facilitating downstream processes (Fig. 2).

Immobilization of photosynthetic cells on thin-layer solid carriers or in hydrogels has been presented as an attractive strategy to overcome several other inherent obstacles specific to suspension cultures. The advantages of solid-state thin biofilms compared to suspension cultures are listed in Tab. 1.

Tab. 1: Comparison of suspension cultures and solid-state thin biofilms from biocatalysis perspective.

	Suspension culture	Solid-state thin artificial biofilms
Cell density	Low (<5 g L^{-1})	High (>50 g L^{-1})
Productivity	Low volumetric productivity	High volumetric productivity (high surface-to-volume ratio)
Biomass formation	Generally unaffected	Diminished (non-growing production cells)
Light availability	Self-shading and low light utilization, requires efficient mixing	Better light distribution and utilization, no mixing
Gas, nutrients, substrate, and product availability	Requires efficient mixing	No mixing, can be negatively affected by matrices, might require tailored matrix
Catalytic activity	Short term (max 1–2 weeks)	Long term (months)
Energy and water consumption	Both high, energy due to intensive mixing and harvesting	Both lower, generally lower volume + no intensive mixing
Downstream processing	Cell-separation necessary	Cell-free effluent from photobioreactor, continuous production mode

The main requirements for solid-state photosynthetic cell factories are mechanical stability, transparency, biocompatibility, tunability and biodegradability of the immobilization matrix.

Algal or cyanobacterial cells have been immobilized within non-toxic biobased polymeric matrices such as alginates, chitin, TEMPO-CNF or synthetic polymeric matrices such as latex, and sol–gels [27–29].

10.4.1 Limitations of conventional immobilization strategies

Still, while raising volumetric cell densities, conventional immobilization of photosynthetic cells on solid matrices generates its own complications such as mass transfer limitations. Low porosity of the applied polymer matrices leads to the accumulation of

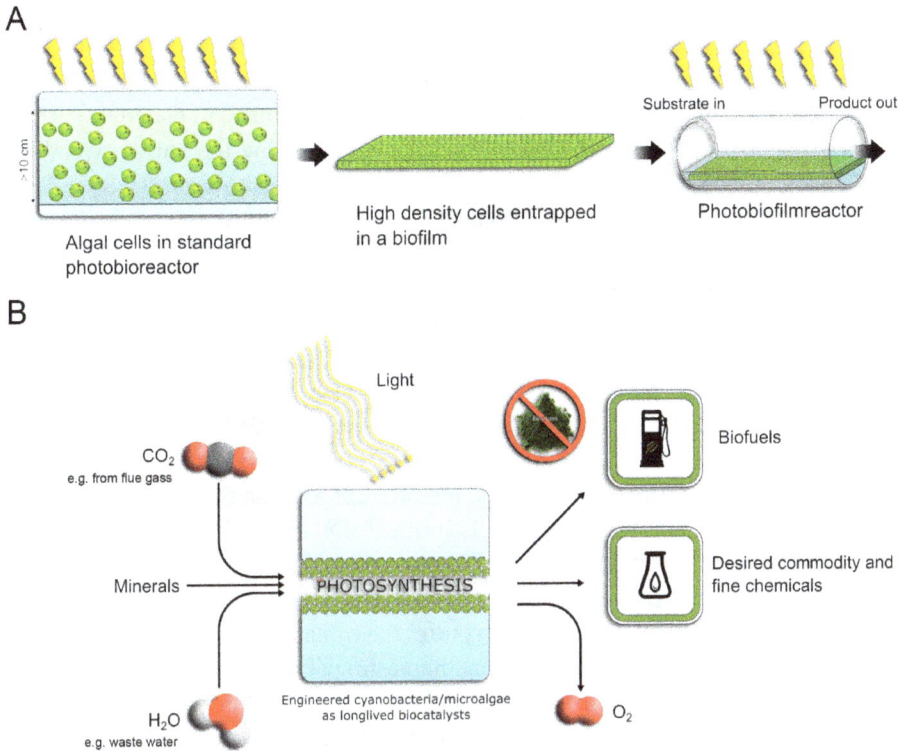

Fig. 2: Biocatalytic production of the targeted chemicals by engineered cyanobacteria or microalgae. The photosynthetic production system includes engineered photosynthetic microbial cell factories harboring the entire metabolic pathway for producing the targeted chemical. This is a two-stage production process. During the first stage, biomass is generated as a feedstock (see Fig. 1), with the majority of absorbed solar energy being directed for biomass build-up. In the second stage, engineered cell factories are entrapped in thin-layer tunable polymer matrices. This solid-state production platform enables direct photosynthetic conversion of CO_2 and light into desired chemicals without wasting energy into excess biomass accumulation.

photosynthetically evolved O_2 inside of the matrix, causing oxidative stress and photo-inhibition of the cells. Moreover, the system limits the diffusion of large substrates or products molecules. Low mechanical stability or rigidity of the immobilization matrix often results in substantial cell leaching. Furthermore, the distribution of cells and materials within conventional biofilm matrices is not uniform, which results in inefficiencies of light utilization, albeit to a lesser extent than in suspension cultures. Finally, conventional immobilization strategies are hampered by limited diffusion rates of substrates, nutrients, gases and products of biocatalysis [30].

10.4.2 Nanocelluloses as tailored matrices

Cellulose is the most widespread sustainable and renewable biopolymer on earth. Wood-/plant-based cellulose fibers have been used for centuries in materials such as papers and textiles. Within the past decades, understanding the native crystalline structure, morphology and assembly, water responsivity of nano-sized cellulose fibers have undergone substantial advancements [30, 31]. Nanocelluloses are colloidal fibril structures with lateral dimensions between 3 and 10 nm and length ranging from approximately 100 nm of cellulose nanocrystals to several micrometers of cellulose nanofibers (CNFs). TEMPO-oxidized CNF is anionically charged nanofibrillated cellulose forming a viscous and transparent gel with pronounced water-binding ability. The presence of COOH groups in TEMPO CNF allows the covalent cross-linking with polyvinyl alcohol (PVA), significantly enhancing the wet strength of TEMPO CNF thin films [32].

In contrast to conventional polymers such as alginate, colloidal nanocellulose allows the creation of a robust, porous, hierarchical and readily adjustable matrix for cell immobilization. Nanocellulose is intrinsically amphiphilic, which is a crucial characteristic that enables their highly adjustable water binding and water conveyance capacities [33]. Moreover, by introducing tailored porosity gradients to the nanocellulose designs, it is possible to control the transportation and storage of nutrients and to facilitate the separation and harvesting of end-products.

In summary, by maintaining more efficient carbon fixation and enhancing cell viability, solid-state photosynthetic cell factory (biocatalysts) designs should outperform suspension photosynthetic cell factories in bioproduction of a wide variety of desired compounds, ranging from active pharmaceutical ingredients (APIs) and fine chemicals to commodity chemicals as well as biofuels (Fig. 2).

10.5 Photosynthetic bioproduction – engineering electron flux

Although the theoretical light-to-product conversion efficiency of natural photosynthesis is about 10–13%, in practice, efficiency is below 1% because of metabolic and technical limitations. Therefore, identification of limiting factors of photosynthesis and overcoming these challenges are the key factors for efficient biocatalytic production.

Oxygenic photosynthesis is a highly tuned series of reactions in plants, algae and cyanobacteria, where the energy of sunlight is harnessed to extract reducing power from water and convert it to the chemical energy of hydrocarbon compounds. These reactions take place on the thylakoid membrane and involve the transport of reducing equivalents from photosystem 2 (PS2) to plastoquinone (PQ), cytochrome b_6f complex (Cyt b_6f), plastocyanin, photosystem 1 (PS1) and finally to ferredoxin (Fd), which functions as a distribution hub of reducing power in the cell or chloroplast.

Importantly, these electron transport reactions generate a difference in electrochemical potential over the thylakoid membrane, the proton motive force (*pmf*), which drives the synthesis of the universal energy currency molecule ATP by the ATP synthase (see Fig. 3).

Alternative electron transport (AET) pathways. In optimal conditions, most reducing power is directed from Fd to ferredoxin:NADP$^+$:oxidoreductase (FNR), generating NADPH, another energy carrier molecule that powers fixation of CO_2 in the Calvin–Benson–Bassham (CBB) cycle as well as other metabolic processes in the cell. Reduced Fd also powers nitrogen assimilation via the nitrate and nitrite reductase enzymes (NaR and NiR, respectively) [34]. In the natural environments of photosynthetic organisms, however, fluctuations in the availability of light and nutrients, or in temperature, for example, result in transient states of imbalance between the electron transport and sink capacities of the photosynthetic electron transport chain (PETC) and the downstream reactions, respectively. Such imbalance can cause overreduction of the PETC, which is hazardous for the photosynthetic machinery. In order to avoid damage to the photosystems, evolution has equipped photosynthetic cells with several AET pathways as photoprotective mechanisms.

In oxygenic photosynthetic organisms apart from flowering plants, a key mechanism to dissipate excessive reducing power in the PETC upon sudden increases in light intensity involves flavodiiron proteins (FDPs) [35–37]. FDPs are modular enzymes that light-dependently catalyze the reduction of O_2 to H_2O using electrons from the PETC, most likely from Fd, providing a release valve for excessive reducing power in the PETC [38–40]. FDPs thus mediate a water–water cycle (electrons extracted from H_2O are returned to H_2O) similarly to the Mehler reaction. However, differently from the Mehler reaction, FDPs catalyze a 4-electron reaction without concomitant production of reactive oxygen species. Therefore, the light-dependent reduction of O_2 by FDPs is often referred to as the Mehler-like reaction [35]. Please see chapter 1.

Cyclic electron transport (CET) around PS1 forms another crucial AET pathway. In cyanobacteria, CET, that is, oxidation of Fd to reduce PQ, is mainly mediated by the NADH-dehydrogenase-like complex (NDH-1). CET does not produce NADPH, but as NDH-1 pumps 2H$^+$ to the thylakoid lumen for each electron it transfers [41], CET contributes to the generation of the *pmf*, increasing the production of ATP, and can thus adjust the ATP:NADPH ratio according to the needs of the carbon metabolism. NDH-1 is a highly versatile complex in cyanobacteria because, in addition to its role in CET, it can also function in respiration and convert CO_2 to HCO_3^- in the carbon-concentrating mechanism (CCM) by incorporating CCM-specific subunits [42]. In the unicellular green alga *Chlamydomonas reinhardtii* NDH-1 is missing, but a type II NAD(P)H dehydrogenase, NDA2, plays a role in the reduction of the PQ-pool [43]. A major CET route in *C. reinhardtii* consists of the so-called ferredoxin:plastoquinone-reductase (FQR) pathway, which depends on (but might not directly involve) the proteins proton gradient regulation 5 (PGR5) and PGR5-like 1 (PGRL1) [44, 45]. PGR5 and PGRL1 orthologues are also present in cyanobacteria, but their physiological roles are uncertain [46, 47].

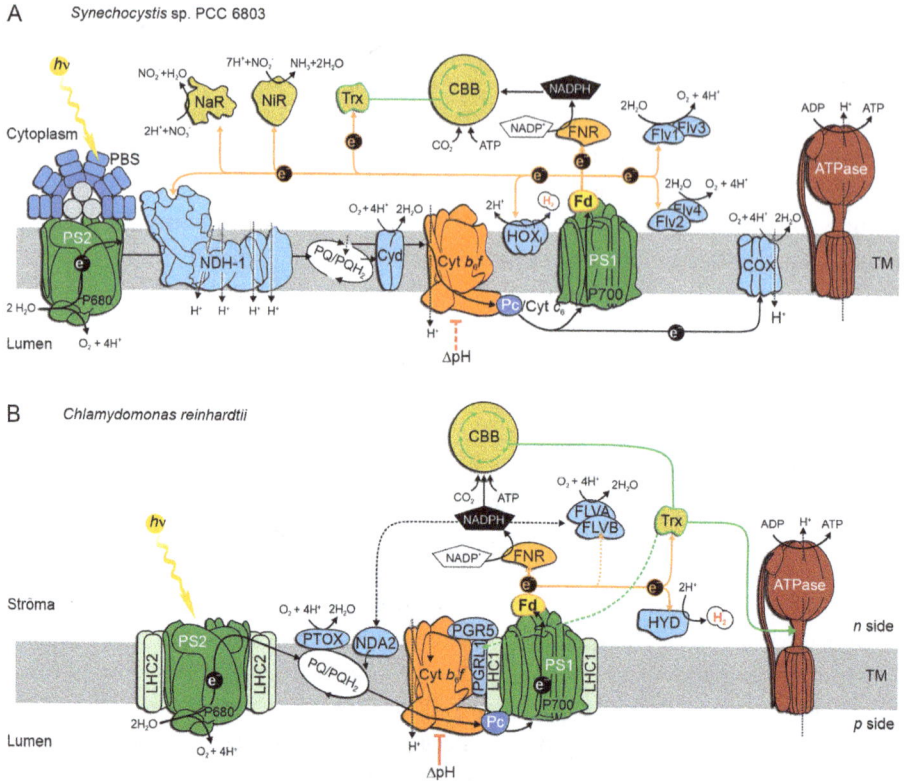

Fig. 3: Photosynthetic electron transport pathways in the model cyanobacterium *Synechocystis* sp. PCC 6803 (A) and in the green alga *Chlamydomonas reinhardtii* (B). Ferredoxin (Fd) functions as a distribution hub of electrons from PS1, and electron transport routes from it are highlighted with orange. In cyanobacteria (A), Fd feeds electrons to FNR, to Flv1/3 and Flv2/4 hetero-oligomers (note that the *flv4-2* operon is not expressed under high CO_2 condition), to NDH-1, the Ni-Fe hydrogenase (HOX), to nitrite and nitrate assimilation by the nitrate reductase (NaR) and nitrite reductase (NiR), respectively, as well as to the thioredoxin system (Trx), whose reduction is pre-requisite for activation of the CBB cycle [55, 56]. In cyanobacteria, light energy is harvested by specific protein-pigment complexes, the phycobilisomes (PBS), attached to the photosynthetic membrane. In *C. reinhardtii* (B), FLVA and FLVB are the only FDP isoforms and NDH-1 is absent. Instead, NDA2 plays a role in the reduction of the PQ-pool. As a major route of CET, electrons from Fd are directed to the FQR pathway depending on the PGR5 and PGRL1 proteins. NiR is omitted from (B) because several laboratory strains of *C. reinhardtii* do not have a functional nitrite assimilation pathway [57]. In *C. reinhardtii* Trx also light dependently activates the ATP synthase [58] and possibly PGRL1/PGR5 [59]. The [FeFe] hydrogenase (HYD) also receives electrons from Fd. The green arrows stand for activatory thiol exchange reactions. Solid black lines depict electron transport, while dotted black lines are for translocation of protons across the thylakoid membrane (TM) or putative electron transport pathways. Acidification of the lumenal space (elevation of ΔpH) causes inhibition of PQH_2 oxidation at the Cyt b_6f complex (photosynthetic control), which is depicted by blunt-ended red arrows. In *C. reinhardtii*, instead of soluble phycobilisomes, photons of light are harvested by the integral membrane protein light-harvesting complexes 1 and 2 (LHC1 and LHC2).

Thylakoid membranes also contain respiratory terminal oxidases (RTOs) that can function as outlets of excessive electrons in the PETC. In cyanobacteria, where the photosynthetic and respiratory electron transport reactions take place in the same membranes [48], the *bd* quinol oxidase (Cyd) and the *aa3*-type cytochrome c oxidase (Cox) take up electrons from the PQ pool and PC or cytochrome c_6, respectively, to reduce O_2 to H_2O [49]. These RTOs function not only as terminal electron acceptors in transport of dark-respiratory electrons from the cytosol to the photosynthetic chain but also as sinks for excessive electrons in the PETC during illumination [50, 51]. RTO activity is low in light but clearly not negligible. In some cyanobacteria, but not in *Synechocystis* sp. PCC 6803, plastid terminal oxidase (PTOX) also resides in the thylakoid membrane and catalyzes oxidation of PQH_2 to reduce O_2 [52]. In green algae, respiration takes place in mitochondria, and Cyd and Cox are absent, but two isoforms of PTOX exist in *C. reinhardtii*, of which PTOX2 is the primary terminal oxidase on thylakoids with an important role in preventing over-reduction of the PQ pool in light [53, 54].

Schematic depictions of photosynthetic electron transport and AET pathways in *Synechocystis* sp. PCC 6803 and *C. reinhardtii* are shown in Fig. 3.

10.6 Strategies for bioengineering of electron transport for enhanced bioproduction platforms

In natural environments with fluctuating conditions, the AET pathways allow the photosynthetic cells to survive by trading a significant proportion of photosynthetic efficiency for protection. In the artificially stable conditions of most photosynthesis-based bioproduction platforms, however, this trade-off constitutes a waste of energy. Therefore, the productivity of various bioproduction systems can be significantly enhanced by several distinct, but mutually non-exclusive, strategies:

(i) *Modification or elimination of the AET pathways*, and thus channeling of photosynthetic electrons to either natural or exogenous desired processes. This approach can be particularly efficient when the bioproduction reaction can utilize reducing power directly from the PETC, for example, in the form of Fd or NADPH.

(ii) *Modification of the PETC* to maximize the supply of the reducing cofactors for bioproduction. Enhancement of electron transport rate and biomass yield of plants was recently achieved via overexpression of the Rieske protein component of Cyt b_6f [60, 61], and similar approaches can potentially be utilized to increase the productivity of cyanobacterial or algal bioproduction systems. Moreover, in natural conditions, a photoprotective mechanism referred to as photosynthetic control is induced by acidification of the thylakoid lumen, and inhibits electron transport at Cyt b_6f [62]. In some bioproduction systems, it may be advantageous to relieve this mechanism in order to maximize the supply of reduced Fd or NADPH. This could be achieved by increasing the proton conductivity of the thylakoid membrane, thus lowering *pmf* and relaxing photosynthetic control, by a) deleting the inhibitory ε-subunit of the ATP synthase [63], or b) overexpressing thylakoid ion channels [64].

(iii) *Introduction of heterologous electron sinks* in cyanobacteria can itself enhance photosynthetic efficiency [65]. Addition of new strong electron sinks downstream of the PETC can alleviate inhibitory feedback regulation that normally occurs as a response to saturation of the electron transport capacity and reduction of the PQ, Fd, and $NADP^+$ pools [62, 66–68]. Similar effects may be achieved by enhancing the activity of endogenous electron sinks, mainly the CBB cycle and other redox-regulated processes, by overexpressing components of the thioredoxin system [69].

(iv) *Immobilization.* Solid-state platforms have several advantages over suspension platforms in terms of photosynthetic efficiency, as discussed above (see also Tab. 1), and can be utilized in combination with other strategies for synergistic effects.

Bioproduction of molecular hydrogen (H_2) is an excellent case study for the validation of improved photosynthetic light reactions and elimination of competing pathways as important strategies for enhancement of product yield. H_2 is considered as a zero-carbon biofuel and energy carrier. The unicellular model cyanobacterium *Synechocystis* sp. PCC 6803 harbors [Ni-Fe]-bidirectional hydrogenase that can oxidize and produce H_2 [70]. Whereas the model green alga *C. reinhardtii*, which is often utilized in H_2 bioproduction platforms, harbors endogenous O_2 sensitive [Fe–Fe] hydrogenase enzymes that oxidize Fd to produce H_2 [71, 72]. Considering the O_2 sensitivity and strong competition between [Fe–Fe] hydrogenases and other electron transport pathways, several strategies have recently been developed to enhance the H_2 photoproduction capacity of *C. reinhardtii*-based platforms. These strategies are channeling of photosynthetic electrons to hydrogenase by fusing the hydrogenase directly to PS1 [73, 74] or to Fd [75] or by increasing the affinity of Fd to hydrogenase by targeted mutations in Fd [76–78]. It is known that in *C. reinhardtii* H_2 production is severely limited by competition for reducing equivalents with CO_2 fixation once the CBB cycle becomes activated in light [79–81]. To overcome this hindrance, a new pulse-illumination protocol was developed. In this protocol, activation of the CBB cycle can be prevented by subjecting the *C. reinhardtii* cells to brief pulses of light on a background of dark or very low irradiance [81, 82]. The pulses are sufficient to produce reduced Fd in the PETC to feed the hydrogenase but insufficient to activate the CBB cycle via redox regulation. Consequently, biomass accumulation is also prevented. Thus, the protocol allows continuous H_2 production under nutrient repellent conditions (Fig. 4). Furthermore, elimination of the competing FDP pathway and application of periodic recovery phases every 3–4 days allowed continuous H_2 photoproduction for at least 18 days [83]. The developed "pulse-illumination" protocol enables photosynthetic microalgae functioning as a real biocatalyst by avoiding "wasting" of absorbed light energy for biomass production, but instead channeling photosynthetic electrons to H_2. Possible application of the "pulse-illumination" protocol also in cyanobacterial cells remains to be investigated.

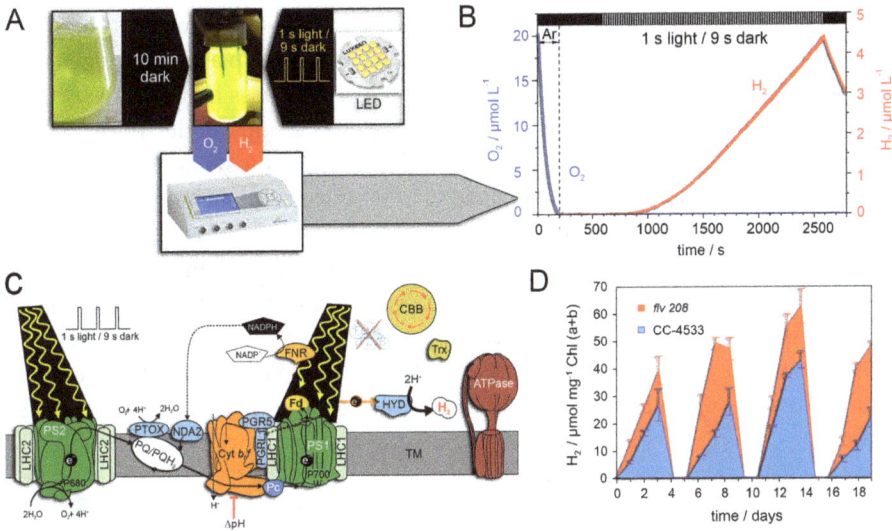

Fig. 4: Photobioproduction of H_2 in *C. reinhardtii* cells by a pulse-illumination protocol. (A) Prior to production, the cultures were flushed with argon for 3 min in darkness in order to decrease O_2 concentration. This is followed by dark acclimation for another 5–10 min in order to inactivate competing electron sink processes, mainly the CBB cycle. H_2 photoproduction was initiated by brief (1 s) pulses of light, flanked by intermittent (9 s) periods of darkness or very low irradiance. (B) A measurement of H_2 and O_2 concentrations during 200 cycles of the pulse-illumination protocol. Strong H_2 photobioproduction is seen after the remaining inhibitory O_2 is consumed. Adapted from [81]. (C) A schematic illustration of the molecular background for the protocol. The intermittent light pulses induce sufficient photosynthetic electron transport to reduce Fd, but not long enough to result in (i) accumulation of photosynthetic O_2 that would outweigh O_2 consumption (mainly by respiration and also FDPs) and presumably (ii) activation of the thioredoxin system (Trx), which would, in turn, result in redox-activation of CBB enzymes and the ATP synthase, or (iii) establishment of sufficient proton motive force for ATP synthesis. Consequently, the CBB cycle is not activated and cannot consume photosynthetic reducing power stored in NADPH by FNR. This results in strong channeling of photosynthetic electrons toward H_2 production by HYD. This effect can be further enhanced by removing FDPs (depicted by the red cross) as another competing electron sink. Difference in H_2 production per chlorophyll (Chl $a + b$) content between WT (CC-4533) and FDP-deficient cells (*flv 208* with the pulse-illumination protocol is shown in (D). A recovery phase consisting of resuspension of cells in fresh medium and incubation under constant illumination for 24 h was done after every 4 days of pulse illumination during the experiment. (B) is adapted from [81] and (D) from [83].

10.7 Light-driven whole-cell biotransformation as a biocatalytic production system

Recently, several proof-of-concept studies have demonstrated the potential of using cyanobacteria and green algae in whole-cell biotransformation platforms where living cells perform transformation of exogenous substrates into desired products via heterologous enzymes. Redox-mediated biocatalytic transformations in photosynthetic microbes involve the heterologous oxidoreductases such as monooxygenases, ene-reductases, imine reductases and alcohol dehydrogenases that selectively catalyze the introduction and modification of functional groups under physiological conditions [84–88]. A particular advantage of utilizing photosynthetic microbes as hosts for biotransformation is their capacity to produce large amounts of the solar-driven renewable cofactors such as NADPH or reduced Fd. Employing cells redesigned with respect to elimination of natural electron sinks [89] and introduction of strong metabolic sinks (oxidoreductases), as well as the profiling of overall metabolic consequences, and finally, combination of photosynthetic whole-cell biotransformation with solid-state systems will enable solar-driven biocatalysts to function in a direct and continuous "substrate in–product out" production mode. See also chapter 3.

10.8 Immobilized heterocysts as N_2-fixing or H_2-producing biocatalytic cell factories

N_2-fixing cyanobacteria possess nitrogenase enzymes that use photosynthetic energy to fix atmospheric N_2 into ammonia. During this process, H_2 is released as a by-product [90]. In the absence of N_2, the nitrogenase enzyme reduces protons to H_2. It is important to note that nitrogenase and hydrogenase enzymes are O_2 sensitive. Cyanobacterial cells have developed different strategies to overcome this problem. In the absence of combined nitrogen in the growth medium, some filamentous cyanobacteria differentiate specific vegetative cells into heterocysts, microoxic cells. Vegetative cells (also called photosynthetic cells) perform oxygenic photosynthesis and CO_2 fixation, while heterocysts do not contain an active O_2 evolving PS2 or CO_2-fixing Rubisco. These specialized cells thus provide microoxic conditions for atmospheric N_2 fixation. Heterocysts and vegetative cells maintain an interdependent symbiosis where vegetative cells provide heterocysts with reduced carbon compounds, for example, glutamate, alanine and sucrose, while heterocysts supply vegetative cells with fixed nitrogen, for example, glutamine and β-aspartyl-arginine dipeptide [91]. Isolated heterocysts are sensitive to oxic environments and remain active only for a few hours in suspensions [92]. A proof of concept for thin-layer immobilization of isolated cyanobacterial heterocysts (Fig. 5) demonstrated improved and prolonged acetylene reduction (a measure of nitrogenase activity) and H_2 photoproduction up to 18 days [92]. Thus, engineered heterocyst biofilms can act as long-lived bio-catalysts for the production of ammonia and/or H_2. Despite the importance of het-erocysts as N_2-fixing anaerobic cell factories, the electron transport pathways in these specialized cells are not well known. It has been shown that N_2-fixing hetero-cystous cyanobacteria contain two extra copies of FDPs named Flv1B and Flv3B, which are exclusively localized in heterocysts [93]. The Flv3B protein actively elimi-nates O_2 inside of the heterocysts, enabling N_2 fixation and diazotrophic growth under light [94]. The Flv1B protein is not involved in O_2 photoreduction, and its exact role in heterocysts remains unknown. It was recently demonstrated that overexpression of the Flv3B protein resulted in increased H_2 photoproduction; however, this was not accompanied with enhanced nitrogenase activity [95]. Further investigations of electron-transport pathways in heterocysts will guide genetic engi-neering strategies to further improve N_2-fixation and/or H_2 photoproduction capacity of immobilized heterocyst cells.

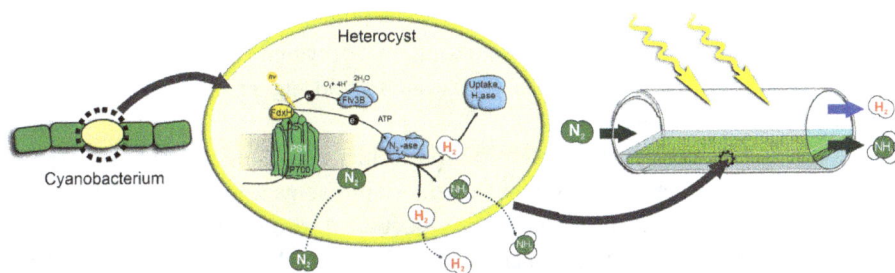

Fig. 5: Heterocysts of N_2-fixing filamentous cyanobacteria as N_2-fixing or H_2-producing cell factories. Heterocyst cells are isolated under anoxic conditions and entrapped in thin-layer polymeric alginate films. The production assay is supplemented with sucrose as a source of organic carbon and reducing power. A specific ferredoxin isoform FdxH feeds electrons from PS1 to the nitrogenase, while Flv3B reduces O_2 to H_2O, thus helping to maintain the microooxic conditions required for nitrogenase activity. Nitrogenase activity can be measured as acetylene reduction or H_2 photoproduction. Immobilization of isolated cyanobacterial heterocysts in thin-later alginate films improves and prolongs nitrogenase activity and H_2 photoproduction up to 18 days [92].

10.9 Biocatalytic photoproduction of carbon-based chemicals in thin-layer artificial cyanobacterial biofilms

Ethylene (C_2H_4) is one of the most important organic commodity chemicals with an annual global demand of more than 150 million tons. It is the main building block in the production of plastics, fibers and other organic materials. Plants produce ethylene in small quantities to control growth, development and stress response. Over the last years, a great effort has been made by applying metabolic engineering and using photosynthetic cyanobacteria as chassis for ethylene production. For example, recombinant cyanobacteria that express the ethylene-forming enzyme (EFE) from the plant pathogen *Pseudomonas syringae* can produce ethylene using solar energy and CO_2 from the air [96–99]. However, the overall efficiency of the available photoproduction systems is still very low for commercial applications. Besides the genetic background, the culture and production conditions have a strong effect on the production yield. Entrapment of ethylene-producing engineered cyanobacterial cells within thin-layer alginate polymer matrix enabled biocatalytic photoproduction of ethylene for up to 40 days, yielding 822 mL m^{-2} ethylene with a light-to-ethylene conversion efficiency of around 1.54%, that is, 3.5-fold higher than in conventional suspension cultures [100].

The production ethylene in the thin-layer artificial biofilms by strongly decreasing nutrient losses to biomass production reduced problems with nutrient deprivation during production cycles. This opens new directions for the further development of efficient solid-state photosynthetic cell factories for biocatalytic production of carbon-compounds.

10.10 Future perspectives

Over the last 3 billion years, evolution has already overcome the challenges faced by suspension-based photosynthetic cell factories via the process of endosymbiosis of an ancient cyanobacterium within a eukaryotic cell, and the subsequent emergence of multicellular, gas- and light-responsive, hierarchical tissue structure of immobilized photosynthetic cell factories known as the plant leaf. By mimicking the functional anatomy of a leaf, it is possible to transfer the advantages of current suspension photosynthetic cell factories to tailored solid-state biocatalyst platforms that would be capable of actively responding to environmental changes. Organized and controlled distribution of cells within nanostructured, hierarchical leaf-inspired matrices should allow optimization of utilization of light energy as well as the distribution and diffusion of water, CO_2 and nutrients to individual cells. Therefore, by maintaining more efficient carbon fixation and enhancing cell viability, such solid-state photosynthetic cell factory designs should outperform suspension cell factories in bioproduction of a wide variety of desired compounds, ranging from APIs and fine chemicals to commodity chemicals as well as biofuels. The productivity of these platforms can be further enhanced by optimizing photosynthetic efficiency via (i) eliminating or modifying wasteful natural electron sinks; (ii) modifying the PETC to maximize production of reducing cofactors; (iii) introducing heterologous strong metabolic sinks to avoid feedback reactions; and (iv) using specialized production conditions that selectively channel electrons to desired processes.

References

[1] Key World Energy Statistics 2020 [internet]. IEA Paris, 2020. (Accessed January 25, 2021 at https://www.iea.org/reports/key-world-energy-statistics-2020).

[2] Saladini F, Patrizi N, Pulselli FM, Marchettini N, Bastianoni S. Guidelines for emergy evaluation of first, second and third generation biofuels. Renew Sustain Energy Rev 2016, 66, 221–7.

[3] Roadmap for the Chemical Industry in Europe towards Bioeconomy – Strategy Document [internet]. RoadToBio, 2019 (Accessed January 25, 2021 https://www.roadtobio.eu/uploads/publications/roadmap/RoadToBio_strategy_document.pdf).

[4] Zhou W, Wang J, Chen P, et al. Bio-mitigation of carbon dioxide using microalgal systems: Advances and perspectives. Renew Sustain Energy Rev 2017, 76, 1163–75.

[5] Chowdhury H, Loganathan B. Third-generation biofuels from microalgae: A review. Curr Opin Green Sustain Chem 2019, 20, 39–44.

[6] Allahverdiyeva Y, Aro EM, van Bavel B, et al. NordAqua, a Nordic Center of Excellence to develop an algae based photosynthetic production platform. Physiologia Plantarum 2021.

[7] Koyande AK, Chew KW, Rambabu K, Tao Y, Chu DT, Show PL. Microalgae: A potential alternative to health supplementation for humans. Food Sci Hum Wellness 2019, 8, 16–24.

[8] Fabris M, Abbriano RM, Pernice M, et al. Emerging technologies in algal biotechnology: Toward the establishment of a sustainable, algae-based bioeconomy. Front Plant Sci 2020, 11, 279.

[9] Jiang L, Wang Y, Yin Q, et al. Phycocyanin: A potential drug for cancer treatment. J Cancer 2017, 8, 3416–29.

[10] Shishido TK, Popin RV, Jokela J, et al. Dereplication of natural products with antimicrobial and anticancer activity from brazilian cyanobacteria. Toxins (Basel) 2019, 12, 12.

[11] Ronga D, Biazzi E, Parati K, Carminati D, Carminati E, Tava A. Microalgal biostimulants and biofertilisers in crop productions. Agronomy 2019, 9, 1–22.

[12] Colla G, Rouphael Y. Microalgae: New source of plant biostimulants. Agronomy 2020, 10, 1–4.

[13] Godlewska K, Michalak I, Pacyga P, Baśladyńska S, Chojnacka K. Potential applications of cyanobacteria: Spirulina platensis filtrates and homogenates in agriculture. World J Microbiol Biotechnol 2019, 35, 1–18.

[14] Costa JAV, Freitas BCB, Cruz CG, Silveira J, Morais MG. Potential of microalgae as biopesticides to contribute to sustainable agriculture and environmental development. J Environ Sci Heal – Part B Pestic Food Contam Agric Wastes 2019, 54, 366–75.

[15] Karan H, Funk C, Grabert M, Oey M, Hankamer B. Green bioplastics as part of a circular bioeconomy. Trends Plant Sci 2019, 24, 237–49.

[16]   Woo HM, Lee HJ. Toward solar biodiesel production from CO2 using engineered cyanobacteria. FEMS Microbiol Lett 2017, 364, 66.

[17]   Slegers PM, Olivieri G, Breitmayer E, et al. Design of value chains for microalgal biorefinery at industrial scale: Process integration and techno-economic analysis. Front Bioeng Biotechnol 2020, 8, 1–17.

[18]   Laurens LML, Markham J, Templeton DW, et al. Development of algae biorefinery concepts for biofuels and bioproducts; a perspective on process-compatible products and their impact on cost-reduction. Energy Environ Sci 2017, 10, 1716–38.

[19]   Lips D, Schuurmans JM, Branco Dos Santos F, Hellingwerf KJ. Many ways towards 'solar fuel': Quantitative analysis of the most promising strategies and the main challenges during scale-up. Energy Environ Sci 2018, 11, 10–22.

[20]   Santos-Merino M, Singh AK, Ducat DC. New applications of synthetic biology tools for cyanobacterial metabolic engineering. Front Bioeng Biotechnol 2019, 7, 33.

[21]   Hitchcock A, Hunter CN, Canniffe DP. Progress and challenges in engineering cyanobacteria as chassis for light-driven biotechnology. Microb Biotechnol 2020, 13, 363–7.

[22]   Miao R, Xie H, Liu X, Lindberg P, Lindblad P. Current processes and future challenges of photoautotrophic production of acetyl-CoA-derived solar fuels and chemicals in cyanobacteria. Curr Opin Chem Biol 2020, 59, 69–76.

[23]   Masojídek J, Sergejevová M, Malapascua JR, Kopecký J. Thin-layer systems for mass cultivation of microalgae: Flat panels and sloping cascades. In: Prokop A, Bajpai R, Zappi M, eds. Algal Biorefineries. Cham, Switzerland, Springer International Publishing, 2015, 237–61.

[24]   Torzillo G, Pushparaj B, Masojidek J, Vonshak A. Biological constraints in algal biotechnology. Biotechnol Bioprocess Eng 2003, 8, 338–48.

[25]   Leino H, Kosourov SN, Saari L, et al. Extended H 2 photoproduction by N 2-fixing cyanobacteria immobilized in thin alginate films. Int J Hydrogen Energy 2012, 37, 151–61.

[26]   Ghirardi ML Algal Systems for Hydrogen Photoproduction [internet]. 2015. Available from https://www.osti.gov/servlets/purl/1222789.

[27]   Moreno-Garrido I. Microalgae immobilization: Current techniques and uses. Bioresour Technol 2008, 99, 3949–64.

[28]   Strieth D, Ulber R, Muffler K. Application of phototrophic biofilms: From fundamentals to processes. Bioprocess Biosyst Eng 2018, 41, 295–312.

[29]   Jämsä M, Kosourov S, Rissanen V, et al. Versatile templates from cellulose nanofibrils for photosynthetic microbial biofuel production. J Mater Chem A 2018, 6, 5825–35.

[30]   Heise K, Kontturi E, Allahverdiyeva Y, et al. Nanocellulose: Recent fundamental advances and emerging biological and biomimicking applications. Adv Mater 2021, 33, 2004349.

[31]   Kontturi E, Laaksonen P, Linder MB, et al. Advanced Materials through Assembly of Nanocelluloses. Adv Mater 2018, 30, 1703779.

[32]   Hakalahti M, Salminen A, Seppälä J, Tammelin T, Hänninen T. Effect of interfibrillar PVA bridging on water stability and mechanical properties of TEMPO/NaClO2 oxidized cellulosic nanofibril films. Carbohydr Polym 2015, 126, 78–82.

[33]   Hakalahti M, Faustini M, Boissière C, Kontturi E, Tammelin T. Interfacial mechanisms of water vapor sorption into cellulose nanofibril films as revealed by quantitative models. Biomacromolecules 2017, 18, 2951–8.

[34]   Flores E, Frías JE, Rubio LM, Herrero A. Photosynthetic nitrate assimilation in cyanobacteria. Photosynth Res 2005, 83, 117–33.

[35]   Allahverdiyeva Y, Isojärvi J, Zhang P, Aro E-M. Cyanobacterial oxygenic photosynthesis is protected by flavodiiron proteins. Life 2015, 5, 716–43.

[36]   Ilík P, Pavlovič A, Kouřil R, et al. Alternative electron transport mediated by flavodiiron proteins is operational in organisms from cyanobacteria up to gymnosperms. New Phytol 2017, 214, 967–72.

[37]   Nikkanen L, Solymosi D, Jokel M, Allahverdiyeva Y. Regulatory electron transport pathways of photosynthesis in cyanobacteria and microalgae: Recent advances and biotechnological prospects. Physiol Plant 2021. [submitted].

[38]   Santana-Sanchez A, Solymosi D, Mustila H, Bersanini L, Aro E-M, Allahverdiyeva Y. Flavodiiron proteins 1-to-4 function in versatile combinations in O2 photoreduction in cyanobacteria. Elife 2019, 8, 1–22.

[39]   Nikkanen L, Santana Sánchez A, Ermakova M, Rögner M, Cournac L, Allahverdiyeva Y. Functional redundancy between flavodiiron proteins and NDH-1 in Synechocystis sp. PCC 6803. Plant J 2020, 103, 1460–76.

[40]   Sétif P, Shimakawa G, Krieger-Liszkay A, Miyake C. Identification of the electron donor to flavodiiron proteins in Synechocystis sp. PCC 6803 by in vivo spectroscopy. Biochim Biophys Acta – Bioenerg 2020, 1861, 148256.

[41]   Schuller JM, Birrell JA, Tanaka H, et al. Structural adaptations of photosynthetic complex I enable ferredoxin-dependent electron transfer. Science (80-) 2019, 363, 257–60.

[42]   Peltier G, Aro E-M, Shikanai T. NDH-1 and NDH-2 plastoquinone reductases in oxygenic photosynthesis. Annu Rev Plant Biol 2016, 67, 55–80.

[43]   Jans F, Mignolet E, Houyoux P-A, et al. A type II NAD(P)H dehydrogenase mediates light-independent plastoquinone reduction in the chloroplast of Chlamydomonas. Proc Natl Acad Sci 2008, 105, 20546–51.

[44]   DalCorso G, Pesaresi P, Masiero S, et al. A complex containing PGRL1 and PGR5 Is involved in the switch between linear and cyclic electron flow in arabidopsis. Cell 2008, 132, 273–85.

[45]   Nawrocki WJ, Bailleul B, Cardol P, Rappaport F, Wollman F-A, Joliot P. Maximal cyclic electron flow rate is independent of PGRL1 in Chlamydomonas. Biochim Biophys Acta – Bioenerg 2019, 1860, 425–32.

[46]   Yeremenko N, Jeanjean R, Prommeenate P, et al. Open reading frame ssr2016 is required for antimycin a-sensitive photosystem i-driven cyclic electron flow in the cyanobacterium synechocystis sp. PCC 6803. Plant Cell Physiol 2005, 46, 1433–6.

[47]   Dann M, Leister D. Evidence that cyanobacterial Sll1217 functions analogously to PGRL1 in enhancing PGR5-dependent cyclic electron flow. Nat Commun 2019, 10, 5299.

[48]   Mullineaux CW. Co-existence of photosynthetic and respiratory activities in cyanobacterial thylakoid membranes. Biochim Biophys Acta – Bioenerg 2014, 1837, 503–11.

[49]   Pils D, Schmetterer G. Characterization of three bioenergetically active respiratory terminal oxidases in the cyanobacterium Synechocystis sp. strain PCC 6803. FEMS Microbiol Lett 2001, 203, 217–22.

[50]   Lea-Smith DJ, Ross N, Zori M, et al. Thylakoid terminal oxidases are essential for the cyanobacterium synechocystis sp. PCC 6803 to survive rapidly changing light intensities. Plant Physiol 2013, 162, 484–95.

[51]   Ermakova M, Huokko T, Richaud P, et al. Distinguishing the roles of thylakoid respiratory terminal oxidases in the cyanobacterium Synechocystis sp. PCC 6803. Plant Physiol 2016, 171,1307–19.

[52]   McDonald AE, Ivanov AG, Bode R, Maxwell DP, Rodermel SR, Hüner NPA. Flexibility in photosynthetic electron transport: The physiological role of plastoquinol terminal oxidase (PTOX). Biochim Biophys Acta – Bioenerg 2011, 1807, 954–67.

[53]   Houille-Vernes L, Rappaport F, Wollman F-A, Alric J, Johnson X. Plastid terminal oxidase 2 (PTOX2) is the major oxidase involved in chlororespiration in Chlamydomonas. Proc Natl Acad Sci 2011, 108, 20820–5.

[54] Nawrocki WJ, Buchert F, Joliot P, Rappaport F, Bailleul B, Wollman F-A. Chlororespiration controls growth under intermittent light. Plant Physiol 2019, 179, 630–9.

[55] Tamoi M, Miyazaki T, Fukamizo T, Shigeoka S. The Calvin cycle in cyanobacteria is regulated by CP12 via the NAD(H)/NADP(H) ratio under light/dark conditions. Plant J 2005, 42, 504–13.

[56] Tsukamoto Y, Fukushima Y, Hara S, Hisabori T. Redox control of the activity of phosphoglycerate kinase in synechocystis sp. PCC6803. Plant Cell Physiol 2013, 54, 484–91.

[57] Fernandez E, Matagne RF. Genetic analysis of nitrate reductase-deficient mutants in Chlamydomonas reinhardii. Curr Genet 1984, 8, 635–40.

[58] Hisabori T, Sunamura E-I, Kim Y, Konno H. The chloroplast ATP synthase features the characteristic redox regulation machinery. Antioxid Redox Signal 2013, 19, 1846–54.

[59] Petroutsos D, Terauchi AM, Busch A, et al. PGRL1 participates in iron-induced remodeling of the photosynthetic apparatus and in energy metabolism in chlamydomonas reinhardtii. J Biol Chem 2009, 284, 32770–81.

[60] Simkin AJ, McAusland L, Lawson T, Raines CA. Overexpression of the RieskeFeS Protein Increases Electron Transport Rates and Biomass Yield. Plant Physiol 2017, 175, 134–45.

[61] Ermakova M, Lopez-Calcagno PE, Raines CA, Furbank RT, Von Caemmerer S. Overexpression of the Rieske FeS protein of the Cytochrome b6f complex increases C4 photosynthesis in Setaria viridis. Commun Biol 2019, 2, 314.

[62] Malone LA, Proctor MS, Hitchcock A, Hunter CN, Johnson MP. Cytochrome b6f – Orchestrator of photosynthetic electron transfer. Biochim Biophys Acta – Bioenerg 2021, 5, 148380.

[63] Imashimizu M, Bernát G, Sunamura E-I, et al. Regulation of F0F1-ATPase from Synechocystis sp. PCC 6803 by γ and ∈ Subunits Is Significant for Light/Dark Adaptation. J Biol Chem 2011, 286, 26595–602.

[64] Correa Galvis V, Strand DD, Messer M, et al. H+ Transport by K+ EXCHANGE ANTIPORTER3 Promotes Photosynthesis and Growth in Chloroplast ATP Synthase Mutants. Plant Physiol 2020, 182, 2126–42.

[65] Grund M, Jakob T, Wilhelm C, Bühler B, Schmid A. Electron balancing under different sink conditions reveals positive effects on photon efficiency and metabolic activity of Synechocystis sp. PCC 6803. Biotechnol Biofuels 2019, 12, 43.

[66] Shaku K, Shimakawa G, Hashiguchi M, Miyake C. Reduction-Induced Suppression of Electron Flow (RISE) in the Photosynthetic Electron Transport System of Synechococcus elongatus PCC 7942. Plant Cell Physiol 2015, 57, 1443–53.

[67] Hald S, Nandha B, Gallois P, Johnson GN. Feedback regulation of photosynthetic electron transport by NADP(H) redox poise. Biochim Biophys Acta – Bioenerg 2008, 1777, 433–40.

[68] Johnson GN. Thiol regulation of the thylakoid electron transport chaina missing link in the regulation of photosynthesis? Biochemistry 2003, 42, 3040–4.

[69] Nikkanen L, Toivola J, Diaz MG, Rintamäki E. Chloroplast thioredoxin systems: Prospects for improving photosynthesis. Philos Trans R Soc B Biol Sci 2017, 372, 20160474.

[70] Gutekunst K, Chen X, Schreiber K, Kaspar U, Makam S, Appel J. The Bidirectional NiFe-hydrogenase in Synechocystis sp. PCC 6803 Is Reduced by Flavodoxin and Ferredoxin and Is Essential under Mixotrophic, Nitrate-limiting Conditions. J Biol Chem 2014, 289, 1930–7.

[71] Winkler M, Kuhlgert S, Hippler M, Happe T. Characterization of the Key Step for Light-driven Hydrogen Evolution in Green Algae. J Biol Chem 2009, 284, 36620–7.

[72] Rumpel S, Siebel JF, Diallo M, Farès C, Reijerse EJ, Lubitz W. Structural insight into the complex of ferredoxin and [FeFe] hydrogenase from chlamydomonas reinhardtii. ChemBioChem 2015, 16, 1663–9.

[73] Kanygin A, Milrad Y, Thummala C, et al. Rewiring photosynthesis: A photosystem I-hydrogenase chimera that makes H$_2$ *in vivo*. Energy Environ Sci 2020, 13, 2903–14.

[74] Appel J, Hueren V, Boehm M, Gutekunst K. Cyanobacterial in vivo solar hydrogen production using a photosystem I–hydrogenase (PsaD-HoxYH) fusion complex. Nat Energy 2020, 5, 458–67.

[75] Eilenberg H, Weiner I, Ben-Zvi O, et al. The dual effect of a ferredoxin-hydrogenase fusion protein in vivo: Successful divergence of the photosynthetic electron flux towards hydrogen production and elevated oxygen tolerance. Biotechnol Biofuels 2016, 9, 182.

[76] Rumpel S, Siebel JF, Farès C, et al. Enhancing hydrogen production of microalgae by redirecting electrons from photosystem I to hydrogenase. Energy Environ Sci 2014, 7, 3296–301.

[77] Wiegand K, Winkler M, Rumpel S, Kannchen D, Rexroth S, Hase T, Fares C, Happe T, Lubitz W, Rögner M. Rational redesign of the ferredoxin-NADP-oxido-reductase/ferredoxin-interaction for photosynthesis-dependent H_2-production. Biochim Biophys Acta 2018, 1859,253–262.

[78] Kannchen D, Zabret J, Oworah-Nkruma R, Dyczmons-Nowaczyk N, Wiegand K, Löbbert P, Frank A, Nowaczyk M, Rexroth S, Rögner M. Remodeling of photosynthetic electron transport in *Synechocystis* sp. PCC 6803 for future hydrogen production from water. Biochim Biophys Acta – Bioenergetics 2020, 1861, 148208.

[79] Milrad Y, Schweitzer S, Feldman Y, Yacoby I. Green algal hydrogenase activity is outcompeted by carbon fixation before inactivation by oxygen takes place. Plant Physiol 2018, 177, 918–26.

[80] Nagy V, Podmaniczki A, Vidal-Meireles A, et al. Water-splitting-based, sustainable and efficient H2 production in green algae as achieved by substrate limitation of the Calvin–Benson–Bassham cycle. Biotechnol Biofuels 2018, 11, 69.

[81] Kosourov S, Jokel M, Aro E-M, Allahverdiyeva Y. A new approach for sustained and efficient H 2 photoproduction by Chlamydomonas reinhardtii. Energy Environ Sci 2018, 11, 1431–6.

[82] Kosourov S, Nagy V, Shevela D, Jokel M, Messinger J, Allahverdiyeva Y. Water oxidation by photosystem II is the primary source of electrons for sustained H 2 photoproduction in nutrient-replete green algae. Proc Natl Acad Sci 2020, 117, 29629–36.

[83] Jokel M, Nagy V, Tóth SZ, Kosourov S, Allahverdiyeva Y. Elimination of the flavodiiron electron sink facilitates long-term H2 photoproduction in green algae. Biotechnol Biofuels 2019, 12, 1–16.

[84] Hoschek A, Heuschkel I, Schmid A, Bühler B, Karande R, Bühler K. Mixed-species biofilms for high-cell-density application of Synechocystis sp. PCC 6803 in capillary reactors for continuous cyclohexane oxidation to cyclohexanol. Bioresour Technol 2019, 282, 171–8.

[85] Velikogne S, Resch V, Dertnig C, Schrittwieser JH, Kroutil W. Sequence-based in-silico discovery, characterisation, and biocatalytic application of a set of imine reductases. ChemCatChem 2018, 10, 3236–46.

[86] Górak M, Żymańczyk-Duda E. Application of cyanobacteria for chiral phosphonate synthesis. Green Chem 2015, 17, 4570–8.

[87] Köninger K, Gómez Baraibar Á, Mügge C, et al. Recombinant cyanobacteria for the asymmetric reduction of C=C bonds fueled by the biocatalytic oxidation of water. Angew Chemie Int Ed 2016, 55, 5582–5.

[88] Böhmer S, Marx C, Gómez-Baraibar Á, et al. Evolutionary diverse Chlamydomonas reinhardtii Old Yellow Enzymes reveal distinctive catalytic properties and potential for whole-cell biotransformations. Algal Res 2020, 50, 101970.

[89] Assil-Companioni L, Büchsenschütz HC, Solymosi D, et al. Engineering of NADPH supply boosts photosynthesis-driven biotransformations. ACS Catal 2020, 10, 11864–77.

[90] Bothe H, Schmitz O, Yates MG, Newton WE. Nitrogen fixation and hydrogen metabolism in cyanobacteria. Microbiol Mol Biol Rev 2010, 74, 529–51.

[91] Herrero A, Stavans J, Flores E. The multicellular nature of filamentous heterocyst-forming cyanobacteria. FEMS Microbiol Rev 2016, 40, 831–54.

[92] Volgusheva A, Kosourov S, Lynch F, Allahverdiyeva Y. Immobilized heterocysts as microbial factories for sustainable nitrogen fixation. J Biotechnol X 2019, 4, 100016.

[93] Ermakova M, Battchikova N, Allahverdiyeva Y, Aro E-M. Novel heterocyst-specific flavodiiron proteins in Anabaena sp. PCC 7120. FEBS Lett 2013, 587, 82–7.

[94] Ermakova M, Battchikova N, Richaud P, et al. Heterocyst-specific flavodiiron protein Flv3B enables oxic diazotrophic growth of the filamentous cyanobacterium Anabaena sp. PCC 7120. Proc Natl Acad Sci 2014, 111, 11205–10.

[95] Roumezi B, Avilan L, Risoul V, Brugna M, Rabouille S, Latifi A. Overproduction of the Flv3B flavodiiron, enhances the photobiological hydrogen production by the nitrogen-fixing cyanobacterium Nostoc PCC 7120. Microb Cell Fact 2020, 19, 65.

[96] Thiel K, Mulaku E, Dandapani H, Nagy C, Aro E-M, Kallio P. Translation efficiency of heterologous proteins is significantly affected by the genetic context of RBS sequences in engineered cyanobacterium Synechocystis sp. PCC 6803. Microb Cell Fact 2018, 17, 34.

[97] Ungerer J, Tao L, Davis M, Ghirardi M, Maness P-C, Yu J. Sustained photosynthetic conversion of CO2 to ethylene in recombinant cyanobacterium Synechocystis 6803. Energy Environ Sci 2012, 5, 8998.

[98] Durall C, Lindberg P, Yu J, Lindblad P. Increased ethylene production by overexpressing phosphoenolpyruvate carboxylase in the cyanobacterium Synechocystis PCC 6803. Biotechnol Biofuels 2020, 13, 16.

[99] Zhu T, Xie X, Li Z, Tan X, Lu X. Enhancing photosynthetic production of ethylene in genetically engineered Synechocystis sp. PCC 6803. Green Chem 2015, 17, 421–34.

[100] Vajravel S, Sirin S, Kosourov S, Allahverdiyeva Y. Towards sustainable ethylene production with cyanobacterial artificial biofilms. Green Chem 2020, 22, 6404–14.

Katja Bühler, Bruno Bühler, Stephan Klähn, Jens O. Krömer,
Christian Dusny, Andreas Schmid

11 Biocatalytic production of white hydrogen from water using cyanobacteria

Abstract: Hydrogen is an important building block in the chemical industry, but over the last decades, several attempts were also made to develop hydrogen as an energy carrier, for example, for fuel cell technology (electricity, mobility, heating). While this had limited success in the past, there is a renewed push for a systematic change to hydrogen as a major energy carrier. This shift is mainly driven by the widely accepted understanding that the observed harsh effects of climate change must be to a great deal connected to the burning of fossil energy carriers.

We introduce a basic concept for producing hydrogen from water using natural photosynthesis applying the toolbox of white biotechnology. We term this "White hydrogen." Water is split to electrons, protons and oxygen by photosystem 2 and hydrogen is subsequently formed. There are basically two approaches to achieve hydrogen formation, both using microbial whole cell biocatalysts. On the one hand, in biophotolysis, electrons liberated in the water oxidizing reaction are delivered via the photosystem(s) to a hydrogenase without a detour through central metabolism, but directly via a redox cofactor like NAD(P)H or ferredoxin. Oxygen is formed in the same cell as a by-product of the water splitting reaction and, as it is the case for hydrogen itself, must be separated. On the other hand, in biophotovoltaics, the photosynthetic electron transport chain is connected to a solid state electrode (anode), driving the reduction of protons to hydrogen on the cathode of a microbial electrolysis cell. Thereby, the formation of oxygen and hydrogen occurs in different reaction chambers, facilitating product recovery. While we expect the future energy mix to comprise bulk hydrogen produced in centralized facilities, for example, via electrolysis driven by electricity derived from photovoltaics or wind power plants, we believe that low-cost and less-resource-intensive solutions will make an important contribution in decentralized, autonomous facilities and applications. This could be small-scale production units for white hydrogen up to a few hundred kg per year that might be directly connected with hydrogen use after short-term storage circumventing complex logistics for large-scale transport and storage. For this purpose, continuous reaction formats are required, for example, the use of phototrophic biofilms as biocatalysts for the production of white hydrogen at biomass concentrations and light supply tuned for optimal hydrogen production efficiency.

This chapter presents and discusses the frame, status, potential and challenges of these two approaches and proposes a concept for the integrated development of the molecular machinery, the biocatalysts and suitable reaction and process formats.

https://doi.org/10.1515/9783110716979-011

11.1 Goal

The biochemical machinery of oxygenic photosynthesis was introduced by nature by cyanobacteria about 3 to 3.5 billion years ago, with the splitting of water into electrons, protons and oxygen being fueled by solar energy. Natural oxygenic photosynthesis is made up of two parts, the light reaction and the dark reaction. The light reaction is yielding oxygen, biochemical energy in the form of ATP and biochemical reduction potential in the form of NADPH, a hydrogen transfer coenzyme stable in the presence of water and molecular oxygen. $NADPH + H^+$ could also be considered biochemically fixed hydrogen. The dark reaction uses the energy stored in ATP and the reduction potential stored in NADPH coenzymes to fix carbon dioxide and to produce glucose as basic synthon for bioorganic chemistry. Nature optimized photosynthesis for this purpose, to store light energy in biological energy and electron carriers, fueling the reduction of carbon dioxide which results in the formation of precursors for all organic compounds in living matter [1].

Utilizing this ancient and highly conserved mechanism for H_2 production is a challenge, as it forces the organism to shuffle its electrons into H_2 production in a kind of "futile cycle" instead of using them to produce biological energy carriers and consequently biomass. Depending on the approach followed, biophotolysis or biophotovoltaics (hereafter BPV), different challenges need to be met. Two key problems of biophotolysis are the nature of the hydrogenases [2] and the oxyhydrogen issue. Many hydrogenases have a pronounced oxygen sensitivity, a strong bias toward H_2 oxidation, and require a strongly negative redox potential to take up electrons. Also, as the water splitting reaction liberates oxygen besides electrons, this will lead to an explosive gas mixture upon accumulation if not effectively separated right from the start. In BPV this problem is not existing, as oxygen evolution and H_2 production occur in separate compartments, which is the major benefit of this approach. However, the electron transfer from the biological cell to the electrode is currently very low and the underlying mechanism not understood; this complicates the development of feasible optimization strategies.

Independent of the approach realized in the end, the ideal biological H_2 production system needs to meet a couple of general conditions to make an application on a larger scale feasible. In our opinion this necessitates a living catalytic system, which can be operated in a continuous mode and is readily regenerating itself. Furthermore, optimal cell-densities need to be accomplished, while at the same time the producing organisms need to stall cell division and shuffle the major part of their electrons into H_2 production instead. Based on the application of white biotechnology the product is called white hydrogen, in line with natural hydrogen already being classified as white hydrogen (https://bdi.eu/artikel/news/wasserstoff-energietraeger-der-zukunft).

Today, different types of hydrogen are classified and distinguished by a color code depending on the technology used for their production (Fig. 1). These classes of hydrogen differ in their sustainability defined by parameters like area efficiency, emission(s) (mainly CO_2) and raw material efficiency, as well as in their storage capacity (storage mitigation), the possibility of decentralized production, and in cost efficiency (Fig. 2). Starting out with a more general discussion on the various possibilities of H_2 production we will look at the pros and cons connected to those technologies. Subsequently, we will address the overall concept of H_2 production via the utilization of the photosynthetic power in cyanobacteria by biophotolysis and BPV. A novel approach for the continuous cultivation of photoautotrophic organisms at optimal cell density will be introduced and the overall process concept will finally be critically discussed.

11.2 Basics

Today, H_2 can be produced by various technologies, utilizing several sources of energy (Fig. 1). Established on a large scale is steam (methane)- and coal-reforming. Both constitute centralized processes relying on fossil resources and are emitting high amounts of CO_2 [3, 4]. In the context of climate friendly fuel production and energy storage, the so-called green H_2 is currently expected to become a key technology. It is based on the electrolysis of water powered by electricity produced from renewable energy (hydro, wind and solar) [5–9]. Acceptable efficiencies have been achieved for this largely CO_2 emission-free technology but costs are still high. Other novel developments utilizing solar energy are photoelectrochemical and thermochemical (powered by a heliostat) water splitting [10–16]. These technologies have a high potential regarding solar-to-H_2 and thus areal efficiency. However, solar power production, as well as the production of the respective technical equipment and infrastructure, is very resource intensive, especially regarding rare and heavy metals [17–19]. Biological H_2 formation is a promising alternative in this context, and is discussed in more detail below [20].

Biological hydrogen production can be achieved in three ways: via fermentation, biophotolysis and microbial electrolysis cells (MECs) [21–28]. All these approaches work because microorganisms are able to use protons (H^+) as an electron sink for two electron equivalents to form H_2. The three approaches differ in the electron donor used, the redox potentials applied to make H_2, and the respective microorganisms capable of performing the necessary reactions.

H_2 formation via fermentation: Fermentative H_2 formation relies on organic substrates such as sugars or organic acids, which serve as electron donors to finally reduce protons. Further, one can differentiate between dark fermentation and anoxygenic photosynthesis both fueled with electrons derived from organic compounds under

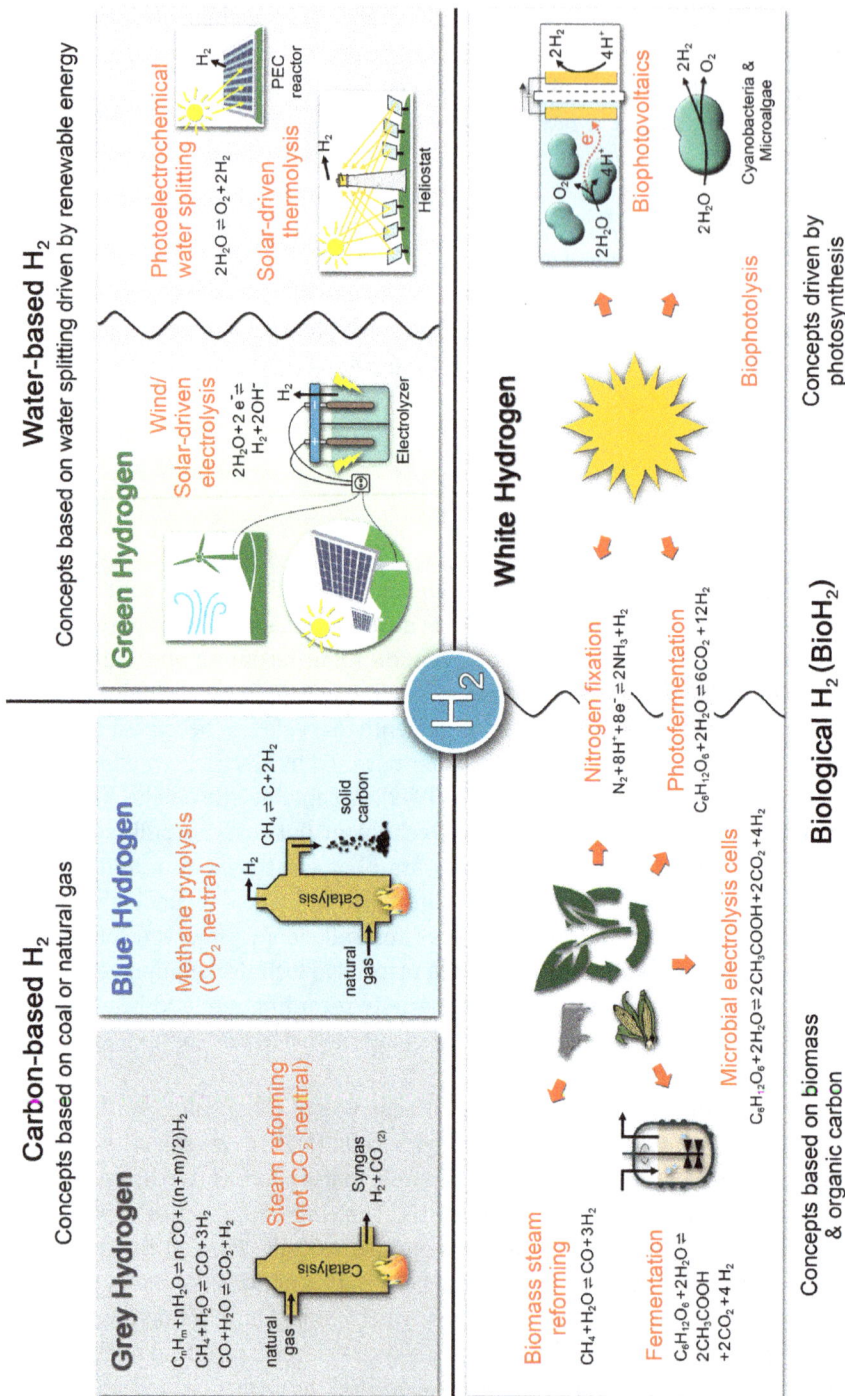

Fig. 1: Overview of the different options to produce H_2.

strictly anaerobic conditions. H_2 production via dark fermentation uses electrons derived from the oxidation either of pyruvate to acetyl-CoA + CO_2 (primarily in strict anaerobes, such as *Clostridium*, *Ethanoligenens* and *Desulfovibrio*), or the oxidation of formate to CO_2 (typically in facultative anaerobes, such as *Enterobacter*, *Citrobacter*, *Klebsiella*, *Escherichia* and *Bacillus*) [29–34]. In the former case, electrons are transferred to ferredoxin, which serves as electron donor for a hydrogenase that catalyzes H_2 formation [29, 35, 36]. In the latter case, H_2 is a direct product of formate oxidation catalyzed by a formate-hydrogen lyase [32, 33, 37]. H_2, however, is not the only electron sink in dark fermentation, and in fact most electrons are used to form reduced organic compounds such as short-chain carboxylic acids and alcohols, which results in low H_2 yields.

H$_2$ formation via anoxygenic photosynthesis can be performed by purple bacteria (e.g., *Rhodobacter*, *Rhodopseudomonas* and *Rhodospirillum*). Thereby, electrons derived from short-chain carboxylic acids are channeled into the photosynthetic electron transport (PET) chain, from which the emerging NADPH is used for H_2 formation – typically by nitrogenases. These enzymes normally catalyze the highly energy- and electron-intensive N_2 fixation process and produce H_2 only as a by-product. This happens at a stoichiometry varying from 1 mol H_2 generated per mol N_2 fixed for the common Mo-containing nitrogenase to 9 mol H_2 per mol N_2 in the case of the highly O_2-sensitive Fe-type nitrogenase [38]. By mutating the uptake hydrogenase to reduce the re-consumption of produced H_2, net H_2 production rates were obtained, despite the unfavorable bioenergetics for nitrogenase-mediated H_2 formation [39]. Integration of dark fermentation with such light-dependent H_2 formation by purple bacteria is a good option to capture electrons diverted to organics during dark fermentation [21]. However, the biggest drawback of fermentative H_2 production is the utilization of organic compounds and thus biomass as electron donor. This results in a very poor light-to-hydrogen efficiency as light energy stored in biomass typically does not exceed 1–2% of total surface irradiation (Fig. 2B). On the other hand, dark fermentation can make use of organic waste streams giving this approach a certain value, albeit being moderate regarding potential production volumes [24].

Photosynthetic H_2 formation via temporal or local decoupling from water oxidation. In this scenario, no organic substrate is added to supply electrons or scavenge O_2 via its respiratory assimilation, and the O_2-sensitive hydrogenases or nitrogenases of cyanobacteria such as *Nostoc* or *Cyanothece* operate in heterocysts as O_2-free compartments or during night [40]. In case of nitrogenases, H_2 is produced as byproduct of N_2 fixation as described above for purple bacteria. The electrons driving H_2 formation (and N_2 fixation) are not directly derived from water oxidation, but enter carbon metabolism to be stored, for example, as glycogen, and are made available again from carbon metabolism under O_2-deprived conditions as provided by heterocysts or during night [41]. Cyanobacteria typically also contain uptake hydrogenases, which recycle the H_2 produced, thereby minimizing energy loss. This activity counteracts H_2 production and needs to be deleted or inactivated, if such organisms are to be utilized for H_2 production. Overall, this approach has been found to enable high H_2 formation rates, especially with nitrogenases, but suffers from the detour via carbon metabolism reducing the light-to-hydrogen efficiency and, in case of nitrogenase-based H_2 formation, from the coupling to nitrogen fixation and thus growth.

H_2 formation via biophotolysis: The holy grail of biological H_2 production is the direct coupling of H_2 formation to the oxygenic photosynthetic light reaction, giving rise to the net reaction of H_2O to H_2 and O_2, with energy provided by light (Fig. 2D) [42]. Thereby, any biomass formation and carbon metabolism is circumvented enabling a high theoretical light-to-hydrogen efficiency. In the oxygenic light reaction, electrons derived from water are channeled into the PET, which, together with the protons released upon water oxidation, generates a proton gradient used for ATP synthesis. Within the PET, two photons drive the transfer of an electron pair from water through photosystem 2, and another two photons are required to drive the electrons further through photosystem 1 to ferredoxin and NADPH, which, in many cyanobacteria and algae, can be used by hydrogenases for H_2 production. Approaches to make efficient use of these reduced electron shuttles or other PET components for hydrogenase-mediated H_2 production will be discussed in Section 4.

Hydrogenases in oxygenic organisms generally are already inactivated at low concentrations of O_2 [43–45]. The energy-yielding transfer of electrons to O_2, that is respiration, makes more sense than "electron loss" to H_2. However, as the use of H_2O as electron donor in photosynthesis obligatorily involves O_2 formation, O_2-tolerant hydrogenases are desirable for biophotolytic H_2 formation. This could be achieved either by making native hydrogenases less O_2-sensitive [46, 47] or by transferring O_2-tolerant hydrogenases from other organisms [48–50] to oxygenic phototrophs. The most common hydrogenases are FeFe hydrogenases and NiFe hydrogenases [51, 52], the former being denatured even at trace concentrations of O_2. NiFe hydrogenases are less O_2 sensitive and generally are reversibly inactivated by O_2 [53] making them more promising for biophotolytic H_2 formation, even though FeFe hydrogenases feature higher k_{cat} values. Furthermore, there are naturally O_2-tolerant NiFe hydrogenases, for example, from *Cupriavidus necator* (formerly *Ralstonia eutropha*) [54–57], which may be introduced into phototrophic microorganisms. This is complicated by the multicomponent nature of these enzymes, which consist of a two- or three-component diaphorase module and a two-component hydrogenase module. The diaphorase module is responsible for electron transfer from NAD(P)H or reduced ferredoxin to the hydrogenase module, which then transfers the electrons to protons by means of the NiFe active site, leading to H_2 formation. Furthermore, hydrogenases are subject to a complex maturation mechanism depending on six to seven additional so-called Hyp enzymes [58, 59].

H_2 production via MEC/BPV: The MEC is an emerging technology that combines bacterial metabolism with electrochemistry. Anode-respiring bacteria, for example

from genera of *Geobacter* [60], *Shewanella* [61, 62], *Pseudomonas* [63], *Clostridium*, *Desulfuromonas*, *Escherichia* [64] and *Klebsiella*, can be attached to an anode, to which they transfer electrons derived from the oxidation of organic compounds [65]. At the cathode, electrons can then be used to produce H_2. This typically requires a power supply to boost the voltage of the electrons, as the standard potential of the electron donor typically is more positive[1] (e.g., −0.28 V for acetate) than for H_2 (−0.41 V). Still, the theoretically required voltage (e.g., −0.13 V) is lower than the voltage typically required for water electrolysis (−1.23 V) as used for green hydrogen. However, energy losses occurring in the MEC currently make it non-competitive. Here, the Holy Grail would again be to use electrons derived from water and energized by light as close as possible to the photosystems in a BPV approach. Thereby, electrodes are used to harvest electrons from the cells and the hydrogen molecule is generated in a separate, abiotic compartment of the reactor producing a clean stream of H_2. This approach has the advantage that depending on the accessible point of electron harvest, an electrical power input much lower than that of a classical electrolyzer is needed, or that a power output might even be generated. This makes a BPV a much more efficient system, than the current way of making "green hydrogen" using an electrolyzer. The challenges at this point are to understand the underlying electron transfer mechanism, to fine tune the point of electron harvest, and to maximize the current density in the BPV system. Especially accessing the photosynthetic electron transfer chain either via genetic engineering or via chemical electron shuttles (so-called mediators) remains a biological challenge [25].

Figure 2 clusters the main approaches for H_2 formation. There are processes relying on fossil energy carriers, processes based on the conversion of biomass or organic carbon sources, non-biological processes making more direct and efficient use of solar energy, and biological processes using microbes as catalysts for the direct use of sunlight for water splitting. Thereby different parameters are assessed in a qualitative matter.

The areal efficiency primarily results from the land use of the respective technology. While fossil technologies rely on organic compounds formed millions of years ago they feature a high area efficiency. In contrast, the area efficiency of technologies relying on present-day sunlight scales with their solar-to-H_2 efficiency. This is a major disadvantage when using biomass/organics [66–68] because solar efficiencies are low. Greenhouse gas emission is low for solar and solar-white H_2, whereas harvesting and transport of biomass increase emissions of the respective processes considerably. Raw material efficiencies are affected by the use of rare earth-, precious-, and heavy- metals, and thus are low for solar technologies, while providing a clear advantage for white H_2 approaches. Due to low technology readiness levels of most evaluated technologies, cost efficiencies are difficult to

---

1 All potentials discussed in this chapter are related to the standard hydrogen electrode (SHE).

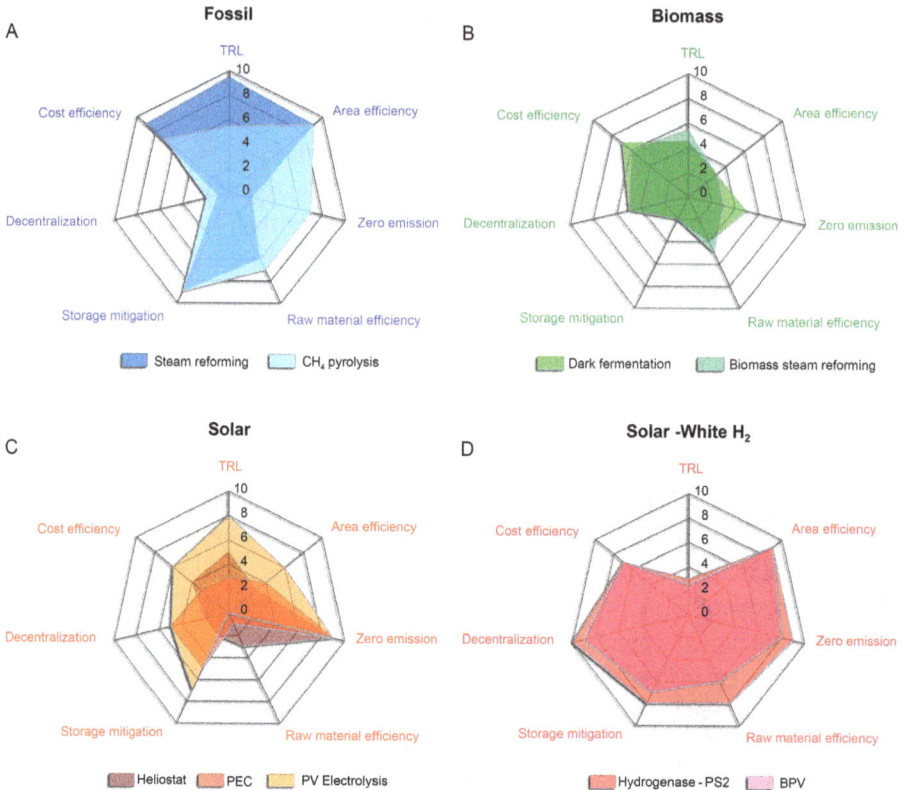

Fig. 2: Comparative analyses of H$_2$-producing technologies, clustered according to the primary energy source. The higher the score, the better the respective technology performs. Abbreviations: TRL, technology readiness level; PEC, photoelectrochemical cell; PV, photovoltaics; BPV, biophotovoltaics.

determine; they are estimated to be similar based on the respective development potentials. This evaluation shows the high attractiveness of direct H$_2$ production from sunlight by means of microbes as catalysts, especially due to the potential to circumvent biomass-derived compounds as an intermediate as well as its high resource efficiency.

11.3 Methods, concepts and state of the art: toward biological H$_2$ production

As described above, there are various ways to produce hydrogen by biological means using solar energy as driver. These strategies differ greatly in efficiency, yield, potential, ecological impact, and development stage (Fig. 2D). All options are in an early development stage and have a high ecological potential, as the primary electron source is water and no or relatively little problematic resources such as rare, precious or heavy metals are required for equipment and catalyst production.

In order to make a significant contribution to future H_2 production, a high energy yield, that is, the efficiency of the conversion of light energy to chemical energy in the form of H_2, is required. A low energy efficiency would lead to an unreasonably high land use. This precludes the use of biomass and biomass-derived organic compounds as electron donors. Whereas the primary electron donor for CO_2 fixation is also water, the light to biomass energy yield is low (typically 1–2%) with a potential to reach 5–8% upon extensive cell engineering [69, 70]. This results in a very low overall light to H_2 energy yield. Due to energy efficiency reasons, also a detour via carbon-based metabolites such as glycogen and glucose within phototrophic microbes should largely be avoided. Hydrogen production thus has to be coupled as close as possible to the primary events of photosynthesis [71]. Thus, the two most promising approaches for biological H_2 production using phototrophs are H_2 formation catalyzed via a hydrogenase operating simultaneously with the photosynthetic light reaction and hydrogen formation at the cathode using a BPV system. Here we will give an overview on strategies and methodologies involved in these concepts, present the current state-of-the-art, and discuss key challenges and possible solutions on the way to a productive, competitive hydrogen producing system.

Hydrogenase-catalyzed H_2 formation driven by electrons derived from O_2-evolving photosynthetic water oxidation in phototrophic microbes (Fig. 3) requires the application of O_2-tolerant hydrogenases and/or a very efficient O_2 removal. Reactive O_2 removal may be an option, but is heavily compromised by the necessity of an additional electron donor for the involved O_2 reduction. A separate substrate functioning as electron donor would increase costs and, considering the low molecular mass of H_2, lead to huge amounts of by-product. Mechanical O_2 removal, for example, via membranes, will not remove O_2 efficiently enough to avoid inhibition/inactivation of typical hydrogenases found in phototrophic microorganisms. Thus, a certain degree of O_2 tolerance can be considered a prerequisite for hydrogenases applied in this approach. As described above, *C. necator* contains promising and well-characterized O_2-tolerant NiFe hydrogenases – prime candidates for an introduction into phototrophic microorganisms, which, however, has not been reported so far. Cyanobacteria constitute promising target hosts for an introduction of O_2-tolerant hydrogenases, considering the extended wavelength range they use by means of phycobilisomes and their ease of manipulation and cultivation. As all hydrogenases, NiFe hydrogenases require additional enzymes for their maturation [58, 59]. Whereas cyanobacteria contain the maturation machinery for their native O_2-sensitive NiFe hydrogenases, efficient maturation of NiFe hydrogenases in the presence of O_2 appears to additionally depend on a formyltetrahydrofolate decarbonylase synthesizing the active site carbon monoxide ligand. Thus, recombinant synthesis of O_2-tolerant NiFe hydrogenases in cyanobacteria may depend on the presence of this enzyme and possibly additional components of their native maturation machinery.

Another challenge is the electron transfer from the PET to these hydrogenases. As described above, photosystem 1 enables the formation of the comparably stable and diffusible electron carriers ferredoxin and NADPH with midpoint potentials of −430 and −320 mV, respectively (H_2/H^+: −413 mV). However, the known O_2-tolerant NiFe hydrogenases, for example, from *Cupriavidus necator*, requires NADH as electron donor. This incompatibility may be solved by transhydrogenases or by protein engineering toward an O_2-tolerant NiFe hydrogenase accepting electrons from NADPH or reduced ferredoxin – similar to native NiFe hydrogenases in cyanobacteria such as *Synechocystis* sp. PCC 6803 [72]. On the other hand, native cyanobacterial NiFe hydrogenases may be engineered toward O_2 tolerance [46, 47]. While the formation of H_2 by NADPH oxidation is energetically unfavorable due to its high midpoint potential, which requires virtually a 100% reduced NADPH pool and low H_2 concentrations [73], reduced ferredoxin is a more suitable electron donor for hydrogenases.

An enhanced coupling of electron transfer from the PET to the hydrogenase system may be achieved by a hydrogenase fusion to photosystem 1 [27, 74–77]. Recently, this has been achieved for the coupling of the native NiFe hydrogenase module from *Synechocystis* sp. PCC 6803 to PS1 within the same strain [27]. Indeed, H_2 formation via direct electron transfer from PS1 to the hydrogenase *in vivo* could be measured, whereas electron transfer from simultaneous water oxidation remained hampered, probably due to the O_2 sensitivity of the native hydrogenase module. Thus, the transfer of an O_2-tolerant hydrogenase into a cyanobacterial strain and the coupling of its hydrogenase module to PS1 remain challenges to be tackled with multiple fusion options being discussed (Fig. 3). Also, a direct coupling of a hydrogenase to PS2 would be of special interest, as the use of only one instead of both photosystems would double the maximum amount of H_2 formed per absorbed quantum. In PS2, pheophytin is the primary electron acceptor, with a midpoint potential between −0.5 and −0.6 V upon reduction, which, in principle, is sufficient to drive H_2 formation [78]. However, reduced pheophytin in photosystem 2 is a single electron transfer system which is quite unstable and therefore requires fast electron transfer to the hydrogenase. Protein engineering attempts to merge appropriate parts of photosystem 2 and a hydrogenase have not been reported so far (for more details see also Chapter 1). In all cases, the competition of hydrogen formation and cellular metabolism, based on a still not fully understood regulation of electron flux and transfer, has to be considered. A first metabolic and regulatory engineering target thereby is to establish high rates of electron transfer to the hydrogenase while preserving a necessary minimum of maintenance metabolism to keep the cells stable under minimal growth. This may be achieved by cultivating the photo-biocatalysts as a biofilm, which is described in more detail below.

Besides establishing an O_2-tolerant hydrogenase within a cyanobacterium combined with its efficient coupling to the cyanobacterial redox metabolism, the

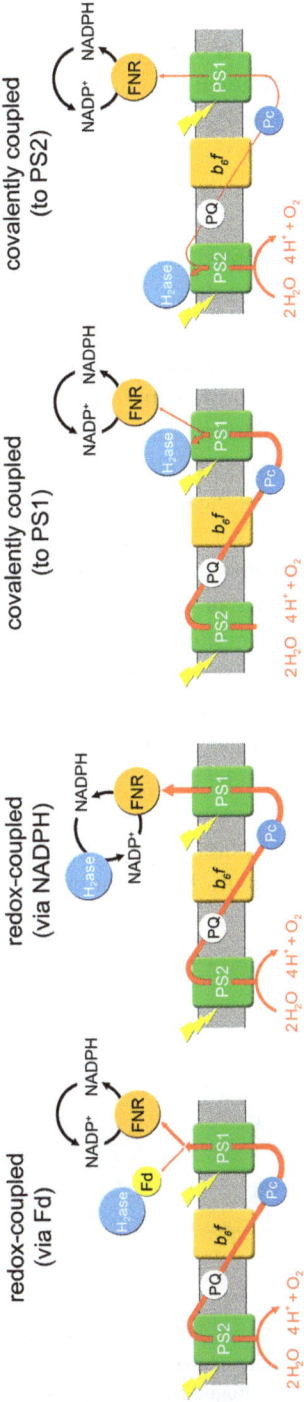

Fig. 3: Various ways to integrate hydrogenase based H₂ production into the cellular machinery. For details please see text.

simultaneous formation of H_2 and O_2 is a major challenge in this approach. First, high H_2 levels in the surrounding of the cells are prone to hamper the possibly reversible H_2 formation; second, high O_2 levels are well known to inhibit cyano-bacterial growth and metabolism; and third, simultaneous H_2 and O_2 accumula-tion leads to an explosive oxyhydrogen mixture. Thus, a reaction engineering approach to remove those gases *in situ* appears to be mandatory and is discussed in more detail at the end of this section.

H_2 formation via BPV: In the BPV approach, a MEC is adopted to use sunlight as the energy source for electron generation. A MEC typically consists of a single or dual chamber device with a working electrode, a counter electrode and sometimes a reference electrode. The latter is needed, if potentiostatic control is desired to main-tain a constant working electrode potential. Microbes are located in the working electrode chamber and generate electrons through the oxidation of metabolites. In the case of a BPV the microbes are capable of using light energy to elevate the elec-trons to a higher energy level [79, 80] – for instance in purple bacteria – or are able to even liberate electrons through water oxidation – for instance in micro algae or cyanobacteria. For the production of hydrogen, a two-chamber three-electrode sys-tem is preferable, because in such a system a cation exchange membrane separates the working electrode chamber (Fig. 4) [25]. Although this complicates the set-up, it enables the generation of oxygen and hydrogen in separate compartments. While the working electrode chamber might generate a stream of CO_2 and O_2, depending on the metabolism of the respective used organisms, the counter electrode chamber will provide a clean stream of H_2. Application of a reference electrode allows using different electrical bias to control the energy level of the electrons supplied to the anode. As electron acceptors require a more positive redox potential than electron donors, tuning toward an accessible point of electron harvest in the PET or broader

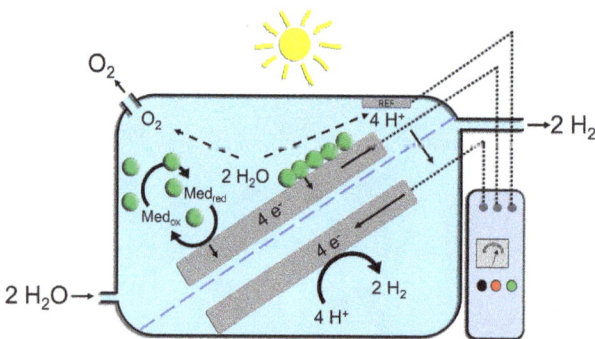

Fig. 4: Schematic view of a BPV system. Cells could form a biofilm on the anode or be dispersed in the anode compartment. Electron transfer could be facilitated via mediator molecules or direct electron transfer. The two compartment set-up enables separation of hydrogen and oxygen in a single device, while the use of a potentiostat maintains the thermodynamic driving force.

metabolism is possible; also potential differences can be used to manipulate the rate of electron harvest (i.e., the current) [25]. In comparison with a standard electrolyzer, the required electrical power input for water splitting in the BPV is minimized, resulting (theoretically) in a much more efficient system. Challenge here is to understand the underlying electron transfer mechanism, to fine tune the point of electron harvest, and to maximize the current density in the BPV system.

In two-chamber systems the biocatalysts, microalgae or cyanobacteria, are incubated in the anode compartment. These microbes contain both photosystems 1 and 2 and hence are able to generate protons and electrons from water splitting with sunlight as the primary source of energy. Key challenges here are the harvesting of electrons from the cellular redox metabolism and their transfer from the cell to the anode. In principle, there are two distinct processes for electroactive microbes to achieve this: Either indirect (or mediated) electron transfer (IET) using low molecular weight compounds as mediators or direct electron transfer (DET) facilitated by conductive membrane components or appendages of the cell, such as pili or cytochrome stacks [25]. To date, it is unclear how the electron transfer in cyanobacteria works, but several publications have shown that IET is more likely in model strains, such as *Synechocystis* sp. PCC 6803. If IET is chosen as the mode of electron transfer, it is important to maintain the potential of the working electrode, which requires a potentiostat and a reference electrode. This ensures the efficient recycling of shuttle molecules or mediators, which are reduced by the cells and oxidized at the anode. If the working electrode potential would be allowed to drift, the potential difference between the poised electrode and the oxidation potential of the mediator could become suboptimal and lower the current output of the system.

Although the use of cyanobacteria in BPV systems is still a young field dominated by descriptive papers, a range of cyanobacteria has been tested for their photo-current output. To date, the systems achieved power densities of up to 0.5 W m^{-2} [81, 82]. Organisms include *Synechococcus* and *Synechocystis* sp. PCC 6803 [83], *Nostoc* [84, 85], *Anabaena variabilis* M-2 [86, 87], *Oscillatoria limnetica* [88] and *Lyngbya* [84, 89]. The highest power density reported so far was above 100 mW m^{-2} with *Synechocystis* sp. PCC 6803, however, only at microscale (with an anode chamber of 0.4 µL) [90]. If a potential of +0.5 V is applied, corresponding to 0.2 A m^{-2}, this is equivalent to the generation of 2.0728 × 10^{-6} mol electrons per second per m^2, yielding 3.7 mmol hydrogen per m^2 per hour. Of course this current density needs to be improved, for instance by engineering the microbes to increase power output [91]. Alternatively, the used mediator molecules may increase the H$_2$ production circumventing genetically modified organisms and their legal implications. This indicates the potential of BPV as an important part of the puzzle for H$_2$ production. Especially in settings where genetically engineered microbes might be unacceptable, BPV in combination with highly designed chemical mediators (e.g., membrane intercalating metal complexes of desired redox potential) might open a path to phototrophic hydrogen production.

Despite some efforts, the breakthrough to enable continuous, light-driven H_2 production from water in cyanobacteria has not been achieved yet. As pointed out above, the current concepts are very interesting and have a high potential for applications; however, there are still huge challenges to overcome. Moreover, there is still a lack of knowledge on photosynthetic organisms, for example, regulation of carbon fluxes, which constitute major competitors for reduction equivalents and thus are important to understand in order to streamline electron flow toward hydrogen production. Metabolically, there are two major challenges in producing hydrogen: On the one hand, maximization of water splitting and electron supply is required to maximize photon efficiency. This applies to both the hydrogenase and the BPV approach and requires systems metabolic engineering for the overall maximization of electron fluxes from photosystem 2 into the PET. On the other hand, electron transfer for hydrogen formation is a metabolic challenge, with BPV and hydrogenase approaches differing in the terminal point of electron harvest. Nevertheless, both approaches are in principle tunable to a desired point of electron withdrawal to maximize thermodynamic driving force and energy content of the electrons.

Screening options: A reliable and sensitive screening method is a prerequisite for effective catalyst engineering in order to identify promising candidate strains for hydrogen production and/or beneficial mutations within these organisms. One potential approach is the use of biological hydrogen sensors coupled to a reporter gene allowing optical monitoring of H_2 production within bacterial cultures or even in a single cell. Interesting is the exploitation of natural H_2-responsive gene regulatory circuits, which have been reported for several bacteria such as the aerobic H_2-oxidizing (*Knallgas*) bacterium *Cupriavidus necator (Ralstonia eutropha)* [92, 93] or the purple non-sulfur bacterium *Rhodobacter capsulatus* [94, 95]. Upon occurrence of H_2, these bacteria induce the expression of an uptake hydrogenase that oxidizes H_2. Their molecular H_2 sensors are composed of another type of hydrogenase with high H_2 affinity but low H_2 oxidation activity [96]. These sensory hydrogenases perceive the presence of H_2 and transfer the signal via a histidine kinase to a transcriptional regulator, which finally activates expression of the uptake hydrogenase genes.

Engineering of the H_2-responsive gene regulatory circuit has been reported for *Rhodobacter*. A gene variant for the green fluorescent protein (GFP)) was fused to the promoter of the gene encoding the uptake hydrogenase, which resulted in a strain showing H_2-dependent GFP fluorescence. The generated reporter strain has not only been used as biosensor to screen and optimize H_2 production in *Rhodobacter* itself [97, 98] but also to follow H_2 production in co-cultivated green algae [98]. For cyanobacteria, however, molecular tools to investigate the limitation of H_2-evolution in a systematic manner are lacking. A co-cultivation approach as shown above for *Rhodobacter* may not be possible as most bacterial strains do not (optimally) grow in cyanobacterial growth media; this prevents their function as biosensor. Moreover, H_2 would only be detected externally in the medium and not intracellular. Therefore, the establishment of a similar gene regulatory circuit for the direct monitoring of H_2 at its production site, for example, in a cyanobacterial cell, would be extremely helpful. In combination with a high-throughput screening method, a cyanobacterial H_2-sensing strain has great potential to improve H_2 production (Fig. 5).

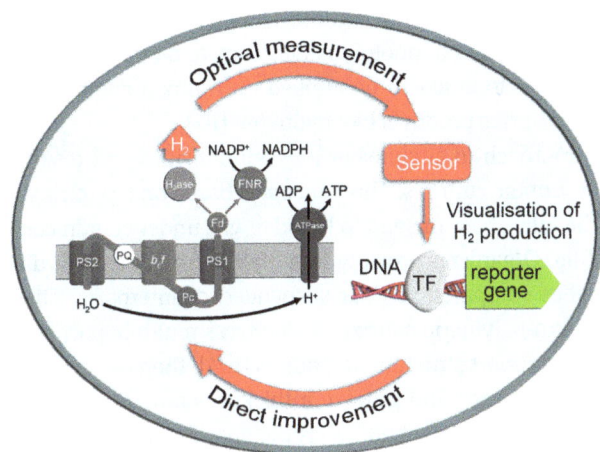

Fig. 5: Modified cyanobacterial strain as platform for the selection of efficient H_2-producers. The strain could express the H_2-sensing regulatory circuit of *Cupriavidus necator* coupled to a reporter gene. Combined with random mutagenesis approaches, the optical detection of H_2 production would allow systematic screening for higher H2-yields as a result of improved hydrogenase properties and/or optimization of the cellular background.

Used as host for random mutagenesis approaches, the obtained mutant library could be directly screened for striking changes in the reporter signal, pointing toward improved H_2-yields. Altogether, it would enable evolutionary approaches to shape hydrogenase properties and/or to identify and engineer metabolic constraints such as high electron fluxes to competing reactions. To finally optimize and enable continuous, light-driven H_2 production from water, the optimized genetic configurations identified by the above mentioned approach could be combined with targeted engineering strategies such as the integration of an O_2-tolerant hydrogenase. The cyanobacterial H_2 biosensor could also help to implement novel, innovative cultivation concepts based on natural [99] and artificial [100] cyanobacterial biofilms (see chapter 10).

Continuous H_2 production using biofilm catalysts – While many research activities focus on photo-biocatalyst engineering, and more and more genetic engineering tools become available for cyanobacteria, reaction engineering is still at an early stage of development. This is reflected in the currently established photo-bioreactors, most of them using either a flat panel or tubular geometry. Only tubular photobioreactors are operated on the industrial scale for the production of biomass or high-priced chemicals (above 10 € kg^{-1}). Examples include a 700 m^3 photobioreactor for the production of biomass for food and feed (Klötze, Germany) and a 25 m^3 photobioreactor for the production of astaxanthin (Hawaii, USA) [101]. However, key problems are not yet solved. To date, there are no scalable bioreactor systems which allow cell densities above 4 g_{CDW} L^{-1} (approximately OD$_{730}$ = 15) [102]. This is too low for a productive bioprocess (for comparison: commercial bioprocesses utilizing heterotrophic microorganisms like *E. coli* easily reach cell densities of 50 g_{CDW} L^{-1} or higher). Thus, photobioreactor development is one of the key limiting factors of photobiotechnology and one of the biggest hurdles to overcome if cyanobacteria are to be applied on a broad commercial scale. This is especially true for low-price products like hydrogen [103].

Here, we describe a novel approach to circumvent low cell densities and insufficient catalytic stabilities in phototrophic cultures. This concept is based on the intrinsic ability of bacteria to attach to the interphase of solid to liquid or gas under certain conditions, thereby forming a biofilm. Organisms growing in a biofilm format exhibit different physiological characteristics compared to their suspended counterparts. They are encased in a self-produced extra-polymeric matrix, which serves multiple purposes like protection, nutrient reservoir, support structure and others [104], thereby these organisms form three-dimensional structures and grow to a thickness ranging between some micrometers and several centimeters, depending on the specific organisms and the respective environment. The metabolic activity of the individual organism in the biofilm is mainly governed by the mass transfer of compounds leaving as well as of nutrients entering the biofilm. This transport in and out of the structure is always passive and is driven by concentration gradients throughout the film [105].

Cultivating photoautotrophic organisms as biofilms comes along with several constraints. First, the cultivation vessel needs to fulfil a certain geometry to allow sufficient light penetration and thus light availability, especially as the culture is not mixed but attached to a surface. Second, sufficient carbon supply needs to be ensured, either as carbonate or as CO_2. Finally, the oxygen liberated during the water splitting reaction needs to be removed effectively, as high oxygen levels will toxify the organisms. In the concept introduced here, these constraints were solved by using a capillary biofilm reactor run in a segmented flow fashion (Fig. 6A) [106, 107]. In such a system, the biofilm is growing inside 3 mm thick capillaries made out of transparent material. The low tube diameter ensures a very high surface area to volume ratio and a short light path, while the air segments, which are constantly passed through the capillary together with the medium flow, serve as 'vehicles'. They carry gaseous substrates like CO_2 into, and waste products like oxygen out of the biofilm and thereby ensure a low oxygen tension in the overall system. Furthermore, they contribute to the mass transfer by the Tylor forces occurring at the interface of the different phases [108], and prevent clogging of the capillary system. Attachment of the cyanobacterial strain, for example, *Synechocystis* sp. PCC 6803, to the capillary wall is promoted by the presence of a chemoheterotrophic organism, *Pseudomonas taiwanensis* VLB120, which is known to be an excellent biofilm former [109]. This artificial consortium follows the concept of prototrophy, where both organisms profit from each other without being dependent on the presence of the partner (Fig. 6B). *Synechocystis* sp. PCC 6803 is delivering oxygen and reduced organic compounds for *P. taiwanensis,* which in turn is respiring the oxygen and providing a conditioning film facilitating the biofilm formation of *Synechocystis* sp. PCC 6803 [99]. This cultivation concept resulted in a homogenous biofilm of 100% capillary surface coverage and extraordinary high cell densities of up to 50 g_{CDW} L^{-1}. Challenging reactions like the conversion of cyclohexane to the corresponding alcohol, representing a huge burden for the microbes due to the toxicity of the reactants, have shown that the biofilm system is capable of maintaining stable reaction activities due to adaptation and constant regeneration sp. PCC 6803[99]. We believe that these findings can be transferred to H_2 production, enabling a continuous mode of operation, very low growth rates and stable activities. In case of H_2 production (with no net electron generation), not all O_2 may be respired; however, the presence of the heterotroph will lower the oxygen tension below critical levels, recycle organic by-products back to CO_2 and stabilize the biofilm.

A

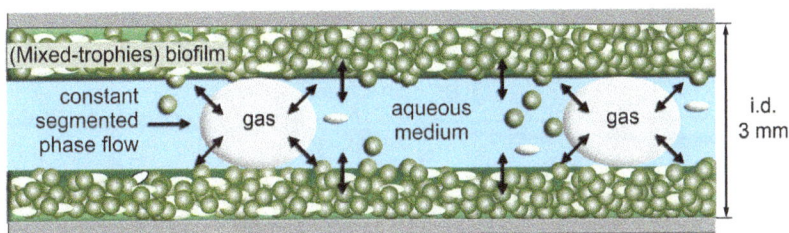

B

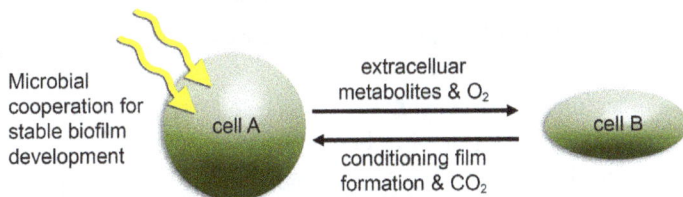

Fig. 6: (A) Capillary biofilm reactor concept: Biofilm growing on the inside of a capillary under constant flushing with medium and gas, resulting in a segmented flow fashion. (B) Prototrophy in mixed species biofilms composed of *Synechocystis* sp. PCC 6803 (cell A) and *Pseudomonas taiwanensis* (cell B) as prerequisite to establish stable *Synechocystis*-biofilms in CBRs. Phototrophic strain providing organic carbon sources and oxygen for the chemolithotrophic organism, respiring the oxygen and providing a conditioning film and CO_2, thereby promoting biofilm formation by *Synechocystis* sp. PCC 6803.

Integrated process design and key constraints – The prospect of producing energy carriers with cyanobacteria received widespread attention in recent years, but their implementation is often restricted by process and economic feasibilities [110]. Catalyst, reaction and process engineering need to be considered in an integrative way in order to develop efficient photobioprocesses that can compete with established H_2 production technologies. However, this requires the application of rational heuristics that guide all stages of process development and scale-up. In general, bioprocess feasibility is constrained by system and process-specific parameters that span the feasible space of operation. Achievable product concentrations and titers, as well as biocatalyst stability and product formation kinetics are the key process constraints and set the quantitative frame for the integrated process design (Fig. 7) [111]. This concept can be transferred to evaluate the theoretical boundaries for BPV concepts and hydrogenase-catalyzed production of white hydrogen. In the following, we discuss the potential efficiency of these processes in decentralized continuous, biofilm-based reaction formats.

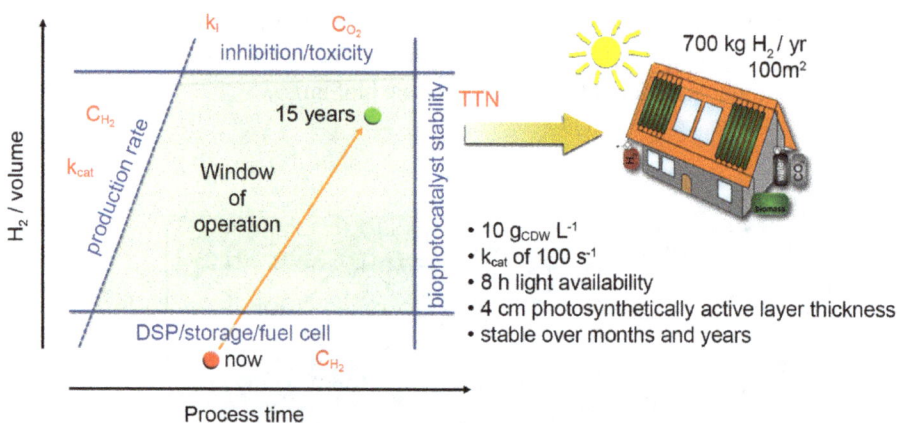

Fig. 7: Window of operation for producing white hydrogen with cyanobacteria (see text for more details).

Cyanobacterial processes promise a sustainable and carbon-neutral chemical production of energy carriers, but are primarily limited by production kinetics of the photobiocatalyst. Slow specific production rates, low biomass concentrations and poor space-time yields (STY) result in typical product concentrations which rarely exceed the mg L^{-1} mark [112]; they are mainly responsible for the high area requirements of state-of-the-art photo-bioprocesses. Novel reaction formats, such as capillary-based biofilm reactors, can remedy this with high biocatalysts densities of up to 50 g_{CDW} L^{-1}, corresponding to a tenfold increase in active biomass per volume compared to common photo-bioreactor setups with suspended cells [99, 113]. However, concerning the production of carbon-based energy carriers, the inherent CO_2 fixation capacity of cyanobacteria and the corresponding metabolic flux toward the product set the upper boundary [114].

Thus, the theoretical kinetic potential for producing hydrogen with cyanobacteria can be considered higher than for products derived from the carbon metabolism [2]. Although reported rates for phototrophic hydrogen production are too low to exceed the minimum process boundaries, the water-splitting reaction at PS2 achieves much higher turnover numbers of up to 250 s^{-1} and can deliver electrons for proton reduction or to the electrode in BPV at the respective rate [27, 115]. With a conservative estimate of an electron delivery rate of 100 s^{-1} $PS2^{-1}$ for hydrogen production processes at a cell density of 10 g_{CDW} L^{-1}, more than 700 kg H_2 a^{-1} could be theoretically produced on an area of 100 m^2. The containing energy is sufficient to cover the energy demand of a four-person household in a modern energy-efficient home.

Next to the kinetic constraints, inhibitory effects of products have to be considered, marking the upper boundary of the window of operation. H_2 accumulation can unfavorably shift chemical equilibria of H_2 formation reactions [116]. Even more critical is the oxygen that is inevitably liberated during water photolysis, as it may have inhibitory effects on the whole cell and especially on the hydrogenases. While this can be handled by the implementation of oxygen-tolerant hydrogenases, high O_2 levels (especially those above air saturation) can also impair these enzymes [117, 118]. At high cell densities, photosynthetic activity can quickly increase oxygen partial pressures above 25% which – besides hydrogenases – also inhibit cell function and biofilm stability [113]. Next to the inhibitory effects of oxygen on the photobiocatalysts, the formation of explosive oxyhydrogen gas at 4% to 95% (v/v) hydrogen poses a significant threat for safe process operation. While oxygen and hydrogen synthesis can be spatially separated in BPV approaches, in situ removal of oxygen and hydrogen from the reaction space is mandatory for hydrogenase-based approaches [119]. As mentioned before, there are several ways to biologically or chemically capture the oxygen liberated in photosynthesis. However, these approaches are neither from an economic, nor from an energetic point of view, feasible. For process feasibility, it is recommendable to integrate downstream processing (DSP) of the gaseous products into the process for avoiding biocatalyst inhibition and oxyhydrogen. Considering the diffusivity of O_2 and H_2, a membrane may enable the *in situ* removal of both gases in a first step, which then is followed by a second, possibly also membrane-mediated step for gas separation. Thereby, the system has to ensure a good light penetration toward the cells. Conceivable strategies involve the implementation of membranes with defined hydrogen and oxygen-selectivities (e.g., modified PDMS or PTFE membranes) directly into the microcapillary reactors or the application of advanced materials such as metal hydrides for hydrogen adsorption [120]. However, critical threshold concentrations of hydrogen have to be reached in the continuous reaction format to realize in situ DSP necessitating the application of hydrogenases with some O_2-tolerance in the biophotolysis approach. Moreover, mass transfer properties of membranes and H_2 adsorption kinetics need to be carefully studied and adjusted to the systems-specific production rates for optimizing reaction conditions.

Process stability is the last key constraint that has to be carefully considered for the integrated design of a competitive white hydrogen production process. While cyanobacterial biofilms are naturally more robust against environmental influences than their suspended counterparts, more sophisticated approaches are needed to extend process uptimes under outdoor conditions [121, 122]. Decentralized H_2 production units are subject to ever-changing conditions due to day/night cycles, weather conditions and drastic seasonal changes in temperature, light quality and irradiance [123]. Therefore, the system has to be robust against such abiotic factors. Survival and growth are the key objective functions of microbial systems in nature. Their underlying principles for robustness against environmental influences can be used to stabilize the process via ecological principles [124]. Future process concepts for white H_2 production under outdoor conditions will focus on the colonization of biofilm microreactors with consortia, whose composition is optimized to fulfil the natural objectives of growth and survival, while maintaining hydrogen production as the main process objective function.

11.4 Outlook and future perspectives

Exploiting natural oxygenic photosynthesis in cyanobacteria for the sustainable production of white H_2 is an attractive complementation to other more advanced H_2 production technologies. With the theoretical solar to H_2 yield of up to 24% [125], photosynthesis-driven H_2 production technologies potentially outperform other solar-based ones. Moreover, the application of living cyanobacterial cells offers the advantage of a self-regenerating photobiocatalytic system enabling a continuous process format. However, the challenges involved in scaling processes based on phototrophic organisms make a large-scale implementation of white H_2 production unlikely. Yet, it holds a huge potential for applications in small, decentralized autonomous units for the production of several hundred kg per year. As outlined in this chapter, the hurdles for developing such processes are still significant and span across all development disciplines from molecular to process engineering. Cyanobacteria are genetically accessible and are ideal host systems for the development of biocatalysts for hydrogen production. First, cellular design targets were defined and met by Rögner et al. and others [2]. Especially during the past few years, new concepts for efficient and theoretically scalable biofilm reactor formats have been developed that enable high cell densities and continuous processing. Nevertheless, significant key challenges remain. White H_2 still suffers from low specific production rates and production stability, which currently poses the main challenge for economically feasible process implementation. This fact requires advanced molecular strategies for catalyst engineering to increase the electron fluxes toward electrodes in BPV approaches, or hydrogenases in biophotolysis. Increase in H_2 production rates would directly translate into the improved economic feasibility of the process. Latest developments, such as the successful fusion of functional hydrogenases directly to the PS1 complex, are promising and point the way for significant future improvements in the photobiocatalysts reactivity toward hydrogen. However, not only efficient molecular configurations for H_2 synthesis itself have to be established, but also a targeted stabilization of cell metabolism via regulatory modifications is required for a balanced division of electrons between the H_2 formation reaction and the demands of cellular metabolisms [126, 127].

Reaction and process concepts must be engineered for enhanced process efficiency and stability, next to the critical steps in strain development. Phototrophic biofilms in capillary-based microreactors are the key reaction format for white H_2 as it enables continuous and stable cultivation of cyanobacteria at optimally high cell densities and hence can provide high STYs. Reaction engineering must focus on optimizing process parameters, including concepts for the efficient in situ removal of O_2 and H_2. Therefore, it is required to design white H_2 processes in development cycles that integrate strain, reaction and process engineering based on the critical process constraints – efficiency, STY, kinetics, product titer and stability.

Overall, the concept of white H_2 production offers a promising source for sustainable bioenergy in decentralized, autonomous applications formats. With future advancements in strain and process development, white H_2 might become a key element in the energy mix of a future bioeconomy.

References

[1] Noor E, Eden E, Milo R, Alon U. Central carbon metabolism as a minimal biochemical walk between precursors for biomass and energy. Mol Cell 2010, 39, 809–20.
[2] Rexroth S, Wiegand K, Rögner M, eds. Cyanobacterial design cell for the production of hydrogen from water. In: Rögner M, eds. Biohydrogen. Berlin, Boston, MA, Walter de Gruyter GmbH & Co. KG, 2015, 61–88.
[3] Chen S, Pei CL, Gong JL. Insights into interface engineering in steam reforming reactions for hydrogen production. Energy Environ Sci 2019, 12, 3473–95.
[4] Gordon M, McFarland E, Metiu H. Catalytic methane pyrolysis for hydrogen production. Abstr Pap Am Chem S 2019, 257.
[5] Shiva Kumar S, Himabindu V. Hydrogen production by PEM water electrolysis – A review. Mater Sci Energy Technol 2019, 2, 442–54.
[6] Ju H, Badwal S, Giddey S. A comprehensive review of carbon and hydrocarbon assisted water electrolysis for hydrogen production. Appl Energ 2018, 231, 502–33.
[7] Diaz-Abad S, Millan M, Rodrigo MA, Lobato J. Review of anodic catalysts for SO_2 depolarized electrolysis for "green hydrogen" production. Catalysts 2019, 9, 63–79.
[8] Shandarr R, Trudewind CA, Zapp P. Life cycle assessment of hydrogen production via electrolysis – a review. J Clean Prod 2014, 85, 151–63.
[9] Vincent I, Bessarabov D. Low cost hydrogen production by anion exchange membrane electrolysis: A review. Renew Sust Energ Rev 2018, 81, 1690–704.
[10] Ratlamwala TAH, Dincer I, Aydin M. Energy and exergy analyses and optimization study of an integrated solar heliostat field system for hydrogen production. Int J Hydrog Energy 2012, 37, 18704–12.

[11] Roca L, De La Calle A, Yebra LJ. Heliostat-field gain-scheduling control applied to a two-step solar hydrogen production plant. Appl Energ 2013, 103, 298–305.

[12] Shukla PK, Karn RK, Singh AK, Srivastava ON. Studies on PV assisted PEC solar cells for hydrogen production through photoelectrolysis of water. Int J Hydrog Energy 2002, 27, 135–41.

[13] Moser M, Pecchi M, Fend T. Techno-economic assessment of solar hydrogen production by means of thermo-chemical cycles. Energies 2019, 12.

[14] Muhich CL, Ehrhart BD, Al-Shankiti I, Ward BJ, Musgrave CB, Weimer AW. A review and perspective of efficient hydrogen generation via solar thermal water splitting. Wires Energy Environ 2016, 5, 261–87.

[15] Assaf J, Shabani B. Experimental study of a novel hybrid solar-thermal/PV-hydrogen system: Towards 100% renewable heat and power supply to standalone applications. Energy 2018, 157, 862–76.

[16] Villafán-Vidales HI, Arancibia-Bulnes CA, Valades-Pelayo PJ, Romero-Paredes H, Cuentas-Gallegos AK, Arreola-Ramos CE. Hydrogen from solar thermal energy, In: Calise F, et al., eds. Solar Hydrogen Production,Academic Press, 2019, 319–63.

[17] Oishi T, Konishi H, Nohira T, Tanaka M, Usui T. Separation and recovery of rare earth metals by molten salt electrolysis using alloy diaphragm. Kagaku Kogaku Ronbun 2010, 36, 299–303.

[18] Vahidi E, Zhao F. Assessing the environmental footprint of the production of rare earth metals and alloys via molten salt electrolysis. Resour Conserv Recy 2018, 139, 178–87.

[19] Yao B, Cai BF, Kou F, Yang YM, Chen XP, Wong DS, Liu LS, Fang SX, Liu HL, Wang HY, Zhang LZ, Li JZ, Kuang GC. Estimating direct CO_2 and CO emission factors for industrial rare earth metal electrolysis. Resour Conserv Recy 2019, 145, 261–7.

[20] Michel H. Editorial: The nonsense of biofuels. Angew Chem Int Ed Engl 2012, 51, 2516–8.

[21] Lee H-S, Vermaas WFJ, Rittmann BE. Biological hydrogen production: Prospects and challenges. Trends Biotechnol 2010, 28, 262–71.

[22] Zhu GF, Wu TT, Jha AK, Zou R, Liu L, Huang X, Liu CX. Review of bio-hydrogen production and new application in the pollution control via microbial electrolysis cell. Desalin Water Treat 2014, 52, 5413–21.

[23] Kadier A, Simayi Y, Abdeshahian P, Azman NF, Chandrasekhar K, Kalil MS. A comprehensive review of microbial electrolysis cells (MEC) reactor designs and configurations for sustainable hydrogen gas production. Alex Eng J 2016, 55, 427–43.

[24] Lukajtis R, Holowacz I, Kucharska K, Glinka M, Rybarczyk P, Przyjazny A, Kaminski M. Hydrogen production from biomass using dark fermentation. Renew Sust Energ Rev 2018, 91, 665–94.

[25] Tschörtner J, Lai B, Krömer JO. Biophotovoltaics: Green power generation from sunlight and water. Front Microbiol 2019, 10, 866.

[26] Zhang C, Ma SS, Wang GH, Guo YP. Enhancing continuous hydrogen production by photosynthetic bacterial biofilm formation within an alveolar panel photobioreactor. Int J Hydrog Energy 2019, 44, 27248–58.

[27] Appel J, Hueren V, Boehm M, Gutekunst K. Cyanobacterial in vivo solar hydrogen production using a photosystem I–hydrogenase (PsaD-HoxYH) fusion complex. Nat Energy 2020, 5, 458–67.

[28] Petrova EV, Kukarskikh GP, Krendeleva TE, Antal TK. The mechanisms and role of photosynthetic hydrogen production by green microalgae. Microbiology+ 2020, 89, 251–65.

[29] Thauer RK, Jungermann K, Decker K. Energy conservation in chemotrophic anaerobic bacteria. Bacteriol Rev 1977, 41, 100–80.

[30]   Reith JH, Wijffels RH. Bio-Methane and Bio-Hydrogen: Status and Perspectives of Biological Methane and Hydrogen Production, Barten H, ed. Dutch Biological Hydrogen Foundation, 2003.

[31]   Lee HS, Krajmalinik-Brown R, Zhang H, Rittmann BE. An electron-flow model can predict complex redox reactions in mixed-culture fermentative bioH$_2$: Microbial ecology evidence. Biotechnol Bioeng 2009, 104, 687–97.

[32]   Axley MJ, Grahame DA, Stadtman TC. *Escherichia coli* formate-hydrogen lyase. J Biol Chem 1990, 265, 18213–18.

[33]   Yoshida A, Nishimura T, Kawaguchi H, Inui M, Yukawa H. Enhanced hydrogen production from glucose using *ldh*- and *frd*-inactivated *Escherichia coli* strains. Appl Microbiol Biotechnol 2006, 73, 67–72.

[34]   Ren N, Xing D, Rittmann BE, Zhao L, Xie T, Zhao X. Microbial community structure of ethanol type fermentation in bio-hydrogen production. Environ Microbiol 2007, 9, 1112–25.

[35]   Jungermann K, Thauer RK, Leimenstoll G, Decker K. Function of reduced pyridine nucleotide-ferredoxin oxidoreductases in saccharolytic *Clostridia*. Biochim Biophys Acta, Bioenerg 1973, 305, 268–80.

[36]   Petitdemange H, Cherrier C, Raval G, Gay R. Regulation of the NADH and NADPH-ferredoxin oxidoreductases in *Clostridia* of the butyric group. Biochim Biophys Acta Gen Subjects 1976, 421, 334–47.

[37]   Shin J-H, Hyun Yoon J, Eun Kyoung A, Kim M-S, Jun Sim S, Park TH. Fermentative hydrogen production by the newly isolated *Enterobacter asburiae* SNU-1. Int J Hydrog Energy 2007, 32, 192–9.

[38]   Masepohl B, Schneider KK, Drepper TT, Müller A, Klippa W. Alternative hydrogenases. In: Leigh GJ, ed. Nitrogen Fixation at the Millennium. Elsevier, 2002, 191–222.

[39]   Harwood CS. Nitrogenase-catalyzed hydrogen production by purple nonsulfur photosynthetic bacteria. In: Wall JD, Harwood CS, Demain A, eds. Bioenergy. ASM Press, 2008, 259–71.

[40]   Bothe H, Schmitz O, Yates MG, Newton WE. Nitrogen fixation and hydrogen metabolism in cyanobacteria. Microbiol Mol Biol Rev 2010, 74, 529–51.

[41]   Min H, Sherman LA. Hydrogen production by the unicellular, diazotrophic cyanobacterium *Cyanothece* sp. strain ATCC 51142 under conditions of continuous light. Appl Environ Microbiol 2010, 76, 4293.

[42]   Michel H. Die natürliche Photosynthese: Ihre Effizienz und die Konsequenzen, In: Gruss P, Schüth F, eds. Die Zukunft der Energie. C. H. Beck, 2008, 80–1.

[43]   Ghirardi ML, Dubini A, Yu J, Maness P-C. Photobiological hydrogen-producing systems. Chem Soc Rev 2009, 38, 52–61.

[44]   Ghirardi ML, Posewitz MC, Maness P-C, Dubini A, Yu J, Seibert M. Hydrogenases and hydrogen photoproduction in oxygenic photosynthetic organisms. Annu Rev Plant Biol 2007, 58, 71–91.

[45]   Stripp ST, Goldet G, Brandmayr C, Sanganas O, Vincent KA, Haumann M, Armstrong FA, Happe T. How oxygen attacks [FeFe] hydrogenases from photosynthetic organisms. Proc Natl Acad Sci U S A 2009, 106, 17331.

[46]   Ghirardi ML, King PW, Posewitz MC, Maness PC, Fedorov A, Kim K, Cohen J, Schulten K, Seibert M. Approaches to developing biological H$_2$-photoproducing organisms and processes. Biochem Soc Trans 2005, 33, 70–2.

[47]   Koo J, Swartz JR. System analysis and improved [FeFe] hydrogenase O$_2$ tolerance suggest feasibility for photosynthetic H$_2$ production. Metab Eng 2018, 49, 21–7.

[48]   Nishimura H, Sako Y. Purification and characterization of the oxygen-thermostable hydrogenase from the aerobic hyperthermophilic archaeon *Aeropyrum camini*. J Biosci Bioeng 2009, 108, 299–303.

[49] Lenz O, Lauterbach L, Frielingsdorf S, Friedrich B. Oxygen-tolerant hydrogenases and their biotechnological potential, In: Rögner M, ed. Biohydrogen. Berlin, Boston, MA, Walter de Gruyter GmbH & Co. KG, 2015, 61–88.

[50] Fritsch J, Lenz O, Friedrich B. Structure, function and biosynthesis of O₂-tolerant hydrogenases. Nat Rev Microbiol 2013, 11, 106–14.

[51] Vignais PM, Billoud B. Occurrence, classification, and biological function of hydrogenases: An overview. Chem Rev 2007, 107, 4206–72.

[52] Lubitz W, Ogata H, Rüdiger O, Reijerse E. Hydrogenases. Chem Rev 2014, 114, 4081–148.

[53] Armstrong FA. Dynamic electrochemical experiments on hydrogenases. Photosynth Res 2009, 102, 541–50.

[54] Burgdorf T, Lenz O, Buhrke T, Van Der Linden E, Jones AK, Albracht SP, Friedrich B. [NiFe]-hydrogenases of *Ralstonia eutropha* H16: Modular enzymes for oxygen-tolerant biological hydrogen oxidation. J Mol Microbiol Biotechnol 2005, 10, 181–96.

[55] Saggu M, Zebger I, Ludwig M, Lenz O, Friedrich B, Hildebrandt P, Lendzian F. Spectroscopic insights into the oxygen-tolerant membrane-associated [NiFe] hydrogenase of *Ralstonia eutropha* H16. J Biol Chem 2009, 284, 16264–76.

[56] Fritsch J, Scheerer P, Frielingsdorf S, Kroschinsky S, Friedrich B, Lenz O, Spahn CM. The crystal structure of an oxygen-tolerant hydrogenase uncovers a novel iron-sulphur centre. Nature 2011, 479, 249–52.

[57] Hartmann S, Frielingsdorf S, Ciaccafava A, Lorent C, Fritsch J, Siebert E, Priebe J, Haumann M, Zebger I, Lenz O. O₂-Tolerant H₂ activation by an isolated large subunit of a [NiFe] hydrogenase. Biochemistry 2018, 57, 5339–49.

[58] Schulz A-C, Frielingsdorf S, Pommerening P, Lauterbach L, Bistoni G, Neese F, Oestreich M, Lenz O. Formyltetrahydrofolate decarbonylase synthesizes the active site CO ligand of O₂-tolerant [NiFe] hydrogenase. J Am Chem Soc 2020, 142, 1457–64.

[59] Bürstel I, Siebert E, Frielingsdorf S, Zebger I, Friedrich B, Lenz O. CO synthesized from the central one-carbon pool as source for the iron carbonyl in O₂-tolerant [NiFe]-hydrogenase. Proc Natl Acad Sci U S A 2016, 113, 14722–6.

[60] Bond DR, Lovley DR. Electricity production by *Geobacter sulfurreducens* attached to electrodes. Appl Environ Microbiol 2003, 69, 1548–55.

[61] Kim HJ, Park HS, Hyun MS, Chang IS, Kim M, Kim BH. A mediator-less microbial fuel cell using a metal reducing bacterium, *Shewanella putrefaciens*. Enzyme Microb Technol 2002, 30, 145–52.

[62] Gorby YA, Yanina S, McLean JS, Rosso KM, Moyles D, Dohnalkova A, Beveridge TJ, Chang IS, Kim BH, Kim KS, Culley DE, Reed SB, Romine MF, Saffarini DA, Hill EA, Shi L, Elias DA, Kennedy DW, Pinchuk G, Watanabe K, Ishii S, Logan B, Nealson KH, Fredrickson JK. Electrically conductive bacterial nanowires produced by *Shewanella oneidensis* strain MR-1 and other microorganisms. Proc Natl Acad Sci U S A 2006, 103, 11358.

[63] Pham TH, Boon N, Aelterman P, Clauwaert P, De Schamphelaire L, Vanhaecke L, De Maeyer K, Höfte M, Verstraete W, Rabaey K. Metabolites produced by *Pseudomonas* sp. enable a Gram-positive bacterium to achieve extracellular electron transfer. Appl Microbiol Biotechnol 2008, 77, 1119–29.

[64] Qiao Y, Bao S-J, Li CM. Electrocatalysis in microbial fuel cells – from electrode material to direct electrochemistry. Energy Environ Sci 2010, 3, 544–53.

[65] Reguera G, McCarthy KD, Mehta T, Nicoll JS, Tuominen MT, Lovley DR. Extracellular electron transfer via microbial nanowires. Nature 2005, 435, 1098–101.

[66] Cao LC, Yu IKM, Xiong XN, Tsang DCW, Zhang SC, Clark JH, Hu CW, Ng YH, Shang J, Ok YS. Biorenewable hydrogen production through biomass gasification: A review and future prospects. Environ Res 2020, 186, 109547–55.

[67] Kalinci Y, Hepbasli A, Dincer I. Biomass-based hydrogen production: A review and analysis. Int J Hydrog Energy 2009, 34, 8799–817.

[68] Takagi D, Okamura S, Tanaka K, Ikenaga N, Iwashima M, Haghparast SMA, Tanaka N, Miyake J. Characterization of hydrogen production by the co-culture of dark-fermentative and photosynthetic bacteria. Res Chem Intermediat 2016, 42, 7713–22.

[69] Barber J. Photosynthetic energy conversion: Natural and artificial. Chem Soc Rev 2009, 38, 185–96.

[70] Zhu XG, Long SP, Ort DR. What is the maximum efficiency with which photosynthesis can convert solar energy into biomass?. Curr Opin Biotechnol 2008, 19, 153–9.

[71] Happe T, Lambertz C, Kwon JH, Rexroth S, Rögner M. Hydrogen production by natural and semiartificial systems, In: Posten C, Walter C, eds. Microalgal Biotechnology: Integration and Economy. De Gruyter, 2013, 118.

[72] Gutekunst K, Chen X, Schreiber K, Kaspar U, Makam S, Appel J. The bidirectional NiFe-hydrogenase in *Synechocystis* sp. PCC 6803 is reduced by flavodoxin and ferredoxin and is essential under mixotrophic, nitrate-limiting conditions. J Biol Chem 2014, 289, 1930–7.

[73] Cournac L, Guedeney G, Peltier G, Vignais PM. Sustained photoevolution of molecular hydrogen in a mutant of *Synechocystis* sp. strain PCC 6803 deficient in the type I NADPH-dehydrogenase complex. J Bacteriol 2004, 186, 1737–46.

[74] Ihara M, Nishihara H, Yoon KS, Lenz O, Friedrich B, Nakamoto H, Kojima K, Honma D, Kamachi T, Okura I. Light-driven hydrogen production by a hybrid complex of a [NiFe]-hydrogenase and the cyanobacterial photosystem I. Photochem Photobiol 2006, 82, 676–82.

[75] Ihara M, Nakamoto H, Kamachi T, Okura I, Maedal M. Photoinduced hydrogen production by direct electron transfer from photosystem I crossl-linked with cytochrome C_3 to [NiFe]-hydrogenase. Photochem Photobiol 2006, 82, 1677–85.

[76] Krassen H, Schwarze A, Friedrich B, Ataka K, Lenz O, Heberle J. Photosynthetic hydrogen production by a hybrid complex of photosystem I and [NiFe]-hydrogenase. ACS Nano 2009, 3, 4055–61.

[77] Lubner CE, Applegate AM, Knörzer P, Ganago A, Bryant DA, Happe T, Golbeck JH. Solar hydrogen-producing bionanodevice outperforms natural photosynthesis. Proc Natl Acad Sci U S A 2011, 108, 20988.

[78] Allakhverdiev SI, Tomo T, Shimada Y, Kindo H, Nagao R, Klimov VV, Mimuro M. Redox potential of pheophytin *a* in photosystem II of two cyanobacteria having the different special pair chlorophylls. Proc Natl Acad Sci U S A 2010, 107, 3924–30.

[79] McCormick AJ, Bombelli P, Bradley RW, Thorne R, Wenzel T, Howe CJ. Biophotovoltaics: Oxygenic photosynthetic organisms in the world of bioelectrochemical systems. Energy Environ Sci 2015, 8, 1092–109.

[80] Bradley Robert W, Bombelli P, Rowden Stephen JL, Howe Christopher J. Biological photovoltaics: Intra- and extra-cellular electron transport by cyanobacteria. Biochem Soc Trans 2012, 40, 1302–7.

[81] Wey LT, Bombelli P, Chen X, Lawrence JM, Rabideau CM, Rowden SJL, Zhang JZ, Howe CJ. The Development of Biophotovoltaic Systems for Power Generation and Biological Analysis. ChemElectroChem 2019, 6, 5375–86.

[82] Saar KL, Bombelli P, Lea-Smith DJ, Call T, Aro E-M, Müller T, Howe CJ, Knowles TPJ. Enhancing power density of biophotovoltaics by decoupling storage and power delivery. Nat Energy 2018, 3, 75–81.

[83] McCormick AJ, Bombelli P, Scott AM, Philips AJ, Smith AG, Fisher AC, Howe CJ. Photosynthetic biofilms in pure culture harness solar energy in a mediatorless bio-photovoltaic cell (BPV) system. Energy Environ Sci 2011, 4, 4699–709.

[84] Pisciotta JM, Zou Y, Baskakov IV. Light-Dependent Electrogenic Activity of Cyanobacteria. PLOS ONE 2010, 5, e10821.

[85] Wenzel T, Härtter D, Bombelli P, Howe CJ, Steiner U. Porous translucent electrodes enhance current generation from photosynthetic biofilms. Nat Commun 2018, 9, 1299.

[86] Tanaka K, Kashiwagi N, Ogawa T. Effects of light on the electrical output of bioelectrochemical fuel-cells containing *Anabaena variabilis* M-2: Mechanism of the post-illumination burst. J Chem Technol Biotechnol 1988, 42, 235–40.

[87] Tanaka K, Tamamushi R, Ogawa T. Bioelectrochemical fuel-cells operated by the cyanobacterium, *Anabaena variabilis*. J Chem Technol Biotechnol 1985, 35, 191–7.

[88] Bombelli P, Zarrouati M, Thorne RJ, Schneider K, Rowden SJ, Ali A, Yunus K, Cameron PJ, Fisher AC, Ian Wilson D, Howe CJ, McCormick AJ. Surface morphology and surface energy of anode materials influence power outputs in a multi-channel mediatorless bio-photovoltaic (BPV) system. Phys Chem Chem Phys 2012, 14, 12221–9.

[89] Pisciotta JM, Zou Y, Baskakov IV. Role of the photosynthetic electron transfer chain in electrogenic activity of cyanobacteria. Appl Microbiol Biotechnol 2011, 91, 377–85.

[90] Bombelli P, Müller T, Herling TW, Howe CJ, Knowles TPJ. A high power-density, mediator-free, microfluidic biophotovoltaic device for cyanobacterial cells. Adv Energy Mater 2015, 5, 1401299.

[91] Kracke F, Lai B, Yu S, Krömer JO. Balancing cellular redox metabolism in microbial electrosynthesis and electro fermentation – A chance for metabolic engineering. Metab Eng 2018, 45, 109–20.

[92] Lenz O, Friedrich B. A novel multicomponent regulatory system mediates H_2 sensing in *Alcaligenes eutrophus*. Proc Natl Acad Sci U S A 1998, 95, 12474–9.

[93] Lenz O, Strack A, Tran-Betcke A, Friedrich B. A hydrogen-sensing system in transcriptional regulation of hydrogenase gene expression in *Alcaligenes* species. J Bacteriol 1997, 179, 1655.

[94] Dischert W, Vignais PM, Colbeau A. The synthesis of *Rhodobacter capsulatus* HupSL hydrogenase is regulated by the two-component HupT/HupR system. Mol Microbiol 1999, 34, 995–1006.

[95] Elsen S, Duché O, Colbeau A. Interaction between the H_2 sensor HupUV and the histidine kinase HupT controls HupSL hydrogenase synthesis in *Rhodobacter capsulatus*. J Bacteriol 2003, 185, 7111–9.

[96] Bernhard M, Buhrke T, Bleijlevens B, De Lacey AL, Fernandez VM, Albracht SP, Friedrich B. The H_2 sensor of *Ralstonia eutropha*. Biochemical characteristics, spectroscopic properties, and its interaction with a histidine protein kinase. J Biol Chem 2001, 276, 15592–7.

[97] Barahona E, Jiménez-Vicente E, Rubio LM. Hydrogen overproducing nitrogenases obtained by random mutagenesis and high-throughput screening. Sci Rep 2016, 6, 38291–300.

[98] Wecker MS, Beaton SE, Chado RA, Ghirardi ML. Development of a *Rhodobacter capsulatus* self-reporting model system for optimizing light-dependent, [FeFe]-hydrogenase-driven H_2 production. Biotechnol Bioeng 2017, 114, 291–7.

[99] Hoschek A, Heuschkel I, Schmid A, Bühler B, Karande R, Bühler K. Mixed-species biofilms for high-cell-density application of *Synechocystis* sp. PCC 6803 in capillary reactors for continuous cyclohexane oxidation to cyclohexanol. Bioresour Technol 2019, 282, 171–8.

[100] Leino H, Kosourov SN, Saari L, Sivonen K, Tsygankov AA, Aro E-M, Allahverdiyeva Y. Extended H_2 photoproduction by N_2-fixing cyanobacteria immobilized in thin alginate films. Int J Hydrog Energy 2012, 37, 151–61.

[101] Fernandes BD, Mota A, Teixeira JA, Vicente AA. Continuous cultivation of photosynthetic microorganisms: Approaches, applications and future trends. Biotechnol Adv 2015, 33, 1228–45.

[102] Lippi L, Bähr L, Wüstenberg A, Wilde A, Steuer R. Exploring the potential of high-density cultivation of cyanobacteria for the production of cyanophycin. Algal Res 2018, 31, 363–6.

[103] Posten C. Design principles of photo-bioreactors for cultivation of microalgae. Eng Life Sci 2009, 9, 165–77.

[104] Flemming H-C, Wingender J, Szewzyk U, Steinberg P, Rice SA, Kjelleberg S. Biofilms: An emergent form of bacterial life. Nat Rev Microbiol 2016, 14, 563–75.

[105] Stoodley P, Sauer K, Davies DG, Costerton JW. Biofilms as complex differentiated communities. Annu Rev Microbiol 2002, 56, 187–209.

[106] David C, Bühler K, Schmid A. Stabilization of single species *Synechocystis* biofilms by cultivation under segmented flow. J Ind Microbiol Biot 2015, 42, 1083–9.

[107] David C, Heuschkel I, Bühler K, Karande R. Cultivation of productive biofilms in flow reactors and their characterization by CLSM. In: Guisan JM, et al., eds. Immobilization of Enzymes and Cells. Humana Press, 2020, 437–52.

[108] Kashid MN, Gerlach I, Goetz S, Franzke J, Acker JF, Platte F, Agar DW, Turek S. Internal circulation within the liquid slugs of a liquid–liquid slug-flow capillary microreactor. Ind Eng Chem Res 2005, 44, 5003–10.

[109] Gross R, Hauer B, Otto K, Schmid A. Microbial biofilms: New catalysts for maximizing productivity of long-term biotransformations. Biotechnol Bioeng 2007, 98, 1123–34.

[110] Kamravamanesh D, Kiesenhofer D, Fluch S, Lackner M, Herwig C. Scale-up challenges and requirement of technology-transfer for cyanobacterial poly (3-hydroxybutyrate) production in industrial scale. Int J Biobased Plastics 2019, 1, 60–71.

[111] Woodley JM, Titchener-Hooker NJ. The use of windows of operation as a bioprocess design tool. Bioprocess Eng 1996, 14, 263–8.

[112] Knoot CJ, Ungerer J, Wangikar PP, Pakrasi HB. Cyanobacteria: Promising biocatalysts for sustainable chemical production. J Biol Chem 2018, 293, 5044–52.

[113] Heuschkel I, Hoschek A, Schmid A, Buhler B, Karande R, Buhler K. Mixed-trophies biofilm cultivation in capillary reactors. MethodsX 2019, 6, 1822–31.

[114] Burnap RL, Hagemann M, Kaplan A. Regulation of CO_2 Concentrating Mechanism in Cyanobacteria. Life (Basel) 2015, 5, 348–71.

[115] Ananyev G, Dismukes GC. How fast can Photosystem II split water? Kinetic performance at high and low frequencies. Photosynth Res 2005, 84, 355–65.

[116] Winkler M, Senger M, Duan J, Esselborn J, Wittkamp F, Hofmann E, Apfel U-P, Stripp ST, Happe T. Accumulating the hydride state in the catalytic cycle of [FeFe]-hydrogenases. Nat Commun 2017, 8, 16115.

[117] Lu Y, Koo J. O_2 sensitivity and H_2 production activity of hydrogenases-a review. Biotechnol Bioeng 2019, 116, 3124–35.

[118] Liebgott PP, Leroux F, Burlat B, Dementin S, Baffert C, Lautier T, Fourmond V, Ceccaldi P, Cavazza C, Meynial-Salles I, Soucaille P, Fontecilla-Camps JC, Guigliarelli B, Bertrand P, Rousset M, Leger C. Relating diffusion along the substrate tunnel and oxygen sensitivity in hydrogenase. Nat Chem Biol 2010, 6, 63–70.

[119] Saper G, Kallmann D, Conzuelo F, Zhao F, Tóth TN, Liveanu V, Meir S, Szymanski J, Aharoni A, Schuhmann W, Rothschild A, Schuster G, Adir N. Live cyanobacteria produce photocurrent and hydrogen using both the respiratory and photosynthetic systems. Nat Commun 2018, 9, 2168.

[120] Schneemann A, White JL, Kang S, Jeong S, Wan LF, Cho ES, Heo TW, Prendergast D, Urban JJ, Wood BC, Allendorf MD, Stavila V. Nanostructured metal hydrides for hydrogen storage. Chem Rev 2018, 118, 10775–839.

[121] Barnes C, Greene L, Lee J. Assessing the stability and expression of transgenes in genetically engineered cyanobacteria for biofuel production. Abstr Pap Am Chem S 2019, 257.

[122] Mb S, Bhattacharjee A. Structural and functional stability of regenerated cyanobacteria following immobilization. J Appl Phycol 2015, 27, 743–53.

[123] Bertucco A, Beraldi M, Sforza E. Continuous microalgal cultivation in a laboratory-scale photobioreactor under seasonal day-night irradiation: Experiments and simulation. Bioproc Biosyst Eng 2014, 37, 1535–42.

[124] Diez-Montero R, Belohlav V, Ortiz A, Uggetti E, Garcia-Galan MJ, Garcia J. Evaluation of daily and seasonal variations in a semi-closed photobioreactor for microalgae-based bioremediation of agricultural runoff at full-scale. Algal Res 2020, 47, 101859.

[125] Esper B, Badura A, Rögner M. Photosynthesis as a power supply for (bio-)hydrogen production. Trends Plant Sci 2006, 11, 543–9.

[126] Wiegand K, Winkler M, Rumpel S, Kannchen D, Rexroth S, Hase T, Farès C, Happe T, Lubitz W, Rögner M. Rational redesign of the ferredoxin-NADP(+)-oxido-reductase/ferredoxin-interaction for photosynthesis-dependent H(2)-production. Biochim Biophys Acta, Bioenerg 2018, 1859, 253–62.

[127] Kannchen D, Zabret J, Oworah-Nkruma R, Dyczmons-Nowaczyk N, Wiegand K, Löbbert P, Frank A, Nowaczyk MM, Rexroth S, Rögner M. Remodeling of photosynthetic electron transport in *Synechocystis* sp. PCC 6803 for future hydrogen production from water. Biochim Biophys Acta, Bioenerg 2020, 1861, 148208.

Index

https://doi.org/10.1515/9783110716979-012

www.ingramcontent.com/pod-product-compliance
Lightning Source LLC
Chambersburg PA
CBHW080927220326
41598CB00034B/5712